Mycorrhizas - Functional Processes and Ecological Impact

Concepción Azcón-Aguilar • José Miguel Barea
Silvio Gianinazzi • Vivienne Gianinazzi-Pearson
Editors

Mycorrhizas - Functional Processes and Ecological Impact

Springer

Editors
Dr. Concepción Azcón-Aguilar
Soil Microbiology and Symbiotic
Systems Department
Estación Experimental del Zaidín (CSIC)
Profesor Albareda 1
18008 Granada, Spain
e-mail: conchi.azcon@eez.csic.es

Dr. Silvio Gianinazzi
UMR 1088 INRA/CNRS 5184
Université de Bourgogne
Plante-Microbe-Environnement
INRA-CMSE, BP 86510
21065 Dijon, Cédex, France
e-mail: silvio@epoisses.inra.fr

Dr. José Miguel Barea
Soil Microbiology and Symbiotic
Systems Department
Estación Experimental del Zaidín (CSIC)
Profesor Albareda 1
18008 Granada, Spain
e-mail: josemiguel.barea@eez.csic.es

Dr. Vivienne Gianinazzi-Pearson
UMR 1088 INRA/CNRS 5184
Université de Bourgogne
Plante-Microbe-Environnement
INRA-CMSE, BP 86510
21065 Dijon, Cédex, France
e-mail: gianina@dijon.inra.fr

Cover illustration: (From *top* to *bottom*) Courtesy of: W. Jean Mather, Vancouver, Canada (Douglas-fir forest recovering after a low severity fire); J. Palenzuela, Granada, Spain (extraradical hyphae of mycorrhizal *Sorghum vulgare*); K. Benabdellah, Granada, Spain (cellular structures of *Glomus intraradices* extraradical hyphae: wall, *green*; membranes, *yellow*; and nuclei, *blue*) and J. Palenzuela, Granada, Spain (*Salix hastata* ectomycorrhizas).

ISBN: 978-3-540-87977-0 e-ISBN: 978-3-540-87978-7
DOI: 10.1007/978-3-540-87978-7

Library of Congress Control Number: 2008937138

Cover design: WMX Design GmbH, Heidelberg, Germany

Printed on acid-free paper

9 8 7 6 5 4 3 2 1

springer.com

Preface

Research has provided clear indications on how mycorrhizal symbioses are central to the multitrophic interactions that occur underground and how these impact the plants aboveground. Conversely, changes that occur aboveground influence development and functioning of mycorrhizas and their relationships with other soil organisms. However, information is still fragmentary concerning the ecological relevance of such phenomena in ecosystem dynamics and stability. In contrast, substantial progress has been made during the last few years in knowledge about the processes regulating mycorrhiza establishment and their function as mutualistic symbioses, as well as their ecological diversity. This has furthered understanding of: (1) the basic biological mechanisms sustaining compatibility and synergism between the interacting organisms, as well as the genetic and cell programmes involved, (2) the impact of symbiotic relationships in terms of individual fitness in ecosystems, and (3) the role of the symbioses in the evolution of plant life styles.

This book integrates present-day knowledge from well-known research groups on some of the topics which are at the forefront of mycorrhizal research, and in particular those related to cell programmes driving mycorrhiza formation and function, the processes sustaining symbiotic mutualism, stress response mechanisms in mycorrhizal symbionts, and the diversity and ecological impacts of mycorrhizal systems. As a general introduction, the first chapter by D. Atkinson places mycorrhizal research in the global context of the evolution of agricultural polices and practices. Methodologies used for improving food production in intensive agriculture are questioned, and evidence is provided for the need to enhance the role of mycorrhizas in order to improve food quality and reduce environmental damage.

The enigma of mutualism evolution is evoked in the second chapter, from the viewpoint of biological altruism in beneficial microbes. Because of the cellular and molecular convergence that can be found between mycorrhizal and rhizobial symbioses, N.A. Provorov and N.I. Vorobyov take the comprehensively studied legume–rhizobia symbioses as the basis to discuss the role of selective pressures in maintaining genes for beneficial traits in soil microbial populations. For a long time, evolutionary research was focussed mainly on antagonistic symbioses but, as de Bary had already pointed out at the end of the nineteenth century, a gradient should exist between microbial parasites that help their partners and those that destroy them. In this context, early fungal–plant interactions are determinant factors for the outcome of symbiotic interactions and, in the following chapter, V. Gianinazzi-Pearson et al. review recent knowledge about the molecular dialogue involved in the primary steps essential to the symbiotic process.

In established mycorrhizal associations, bidirectional exchange of nutrients and other benefits that occur require the formation of symbiotic interfaces resulting from morphophysiological changes in both plant and fungal tissues. The next two contributions are devoted to ongoing non-targeted research to identify transcripts (M. Arlt et al.) and proteins (G. Recorbet and E. Dumas-Gaudot) involved in mycorrhizal functioning, whilst in the sixth chapter N. Ferrol and J. Perez-Tienda consider in detail the function of specific molecular and cellular modifications in the coordinated nutrient exchange across interfaces between the two symbionts. This is followed by a chapter where E.J. Grace et al. evoke the possibility of exploring functional diversity in plant responsiveness to mycorrhiza at the molecular level, in the light of mechanisms for P uptake via mycorrhizal and direct uptake pathways.

Mycorrhizas play a key role in regulating abiotic and biotic stresses to plants, particularly at the rhizosphere level, and this positively impacts on safe food production and bioremediation programmes. Research on mechanisms underlying such beneficial effects has developed considerably in recent years, and two chapters focus on how mycorrhizal fungi may interfere with heavy metal transport within plant tissues (M. González-Guerrero et al.) or induce plant resistance against pathogens (M.J. Pozo et al.).

Recognition of the Glomeromycota as a separate phylum of true fungi (and the extension of their genetic variability by combining molecular and morphological taxonomy) is an important advance in the last decade. J.B. Morton reviews in his chapter the persisting discord between rRNA gene and morphology-based phylogeny, and discusses how this can be overcome within the framework of a balanced multidisciplinary approach.

The mycorrhizal condition is the state of the large majority of plants under most ecological conditions, and therefore these symbioses are a key element in ecosystem functioning. The importance of a spatial context for the understanding of the ecology and evolution of organisms has become increasingly clear. B.J. Pickles and co-workers examine methodological approaches for spatial analysis of mycorrhizal fungi, and show how their application could yield more information about communities than that obtained from simple species abundance and frequency data. Potential mechanisms responsible for generating and maintaining spatial variation in populations and communities of mycorrhizal fungi at different scale levels (root, community, landscape) are then reviewed by B.E. Wolfe et al., and the difficulty in linking such mechanisms to observed spatial patterns is discussed.

Ecological disturbance, whether natural or through human activity, also impacts on the diversity of mycorrhizal fungi, and ecological resilience of these symbionts is essential for sustaining productivity. S.W. Simard considers response diversity of mycorrhizal fungi to disturbance by wildfire or cutting in forest ecosystems, and the impact on survival and growth rate of regenerating vegetation. Likewise, in their chapter, R.J. Bills and J.C. Stutz monitor species richness of mycorrhizal fungi associated with plants in disturbed urbanized, as compared to natural desert, sites.

Drivers of alterations in community structure of mycorrhizal fungi need to be identified. In the forest ecosystem example, reduction in fungal diversity may be avoided when forest floor and key functional groups of plants are maintained. Such an observation is of primary importance in defining optimal forest management practices. The final chapter by A. Morte and collaborators illustrates how successful management strategies can ensure the introduction and production of the desert truffle, which could represent an interesting model for managing production of other edible mycorrhizal fungi.

In conclusion, interest in mycorrhizal symbioses continues to expand and to encompass an increasing range of disciplines. Plants and mycorrhizal fungi have co-evolved in continuous interaction with their abiotic and biotic environments, developing a wide range of coordinated mechanisms which considerably favour the production of primary biomass. Information about such mechanisms is of outstanding importance for promoting sustainable practices in plant production systems as well as for ecosystem conservation and restoration schemes. In conclusion, the efficient management of mycorrhizal systems has the potential to favour the sustainable production of quality foods while ensuring environmental quality for future generations.

We would like to express our gratitude to the chapter authors for their contributions and also to the anonymous reviewers for their constructive feedback on the different manuscripts. It is hoped that the chapter contents will stimulate further research; the concepts and opinions that are expressed are the exclusive responsibility of the author(s).

November 2008

Concepción Azcón-Aguilar
José Miguel Barea
Silvio Gianinazzi
Vivienne Gianinazzi-Pearson
Editors

Contents

Contributors

Ian J. Alexander
School of Biological Sciences, Plant and Soil Science, University of Aberdeen, St. Machar Drive, Aberdeen, AB24 3UU, UK

Ian C. Anderson
The Macaulay Institute, Craigiebuckler, Aberdeen, AB15 8QH, UK

Matthias Arlt
Institute for Vegetable and Ornamental Crops, Theodor-Echtermeyer-Weg 1, 14979 Grossbeeren, Germany

David Atkinson
SAC, Craibstone Estate, Bucksburn, Aberdeen, AB21 9YA, UK

Concepción Azcón-Aguilar
Departamento de Microbiología del Suelo y Sistemas Simbióticos, Estación Experimental del Zaidín, CSIC, Profesor Albareda 1, 18008 Granada, Spain

Karim Benabdellah
Departamento de Microbiología del Suelo y Sistemas Simbióticos, Estación Experimental del Zaidín, CSIC, Profesor Albareda 1, 18008 Granada, Spain

Robert J. Bills
School of Life Sciences, Arizona State University, Tempe, Arizona, USA

Eliane Dumas-Gaudot
UMR 1088 INRA/CNRS 5184/UB Plante-Microbe-Environnement, INRA/CMSE, BP 86510, 21065 Dijon Cedex, France

Nuria Ferrol
Departamento de Microbiología del Suelo y Sistemas Simbióticos, Estación Experimental del Zaidín, CSIC, Profesor Albareda 1, 18008 Granada, Spain

Philipp Franken
Institute for Vegetable and Ornamental Crops, Theodor-Echtermeyer-Weg 1, 14979 Grossbeeren, Germany

Juan M. García
Departamento de Microbiología del Suelo y Sistemas Simbióticos, Estación Experimental del Zaidín, CSIC, Profesor Albareda 1, 18008 Granada, Spain

Javier García-Andrade
Departamento de Microbiología del Suelo y Sistemas Simbióticos, Estación Experimental del Zaidín, CSIC, Profesor Albareda 1, 18008 Granada, Spain

Monique Gardes
Laboratoire Evolution et Diversité Biologique, Université Paul Sabatier, 118 Route de Narbonne, 31062 Toulouse Cedex 4, France

David Genney
School of Biological Sciences, Plant and Soil Science, University of Aberdeen, St. Machar Drive, Aberdeen, AB24 3UU, UK

Vivienne Gianinazzi-Pearson
CNRS 5184/Université de Bourgogne Plante-Microbe Environnement, BP 86510, 21065 Dijon Cedex, France

Manuel González-Guerrero
Department of Chemistry and Biochemistry, Worcester Polytechnic Institute, 100 Institute Rd., Worcester MA01609, USA

Emily J. Grace
Soil and Land Systems, School of Earth and Environmental Sciences, Waite Campus, The University of Adelaide, Adelaide, South Australia, 5005

Almudena Gutiérrez
Departamento de Biología Vegetal (Botánica), Facultad de Biología, Universidad de Murcia, Campus de Espinardo, 30100 Murcia, Spain

Mario Honrubia
Departamento de Biología Vegetal (Botánica), Facultad de Biología, Universidad de Murcia, Campus de Espinardo, 30100 Murcia, Spain

John N. Klironomos
Department of Integrative Biology, University of Guelph, Guelph, ON, N1G 2W1, Canada

Alexander M. Koch
Department of Integrative Biology, University of Guelph, Guelph, ON, N1G 2W1, Canada

Asunción Morte
Departamento de Biología Vegetal (Botánica), Facultad de Biología, Universidad de Murcia, Campus de Espinardo, 30100 Murcia, Spain

Joseph B. Morton
1090 Agricultural Sciences Building, West Virginia University, Morgantown, WV 26506-6108, USA

Jeri L. Parrent
Department of Integrative Biology, University of Guelph, Guelph, ON, N1G 2W1, Canada

Jacob Pérez-Tienda
Departamento de Microbiología del Suelo y Sistemas Simbióticos, Estación Experimental del Zaidín, CSIC, Profesor Albareda 1, 18008 Granada, Spain

Brian J. Pickles
Ecosystem Science and Management, University of Northern British Columbia, 3333 University Way, Prince George, BC, V2N 4Z9, Canada

María J. Pozo
Departamento de Microbiología del Suelo y Sistemas Simbióticos, Estación Experimental del Zaidín, CSIC, Profesor Albareda 1, 18008 Granada, Spain

Nikolai A. Provorov
All-Russia Research Institute for Agricultural Microbiology, Podbelsky Sh. 3, St.-Petersburg, Pushkin-8, 196608, Russia

Ghislaine Recorbet
UMR 1088 INRA/CNRS 5184/UB Plante-Microbe-Environnement, INRA/CMSE, BP 86510, 21065 Dijon Cedex, France

Dietmar Schwarz
Institute for Vegetable and Ornamental Crops, Theodor-Echtermeyer-Weg 1, 14979 Grossbeeren, Germany

Pascale Seddas
UMR INRA 1088 Université de Bourgogne Plante-Microbe Environnement, BP 86510, 21065 Dijon Cedex, France

Benjamin A. Sikes
Department of Integrative Biology, University of Guelph, Guelph, ON, N1G 2W1, Canada

Suzanne W. Simard
Department of Forest Sciences, University of British Columbia, Vancouver, British Columbia, V6T 1Z4, Canada

F. Andrew Smith
Soil and Land Systems, School of Earth and Environmental Sciences, Waite Campus, The University of Adelaide, Adelaide, South Australia, 5005

Sally E. Smith
Soil and Land Systems, School of Earth and Environmental Sciences, Waite Campus, The University of Adelaide, Adelaide, South Australia, 5005

Jean C. Stutz
Applied Biological Studies, Arizona State University Polytechnic, Mesa, Arizona, USA

Marie Tollot
UMR INRA 1088 Université de Bourgogne Plante-Microbe Environnement, BP 86510, 21065 Dijon Cedex, France

Adriaan Verhage
Departamento de Microbiología del Suelo y Sistemas Simbióticos, Estación Experimental del Zaidín, CSIC, Profesor Albareda 1, 18008 Granada, Spain

Nikolai I. Vorobyov
All-Russia Research Institute for Agricultural Microbiology, Podbelsky Sh. 3, St.-Petersburg, Pushkin-8, 196608, Russia

Benjamin E. Wolfe
Department of Organismic and Evolutionary Biology, Harvard University, 16 Divinity Avenue, Cambridge, MA 02138, USA

Mar Zamora
Departamento de Biología Vegetal (Botánica), Facultad de Biología, Universidad de Murcia, Campus de Espinardo, 30100 Murcia, Spain

Chapter 1
Soil Microbial Resources and Agricultural Policies

David Atkinson

Abstract The current systems of food production used in Europe and North America have been hugely successful in increasing the yields of arable crops and in reducing the proportion of household budgets devoted to food. Despite this, intensive agricultural systems have had adverse effects on the environment, and the resulting diet has been linked to current health issues. Both the EU and the USA Government have subsidised production. The income derived from subsides has often exceeded that from farming. Such subsidies are unsustainable. Farming must become more market orientated. Past attempts have largely related to attempts to increase yields and reduce costs. The introduction of genetically modified crops is an example of this approach. An alternative has been to increase income from farming by producing products selling at a premium price. Recently, the public have come to question how food is produced and the consequences of particular methodologies. Questions related to the ethical values inherent in systems are increasingly raised. Future agriculture will be more eco-efficient. Against this background, the role of mycorrhizal fungi should be enhanced. These issues are examined here.

1.1 Introduction

The proceedings of the 5th International Conference on Mycorrhiza (Anon 2006) indicate that at the current time several lines of research are being followed. There are studies aimed at documenting the genome of the fungi and related work to understand molecular communications between fungi, host plants and other micro-organisms. There are studies which aim to understand how the association works to promote the success of the plant partner. There are an increasing number of studies

D. Atkinson
SAC. Craibstone Estate, Bucksburn, Aberdeen, AB21 9YA, UK
e-mail: atkinson390@btinternet.com

C. Azcón-Aguilar et al. (eds.), *Mycorrhizas - Functional Processes and Ecological Impact*,
DOI: 10/1007-978-3-540-87978-7_1,

aimed at understanding the ecology of the fungal plant association (see Chapter 12 by Wolfe et al.) and, derived from this interest, in inoculating agricultural and horticultural crops with arbuscular mycorrhizal fungi (AMF) so as to promote performance. With the exception of the last of these areas, research is justified on the basis of its contribution to our fundamental understanding of symbiotic relationships between fungi and plants and of the structure and functioning of soil communities. The success of inoculation technology depends on its potential to improve current agricultural practices. However, whether any of this research is put into practice, and the amount of funding which may be made available for even the most fundamental of the research areas, will depend on factors mainly unrelated to mycorrhizal fungi.

The ability for progress in any area of science to have a significant effect on the economics of the area it seeks to influence or even on practice within the technologies to which it relates is far from linear (Kealey 1996). Firstly, the science must deliver real understanding and the ability for that understanding to be developed into either a new technology or product or to result in a significant gain in the delivery of current objectives. In the context of agriculture or horticulture, this would normally be seen as an increase in yield or product quality or in the ability to carry out an action more easily. Secondly, the science must enhance the direction of travel of a particular set of technologies and, as a result, aid a saving in costs or render a key activity doable. It will contribute most if it is able to reduce costs in a major part of the cost structure of the industry or if it enhances the practicality of a distinct approach to production. Thirdly, it must also work with the prevailing political model. In the context of agriculture in Europe it must sit comfortably with the rules of the Common Agricultural Policy (CAP) and the environmental norms of the EU as enacted in the Water Framework Directive (WFD). Fourthly, it must also be considered to be ethical, safe and necessary if it is to be acceptable to society.

Most reviews of progress in particular areas of science and on their potential impact deal primarily with the first of these factors. Issues related to the second are often used as justification within the case for carrying out research in that particular area of science, although analysis of financial impact, after the completion of the research, is seldom carried out in a rigorous manner. The potential for system change is currently topical. Until relatively recently it has been assumed that the political process and the public will always support technological advances if they are apparently safe and if they seem likely to result in cost saving. For much of the twentieth century this was the case. Towards the end of the century, attempts to introduce agricultural biotechnological products, such as genetically modified (GM) crops (Bruce and Bruce 1998; Marsh 2003), indicated that there had been a major change in the mood of the public, at least in Europe. This indicated that public acceptance of technological advances could not be automatically assumed. It became clear that the views of those who did research and the creators of new technological products were likely to be driven by different considerations to the users of products and those who purchase foods (Bruce and Bruce 1998).

The resistance of the public to GM crops is currently a factor in attempts to introduce nanotechnology to agriculture.

This chapter examines the extent to which political and economic influences, changes in farming practice and ethical issues will influence the potential role and contribution to agriculture and related industries of soil microbes in general and AMF in particular in the coming decade.

1.2 The Current State of European Agriculture

1.2.1 The Common Agricultural Policy

For most of the twentieth century, in most European countries, agriculture has been subsidised by Governments, latterly via the Common Agricultural Policy (CAP) (Plum 1998). The basis of subsidy has varied. For much of the period, payments were tied to the production of certain crops or resulted from the willingness of the EU to purchase (intervention) crops at prices which were above the cost of production (North 1987; Atkinson 1990). As a result, production was related to the rules of the CAP rather than being geared to the needs of the consumers. The impact of the CAP, together with decreases in the market prices of major commodities during the 1990s, meant that for farming in a country like Scotland, income from the direct products of farming, e.g. grain or milk, represented, on average, only one-third of the total income of the farming business. The remaining two-thirds of income came from the CAP and related sources. An additional consequence of subsidies was that for many commodities the cost of production was above farm gate prices. For example, milk production costs in 2006 were 19.5 p/l against a selling price of 18 p/l, beef production £2/kg against a selling price of £1.9 and oat production £100/t against a selling price of £70/t (NFU Scotland 2006). The survival of farming with prices at these levels was only possible because of the existence of the CAP (DEFRA 2006). Over the same period, the US Farm Bill had a similar effect on production in the USA. In the UK, the competitive nature of the multiple retailers and the low direct prices received by producers meant that the proportion of household incomes spent on food, <10%, fell to an all-time low (DEFRA 2006; Hampson 2006). The value of primary production in UK decreased to 1.7% of gross domestic product (GDP) although the whole agri-food sector represented 7.7% of GDP. This situation was generally perceived to be unsustainable (Buckwell 1998; Anon 2002). In 2005, the CAP was reformed with the replacement of the whole series of production subsidies by a single farm payment which was given irrespective of the farming enterprises being undertaken. The payment requires farming to be continued and a series of environmental goals to be met. At the same time, the EU saw its largest single enlargement (Marsh 2005).

1.2.2 CAP Reform

The combined effect of the above is that the CAP is due for reform in 2013 (see also Sect. 1.7) with the expectation that the level of subsidy will decrease, perhaps by 50% (Fischer Boel 2007). These changes have resulted in a significant awareness within European agriculture of the need to change, and hence in more interest in novel activities and novel methods than has existed for some time. Were support from the CAP to decrease to 50% of current levels this would increase the significance of income derived from farming. This would need to increase significantly to preserve current average incomes. Increases in farming incomes will need to come from a reduction in costs or from improvements in quality; there is insufficient scope to increase yields to the extent needed to replace lost income. This presents options for soil micro-organisms to be part of a radical reform of farming practice and to increase their impact.

1.3 The Current Farming Model and Its Consequences

1.3.1 The Conventional Model

Agriculture, post-1945, changed so as to increase its emphasis on maximising yields through the optimised/maximised use of nitrogen fertilisers. Pesticides were used to protect these high nitrogen crops from the adverse impact of pests and diseases (Martin 1998). Agriculture moved away from a model where crop rotation, which recycled nutrients between crop and stock enterprises and used biological processes such as nitrogen fixation by legumes, was the norm. It moved to a system which saw the principle role of soil as physically supporting the plant so that it could absorb fertiliser-applied nutrients (Conford 2001). The use of varieties in which the partitioning of assimilates into grain, rather than into vegetative elements such as straw, resulted in the predominance of shorter varieties of cereals which needed herbicides to permit them to compete with weeds. Overall, the combination of cheap fertiliser, a continual stream of new pesticides and increasingly high-yielding varieties of cereals increased average yields to record levels; average yields increased by 1–2% each year for most of the period 1975–2000 (Atkinson 1990).

1.3.2 The Environmental Impact of Agriculture

Production carried out in this way and with these inputs was not without its costs in relation to biodiversity of wildlife and soil and environmental quality (Nortcliff 2006). This has become a factor of increasing importance and now is integral to the design of farming systems. Pretty et al. (2000) estimated that in the UK alone the externalities of agriculture, the real costs not being born by those who incurred them, totalled around £2.3 bn/year. This resulted from negative impacts on water, air, soil and biodiversity. For example, they estimated the cost of cleaning water from contamination with fertiliser and pesticides alone was £231 m/year. A hectare of farmed land which might produce £700 of income for cereals would incur a cost of £250 in respect of environmental damage.

1.3.3 Pesticide Concerns

In addition to their environmental impact, conventional farming methods became associated with concerns over the frequency with which pesticide residues were found in food products. This has become significant for many European consumers. The presence of pesticides in food is related to the farming system used. In a study of apple production in the USA, Baker et al. (2002), found pesticide residues in 82% of fruit from conventional production compared to 49% in fruit from farms using integrated pest management (IPM) and 23% from farms which were using organic production. For vegetable production, the comparable figures were 65%, 45% and 23%, respectively. For vegetables, when residues of materials which were currently not permitted for use on any crop were removed from the analysis, the figures decreased to 61%, 44% and 9%, respectively. For fruit, removing such residues had little effect on the analysis. The high frequency of presence in organic production indicated the prevalence of long-lasting contamination from materials applied many years ago, and the possible contamination of production by applications to conventional systems.

1.4 Alternatives to the Current Agricultural Model

1.4.1 Ways of Reducing Environmental Impact

Concerns about the impact of current agricultural systems on wildlife and the environment, and about the presence of pesticide residues in foods, have led to calls for changes to the way farming is practised so as to achieve reductions in pesticide use and a reduced impact of agriculture on the environment (Baillier et al. 1997; Bax 2002). It has resulted in an increase in the proportion of the market being supplied by production which does not rely, or rely as heavily, on the use of pesticides, such as organic farming (Conford 2001). Organic farming is the most extreme version of a system which aims to use natural processes and cycles as the basis of nutrient supply to crops and for the protection of crops from the adverse impact of other organisms. At the present time, such systems produce yields of the main agricultural commodities which are significantly below those which are produced by systems based on the current predominant agricultural model (Atkinson et al. 2004). A major challenge is thus to increase nutrient supply in such systems through using an enhancement of biological processes to enhance rates of nutrient release (Watson et al. 2002; Atkinson et al. 2005). It is here that there is a clear role for the use of mycorrhizal fungi in the design and management of such systems (Harrier and Watson 2004; Atkinson 2006a; Gianinazzi and Gollotte 2006). Systems such as integrated crop management use contributions from such sources together with a more restricted use of fertilisers and pesticides.

1.4.2 Reduction in Environmental Impact Through Organic Farming

In the past decade, foods produced using organic production methods have been the fastest growing sector of the market which in the UK alone is now worth almost £2 bn/year. Organic farming is based on the premise that, if the reserves of the soil are carefully managed, then there is no requirement for the use of the fertilisers and pesticides, which are needed in conventional systems because of a failure to focus on the soil as the centre of production. Not using pesticides has meant that systems farmed using organic methods have been richer in wildlife. In such systems, the options for soil microbes to be the basis of the supply of nutrients and the source of plant health are maximised while an understanding of and management of soil microbes becomes critical.

1.5 Eco-Efficient Agriculture

Systems of the above types sit comfortably with the concept of eco-efficient farming. Developing this concept questions how agriculture might best provide ecosystem services such as the maintenance of soil microbial resources and biodiversity and more generally clean water and air. The latter has become more important since the heightened emphasis being placed on atmospheric carbon levels and carbon footprints. There are now questions about how science might reduce total resource use and enhance the environment and the ways in which AMF could have a significant role.

1.5.1 Eco-Efficient Agriculture Criteria

The following have been suggested as key criteria.

- It is efficient in its resources use, especially of resources derived from non-renewable sources such as fossil fuels. Nitrogenous fertilisers derived from the Haber-Bosch process and most pesticides require large energy inputs, usually from the oil reserves used during their manufacture (Montague 2000). Systems which depend on the soil reserves created by microbial nitrogen fixation and released through the microbial decomposition are more energy efficient.

- It emphasises the use of renewable resources. All agricultural practice, including organic agriculture, depends on the maintenance of an unstable ecosystem. In nature, monocultures dominated by a single plant species and low presence of micro-organisms and wildlife or insects are unprecedented and at variance with well-established ecological concepts such as the niche (Schoener 1989) and the importance of diversity to stability (Walker 1989). Maintaining such ecosystems requires inputs of energy to move a natural balance in favour of the crop. In conventional systems, most of this energy is needed to manufacture fertilisers and pesticides. Machinery to apply these materials and for cultivation, harvesting, etc. is also a significant energy input (Montague 2000). Nutrients and crop protection can be delivered, although not easily and with significant yield penalties, without some of this energy. This was the rationale for crop rotations. The rotation has phases when fertility is developed and others when it is depleted in crop production. Fertility improvement involved legumes fixing atmospheric nitrogen and grasses benefitting soil structure and providing carbon to soil microbial communities. Green manures removed soil nitrogen at the end of the growing season preventing its loss by leaching. Rotations, although effective at minimising the need for external resources, resulted in saleable product being produced, on a particular land area, in only half of the years, unless the grass–clover phase was used for animal production. Developing the rotation to deliver nutrients other than nitrogen and to improve crop health provides a significant role for AMF (see also Sect. 1.8).
- It does not result in the creation of pollutants or transfer them elsewhere to become externalities (Pretty et al. 2000). The inefficiency with which nitrogen fertilisers are used and the nature of action of pesticides mean that significant quantities of these transfer to ground water. Energy inputs to agriculture and the release of soil carbon during cultivation, of methane as a consequence of rumen fermentation and of nitrogen oxides have meant that agriculture has a disproportionate influence on global warming relative to its economic value. Agriculture produces 7% of greenhouse gases; several times its 1.7% contribution to GDP.
- Eco-efficient farming conserves functional diversity. Agriculture is a major user of land. In Scotland, 75% of the land area is used for agriculture, much of it for extensive animal production. This land must also host wildlife. Extensive farming and rotations meet this need by allowing diversity to exist in parallel with agriculture. Intensive agriculture aims to do the same by minimising its agricultural footprint, i.e. the amount of land required for production, and freeing other areas of land for wildlife conservation. Both approaches represent a compromise. Not all species can co-exist in an agricultural system and not all flourish

on the poorer land which intensive agriculture releases. Both impact on soil functional diversity and on AMF. It is here that the selection of agricultural system will impact most on the ability of agriculture to become a net contributor to natural capital (Atkinson et al. 1994). Microbial activity is at the heart of natural soil fertility and the conservation of soil carbon resources. Much global carbon is stored in soils. In Scotland, soil organic carbon contents are high. How soils are managed and the type of agriculture employed will influence the release of previously stored carbon. How a balance is set between land use for food production and use for carbon storage will depend upon the value accorded by society to these interests. This balance will vary between countries and within countries at different times.

- Eco-efficient farming recognises a social dimension to agriculture and its part in the maintenance of cultural heritage, diet, cuisine and many core values. This is at the heart of current differences in attitudes between the USA and Europe over the wisdom of the use of genetically modified crops (Bruce and Bruce 1998). Approaches to global climate change and the value placed by the populations on the environment are currently at the heart of such debates.

1.6 Current Ethical Challenges

1.6.1 The History of Ethics in Agriculture

In one sense, the ability to take ethically-based decisions in agriculture is a luxury resulting from a guaranteed supply of inexpensive food. The need is not new. In the book of Genesis, in the Hebrew Scriptures (Chapter 41, v. 25–49), Joseph, the head steward of the King of Egypt, had to decide how to deal with a situation in which there were years of high production and years of poor production, and with a situation where Egypt had plentiful supplies of grain when other countries lacked food. Awareness of the importance of the cultivation and systems of animal husbandry led to detailed codes of land management, e.g. Deuteronomy, Chapter 26. Historical records from other ancient civilisations show similar concerns.

1.6.2 Silent Spring and Agriculture

Continuing concerns in this area are illustrated by reactions to the issues raised by *Silent Spring* (Carson 1963). *Silent Spring* appeared when chemical pesticides were first introduced to agriculture on a major scale and when the predominant agricultural model was being revolutionised. Carson predicted that pesticides would be harmful to the environment and to wildlife. She anticipated that chemically-based agriculture would impact on landscape, the rural economy and our relationship with nature. The response of chemical industry was to deny chemicals would have these effects but in parallel to move to products with less impact on non-target organisms and the environment. Producers welcomed pesticides as a means of reducing the variable costs of production, e.g. the bill for inputs, and fixed costs, e.g. labour, and of increasing yields and profitability. Initially, the public welcomed the relative decrease in the price of food. However, when it became clear that this type of agriculture was having a major effect on wildlife, e.g. farmland birds, the appearance of the countryside, e.g. through hedge removal, and on agricultural heritage, sympathy for the thesis of *Silent Spring* increased. This debate continues (Atkinson 2007).

1.6.3 The GM Debate

A consequence of the pesticide debate was public resistance to the introduction of GM crops (Bruce and Bruce 1998). They were seen by many in Europe to represent a further intensification of agriculture, further damage to the countryside and questionable long-term safety. The GM debate indicated the arrival of a new set of criteria for the introduction of change, the views of the public.

1.6.4 The Ethics Inherent in Farming Systems

The different ways in which agriculture is practised involve decisions about values and priorities. The range of types of farming systems can be presented as a continuum but with major variations in their underlying values (Atkinson and Watson 2000). Differences are most extreme in a comparison of intensive arable agriculture and organic farming.

1.6.5 Ethics and Current Agriculture

Intensive agriculture is driven by the values of industrialisation. Its primary aim is to maximise financial returns thorough optimising production and minimising costs, including those related to labour. Management is simplified by the use of deterministic linear models which optimise the use of fertilisers and pesticides. The externalisation of costs, including environmental and social costs, is important. This approach has delivered dependable quantities of food and low prices at the till for the past three decades. Its continued success depends upon the adoption of new technologies such as GM crops to facilitate pesticide use and nanotechnology to increase efficiency.

1.6.6 Ethics and Other Farming Systems

Other conventional systems are built around some of these values but with less focus on profitability. Integrated crop management (ICM) systems sacrifice some yield by restricting pesticide and fertiliser applications in order to deliver environmental goods and reduce environmental impact. Mixed farming systems are more complex than specialised animal or crop production systems. They improve the use of resources and could reduce both the need for fertilisers and the problems related to the disposal of animal wastes. If all the waste produced by poultry and pigs in the UK were to be used as N fertiliser it would eliminate the nitrogen fertiliser requirement for over 250,000 ha of cereal production.

1.6.7 Ethics and Organic Farming

In contrast to the above, organic agriculture focusses on the management of natural cycles and on their ability to provide crop nutrients and protection from adverse effects of other organisms. This cannot be achieved with simple linear models. It requires the management of a series of networks. Creating the conditions which allow microbes to mineralise nitrogen does not guarantee its release when needed. Similarly, maximising mycorrhizal colonisation will not guarantee an appropriate supply of phosphorus or improved water relations, but without colonisation these effects will not occur. Organic production will never match the production of conventional systems because this is not its core objective (Atkinson et al. 2004). Organic agriculture aims to produce high quality food. Through the selection of varieties and systems which encourage the production of secondary metabolites, they produce foods containing chemicals now known to be important to health. This contrast between intensive and organic systems identifies some of the significant ethical issues associated with food production.

1.6.8 Current Areas of Ethical Challenge

- The use of biotechnological methods, especially where crops have been modified using transgenic methodologies. There are questions about how risk should be apportioned between the producer who gains when technology increases profitability and consumers who lose if there are adverse effects on health or the environment.
- The appropriateness of systems of production in which maximising yields is the major design criteria at a time when much farm income is from the CAP. This questions the balance required between production and other needs. It asks how consumers can obtain foods consistent with their values.

- How innovations are introduced. The biotechnology debate has moved to nanotechnology. The risks and benefits are hard to discern, yet without some risk areas of potential benefit will be missed.

This questions how such decisions might be made, the principles which underpin decision making, how intergenerational equity can be represented, how the needs of multi-national businesses can be balanced with those of individuals, and how decisions, which potentially affect the whole world, can be taken by individual nations. Discussion of the above issues indicates that much current debate is partial. Parts of a wider picture are argued by protagonists. Those who support biotechnological methods to enhance crop growth argue that such methods are needed to feed those without adequate supplies of food in the developing countries. They focus not on what current technology can deliver but on future hopes, e.g. drought-resistant varieties or crops with lower needs for water or with more effective photosynthetic potential. Others, however, emphasise the complex reasons underpinning poverty in the developing countries. They also illustrate the difficulties of achieving future aims such as crop plants fixing their own nitrogen. Most plants function without this ability and evolution has not resulted in it being an attribute of the more successful species, perhaps because nitrogen fixation demands around 25% of total photosynthates. These arguments are not opposites but different cases.

1.6.9 The Banner Framework

There is need for principles to decide between cases and to put arguments into perspective. A framework to aid this was suggested by the Banner committee which considered issues related to animal breeding (Anon 1995). They suggested a series of questions to be addressed in sequence. Later questions are answered only if the preceding question was answered in the negative. The initial question covers inherent considerations, the later ones consequential issues. Modifying the framework for use beyond its original context, we have the following three questions:

1. Does the development represent a harm which should never be permitted?

   This asks whether it raises inherent, non-consequential, moral issues which are simply unacceptable. In the GM debate, many objectors felt it did and so were not swayed by discussions on safety or future potential to relieve suffering.

2. Do the projected benefits of the development outweigh the projected costs and disbenefits by a significant margin?

   Answering this needs the identification of the consequential issues which the development raises. This cost–benefit analysis is where scientists argue the importance of a discovery. This is not simple as benefits and disbenefits may be of different kinds. Benefits may be in terms of job creation or additions to national wealth. disbenefits may relate to environmental damage. These can be costed in financial terms, directly or through the use of surrogate values, but the costs and benefits rarely fall equally across stakeholders.

3. Can any significant disbenefits be minimised?

   Most activities will have negative consequences. If the important ones can be ameliorated then the cost–benefit ratio will improve and the development will become more acceptable. Early pesticides had significant impacts on wildlife. Later molecules had much smaller effects and so were acceptable to the public until other issues came to dominate the discussion. Like question 2, the future impact of the development is important to this question.

The success of this type of analysis requires the correct questions to be asked. Context will always be important. Some of those who would inherently object to the use of transgenic technology to modify animals might be less concerned at its use in relation to crops, especially if the modification gave a direct benefit to the consumer, e.g. an improved food product. The acceptability of the technology appears to be influenced by the genetic distance between the donor and the receiver species and the source of the donor genes. The use of human genes to allow new crop products seems less acceptable than the use of microbial genes.

1.7 The Agriculture of the Next Decade

1.7.1 CAP Reform

European agriculture is likely to change in major ways. Some changes can be anticipated as a result of known trends in EU attitudes, legislation and financial support measures. Agriculture has always included those who farmed to maximising production and those who farmed for a wider series of objectives. This division seems likely to intensify. The CAP must reduce its call on the EU budget but also improve market orientation and the competitiveness of European agriculture (Marsh 2005; Fischer Boel 2007). In 2005, the need to produce specific commodities or to produce them in a specific way was removed. Farms received a single farm payment decoupled from production. The basis on which this was paid varied between countries. It resulted in a rebalancing of priorities with higher priority given to the maintenance of rural communities, health and the environment.

1.7.2 The Removal of Subsidies

These objectives were aided by the removal of some of the payment, through modulation or degression, to permit its use for such purposes. Such changes make it easier for the EU to comply with the World Trade Organisation (WTO) objective of opening up world trade through reducing the level of subsidy attracted by agricultural products. In total, predictable changes seem likely to result in increased emphasis on locally produced food, food meeting specific quality criteria, rural development projects, a reduced carbon footprint, the use of renewable energy sources and the positive enhancement of the environment. Agriculture developed against these priorities seems likely to place an increased emphasis on understanding and using soil microbes, especially AMF. The agricultural production methodology in which their use is currently most prevalent is organic farming. It seems appropriate to examine their role in this type of production as this should indicate their potential contribution to other forms of production and more generally.

1.8 AMF and Organic Farming

1.8.1 AMF and the Crop Rotation

In organic farming, the primary focus is on the development and optimisation of soil processes. These are the means through which crops are supplied with the nutrients they need for growth and by which the wider ecosystem is managed to enhance the health of the crop (Pozo et al. 2002; Atkinson et al. 2002). The key means of delivery is the crop rotation. This aims to provide the amounts of soil nutrients, especially nitrogen, needed for crop production, but through biological fixation and controlled mineralisation not fertiliser additions. Using biological processes, however, to increase the rate at which nitrogen become available and matching this to the needs of the crop has to date has been achieved less well. Similarly, rotations have deficiencies in the key aims of influencing the availability and modifying the speciation of nutrients such as phosphorus. If crop rotations are to be of greater importance in agriculture then there is a need for them to be modified so as to influence by design a wider range of aims.

1.8.2 AMF and the Redesigned Rotation

In this, there is a clear role for AMF which influence many of these processes. A comparison of how some of the key attributes of farming systems are achieved in organic farming and intensive arable systems highlights some of the future targets, to which AMF must contribute, if the organic approach is able to increase its output of food and its market share.

Improving the supply of phosphate to plants in P-limited systems is among the best documented roles of AMF (see Chapter 7 by Grace et al.). AMF increase a crop's interface with the soil and consequently its ability to influence the rate of release of phosphorus from poorly soluble sources. This will remain an important attribute in low input agriculture, but there is scope through the selection of more appropriate "crops" to develop the design of the rotation so that its ability to supply nutrients, other than nitrogen, is enhanced. However, the role of AMF extends beyond the ability to increase the supply of phosphorus. The AMF symbiotic relationship is normal for most plants. It has evolved over 500 million years and so it is to be expected that plants without AMF will carry out a range of functions less well. This includes the ability to sense the soil environment and as a result to regulate the loss of water especially under conditions when water supplies are limited (Atkinson et al. 2005), the ability to withstand the effects of infection by a number of pathogenic fungi and, in some circumstances, the ability to prevent infection from occurring by stimulating the production of secondary metabolites (Pozo et al 2002; see also Chapter 9 by Pozo et al.). In addition to these direct effects, AMF influence soil structure making it a better environment for plant growth (Atkinson 2006b). The continuing rise in the demand for "organic" foods suggests that AMF might have a bigger role in agriculture than in the past.

1.8.3 The Need for Inoculation

Many of the soils which will be used for organic, and other low input systems of agriculture, will in the recent past have been used for conventional agriculture; this is an inevitable consequence of the increase in organic production. As many of these soils will have depleted populations of AMF, their mycorrhizal status will require to be increased by appropriate management and by inoculation of AMF (Gianinazzi and Gollotte 2006) so that the AMF contribution is enhanced.

1.9 Conclusions

The next decade seems likely to see larger changes in agriculture than at any time since the prevailing model was established around 50 years ago. Changes will result from a reduction in the funding which farming has received from Governments, and from the need to achieve environmental and social goals as a condition of state support. There will be a continuing rise in demand for products produced without the use of pesticides and for food with zero-pesticide content. There will be greater public interest in how food is produced and, as a result, increased emphasis on public acceptance of new technologies and of how food is produced. If research on AMF demonstrates an ability to aid production with lesser use of energy-demanding external resources then they should become a significant element in eco-efficient farming systems.

References

Anon (1995) Report of the committee to consider the ethical implications of emerging technologies in the breeding of farm animals. HMSO, London

Anon (2002) Farming and food: A sustainable future. The report of the policy commission on the future of farming and food. http://www.cabinet-office.gov.uk/farming

Anon (2006) Mycorrhiza for science and society. Abstracts of Papers Presented at 5th International Conference on Mycorrhiza, July 2006 Estacion Experimental del Zaidin, CSIC, Granada, Spain

Atkinson D (1990) Land: Agricultural resource or wildlife reserve? Reorganisation in the food factory. Aberdeen Univ Rev 183, 218–225

Atkinson D (2006a) Why this needs to be the era of soil biology. Proceedings BCPC International Conference Crop Science and Technology, pp 51–52

Atkinson D (2006b) Arbuscular mycorrhizal fungi and the form and functioning of the root system. In: Mukerji KG, Manoharachary C, Singe J (eds) Microbial activity and the rhizosphere. Springer, Berlin, pp 199–222

Atkinson D (2007) This house believes that Rachel Carson would not today have written Silent Spring: An overview of the issues for debate. Proceeding International Crop Protection Conference, pp 208–209

Atkinson D, Watson CA (2000) The research needs of organic farming: distinct or just the same as other agricultural research? Proceeding BCPC Conference Pest and Diseases, pp 151–159

Atkinson D, Chalmers N, Cooper Crane D, Corcoran K, Crabtree R, Dent B, Watson CA (1994) The sustainability of lowland management systems. In: Sustainable systems of land and water use in Scotland: Four case studies. Scottish Office Agriculture and Fisheries Department, Edinburgh, pp 9–17

Atkinson D, Baddeley JA, Goicoechea N, Green J, Sanchez-Diaz M, Watson CA (2002) Arbuscular mycorrhizal fungi in low input agriculture. In: Gianinazzi S, Schueppe H, Barea JM, Haselwandter K (eds) Mycorrhizal Technology in Agriculture. Birkhauser, Basel, pp 211–222

Atkinson D, Litterick AM, Walker KC, Walker R, Watson CA (2004) Crop protection-what will shape the future picture? Pest Manag Sci 60, 105–112

Atkinson D, Black KE, Dawson LA, Dunsiger Z, Watson CA, Wilson SA (2005) Prospects advantages and limitations of future crop production systems dependant on the management of soil processes. Ann Appl Biol 146, 203–215

Baillier SR, Gregory RD, Siriwardena GM (1997) Farmland bird declines: patterns, processes and prospects. In: Biodiversity and conservation in agriculture. BCPC Symposium 69, pp 65–87

Baker BP, Benbrook CM, Groth E, Benbrook KL (2002) Pesticide residues in conventional, integrated pest management (IPM) grown and organic foods: insights from three US data sets. Food Addit Contam 19, 427–446

Bax W (2002) Consumer attitudes with respect to crop protection. In: The global challenge: Sustainable food production. BCPC Symposium 79, pp 9–14

Bruce D and Bruce A (1998) Engineering Genesis. Earthscan, London

Buckwell A (1998) The future of the common agricultural policy and the implications for crop protection. In: The Bawden Memorial Lectures 1973–1998. BCPC, Farnham, pp 293–308

Carson R (1963) Silent Spring. Hamish Hamilton, London

Conford P (2001) The Origins of the Organic Movement. Floris Books, Edinburgh

DEFRA (2006) Food security: An evidence and analysis paper. http://statistics.defra.gov.uk/esg/reports/foodsecurity/default.asp

Fischer Boel M (2007) A vision for EU agriculture and opportunities for UK farmers. Presentation to the 2007 Oxford Farming Conference http://europa.eu/rapid/pressReleasesAction.do?

Gianinazzi S, Gollotte A (2006) Why and where mycorrhizas should become a significant feature in agriculture. Proceedings BCPC International Conference Crop Science and Technology, pp 53–54

Hampson S (2006) Differentiation: A sustainable future for UK agriculture. Royal Agricultural Society of England, Stoneleigh
Harrier L, Watson CA (2004) The potential role of arbuscular mycorrhizal (AM) fungi in the bioprotection of plants against soil borne pathogens in organic and/or other sustainable farming systems. Pest Manag Sci 60, 149–157
Kealey T (1996) The economic laws of scientific research. Macmillan Press, Basingstoke, UK
Marsh J (2003) The impact of GM on the food and farm sector. In: GM Crops, Ford BJ (ed) Rothay House, Cambridge, pp 29–41
Marsh J (2005) A reformed CAP and its impact on agriculture in an enlarged community. BCPC International Conference Crop Science and Technology, pp 3–16
Martin JS (1998) The use of chemicals in modern farming- a farmers view In: The Bawden Memorial Lectures 1973–1998. BCPC, Farnham, pp 105–114
Montague D (2000) Farming Food and Politics. IAWS, Dublin
NFU Scotland (2006) Farming Facts, http://www.nfus.org.uk
Nortcliff S (2006) Review of management strategies to protect soil and maintain its function for agricultural production. Proceedings BCPC International Conference Crop Science and Technology, pp 32–34
North J (1987) Land use for food production 2015, J RASE 148, 16–20
Plum H (1998) The future of agriculture in the EEC. In: The Bawden Memorial Lectures 1973–1998. BCPC, Farnham, pp 1–8
Pozo MJ, Slezack-Deschaumes S, Dumas-Gaudot E, Gianinazzi S, Azcon-Aguilar C (2002) Plant defence responses induced by arbuscular mycorrhizal fungi. In: Gianinazzi S, Schuepp H, Barea JM, Haselwandter K (eds) Mycorrhizal technology in agriculture. Birkhäuser Verlag, Basel, pp 103–112
Pretty JN, Brett C, Gee D, Hine RE, Mason CF, Morrison JIL, Raven H, Raymond MD, Van der Bijl (2000) An assessment of the total external costs of UK agriculture. Agric Syst 65, 113–136
Schoener TW (1989) The ecological niche. In: Cherrett JM (ed) Ecological Concepts. Blackwell, Oxford, pp 79–114
Walker D (1989) Diversity and stability. In:Cherrett JM (ed) Ecological Concepts. Blackwell, Oxford, pp 115–146
Watson CA, Atkinson D, Gosling P, Jackson LR, Rayns FW (2002) Managing soil fertility in organic farming systems. Soil Use Manag 18, 239–247

Chapter 2
Interspecies Altruism in Plant–Microbe Symbioses: Use of Group Selection Models to Resolve the Evolutionary Paradoxes

Nikolai A. Provorov and Nikolai I. Vorobyov

Abstract The mutualistic plant–microbe symbioses are represented as the products of interspecies (reciprocal) altruism which resemble to some extent pathogenic (antagonistic) interactions at the molecular and cytological levels. The evolutionary interconversions (direct filiations) between these types of plant–microbe relationships are evident from the side of the plant hosts which possess gene networks for regulating beneficial microbes and for defending against pathogenic ones. From the microbial side, the direct filiations between mutualism and antagonism are restricted to defensive interactions, while for nutritional interactions (mycorrhiza, N_2-fixing symbioses) the evolutionary relatedness of microsymbionts to saprophytic microbes is more evident than to pathogenic ones. We suggest that evolutionary divergence of beneficial and deleterious plant symbionts may result from specific selective pressures (different forms of group selection) maintaining genes for beneficial traits in microbial populations. Using the legume–rhizobia symbiosis as a model, we argue that this selection may be based on (1) positive feedbacks operating between partners at the levels of symbiosis metabolism (exchange with C and N compounds) and of partners' co-evolution (synergistic selective pressures supporting mutualism in plants and in microbes); and (2) clonal structures of the *in planta* microbial populations resulting in kin and interdeme selection in favor of symbiotically efficient strains. Some of the revealed rules may be valid for a broad spectrum of mutualistic symbioses including mycorrhiza.

2.1 Introduction

The ability to integrate into symbiotic systems represents a fundamental property of organisms expressed at all levels of biological organization. According to the concept of symbiosis as "living together of differently named organisms" (de Bary 1879), the

N.A. Provorov (✉) and N.I. Vorobyov
All-Russia Research Institute for Agricultural Microbiology, Podbelsky Sh. 3, St.-Petersburg, Pushkin-8, 196608, Russia
e-mail: provorov@newmail.ru

C. Azcón-Aguilar et al. (eds.), *Mycorrhizas - Functional Processes and Ecological Impact*,
DOI: 10/1007-978-3-540-87978-7_2, © Springer-Verlag Berlin Heidelberg 2009

symbioses may be classified into mutualistic or antagonistic (parasitic) types: the first one is beneficial for both partners while the second one is beneficial for one partner (usually for the microsymbiont) but deleterious for the other (usually for the host plant). These types of symbioses were for a long time studied separately or were even opposed one against the other. However, it is now clear that plant–microbe mutualistic and pathogenic interactions are based on some common molecular and cellular mechanisms. For example, arbuscules developed by glomeromycotan fungi inside mycorrhizal root cells are structurally and functionally close to haustoria formed by biotrophic fungal pathogens (Hahn and Mendgen 2001). In plant-associated bacteria, there are many common infection strategies (Soto et al. 2006), for example, Type Three Secretion Systems which transfer microbial effecter proteins into host cells to reorganize their metabolism for symbiotic purposes (Buttner and Bonas 2003; Abe et al. 2005).

For a long time, evolutionary research was focused mainly on antagonistic symbioses, the dynamics of which may be described using Darwinian selection models. To stress the enigmatic nature of mutualism evolution, Maynard Smith (1989) posed the question: "Why should a symbiont benefit its host if it gains an immediate advantage by injuring it?" In this chapter, we shall address this question from the viewpoint of biological altruism, the rules for which base our arguments mainly on the N_2-fixing symbiosis formed by leguminous plants (Fabaceae) and nodule bacteria (*Rhizobium, Bradyrhizobium, Sinorhizobium, Mesorhizobium, Azorhizobium*; collectively called "rhizobia") which are studied comprehensively at the molecular and population levels (Spaink et al. 1998; Tikhonovich and Provorov 2007). However, at least some of the revealed rules may be valid for a broad spectrum of mutualistic symbioses including mycorrhiza.

2.2 Mutualism and Parasitism: Evolutionary Continuum or Discontinuity?

The structural/functional similarities between mutualistic and pathogenic plant–microbe symbioses were discussed over a long period in terms of fluent interconversions (direct filiations) based on the changes of nutritional types (saprotrophy→ necrotrophy→biotrophy) in microbial partners (Lewis 1974). It was often considered that parasitism universally evolves into mutualism since "any genetic changes which increase the benefits or diminish the harm to either associated species will be adaptively advantageous and hence will be encouraged by natural selection" (Dobzhansky 1951). However, recent molecular phylogenetic data suggest that these ideas are often not valid from the microbial side. For example, arbuscular mycorrhizal (AM) fungi, the most widespread beneficial plant symbionts, have no known relatives among plant pathogens (Schußler 2002). The phylogenetic analysis of the Homobasidiomycetes suggests that ectomycorrhizal fungi are readily interconverted with saprotrophs not with pathogens (Hibbett et al. 2000). The same is true for plant–bacteria interactions: in spite of close taxonomic relatedness between rhizobia and agrobacteria, the genetic systems involved in nodulation (N_2-fixing symbioses) or oncogenic transformation are completely different (Spaink et al. 1998), suggesting an evolutionary separation of these interactions.

Direct filiations between mutualistic and pathogenic forms are evident for those microbes forming defensive symbioses with plants, based on the synthesis of toxins against herbivores or pathogens, as in the case of Clavicipitales ergot fungi (Schardl et al. 1997), *Clavibacter* endophytes (Metzler et al. 1997) or rhizospheric *Pseudomonas corrugata* strains (Catara 2007). For example, some ergot fungi (*Neotyphodium*, *Cordyceps*) do not exhibit pathogenic symptoms in hosts (mutualism) while plant interactions with *Claviceps*, *Epichloë* or *Balansia* combine finely balanced beneficial and deleterious effects (White et al. 2002). However, Clavicipitales represent a specific group of Ascomycota with restricted host ranges (Gramineae and some Juncaceae). Interestingly, different co-evolutionary relationships with hosts may be observed in these defensive microsymbionts; they are evident from a good congruence of partners' phylogenies for the mutualistic (vertically transmitted) *Neotyphodium* strains but poor congruence in the pathogenic (horizontally transmitted) *Epichloë* strains (Schardl et al. 1997; Gundel et al. 2008).

In spite of evolutionary discontinuity between mutualism and antagonism observed on the microbial side in nutritional plant–microbe interactions, similar reactions from the host side have been revealed after penetration of roots by symbionts of these two types. The defence-like responses within the root epidermis and cortex during mycorrhiza development include cell wall modifications, phytoalexin synthesis, and accumulation of callose and of some pathogen-regulated proteins, including peroxidases and lytic enzymes (Gianinazzi-Pearson et al. 1996; Kapulnik et al. 1996; Bidartondo 2005; Garmendia et al. 2006; see also Chapter 9 by Pozo et al.). Therefore, plants possess the universal systems for hosting/regulation of different microbial invasions, which may be beneficial or deleterious depending on the partners' genotypes and on environmental conditions. Universality of these regulatory systems is illustrated at the level of legume genes controlling both AM establishment and rhizobia-induced nodulation (Parniske 2004). Since AM fungi appeared together with the first land plants (400–500 million years ago) while nodulation appeared much later (60–80 million years ago), together with the Rosid 1 dictos (La Rue and Weeden 1994; Kistner and Parniske 2002; Provorov et al. 2002), it is logical to suppose that many plant genes involved in legume nodulation have been recruited from the more ancient AM system.

All these data indicate that the evolutionary continuum proposed for different types of plant–microbe symbioses is valid from the host side but not from the microbial side. The phylogenetic separation of mutualistic and antagonistic microbes suggests that some important mechanisms involved in microbial evolution should be different. It would be difficult to reveal the nature of this difference by comparing the symbiotic genes in such taxonomically diverse groups as mutualistic and pathogenic microbes. For a more valid comparison, we shall consider nodule bacteria which combine mutualistic and pathogenic effects in their interactions with the same legume hosts.

2.3 Legume–Rhizobia Symbioses: Mutualism and Antagonism Together

The legume–rhizobia symbioses develop and function through tight co-operation between the partners' gene systems, based on their cross-regulation by specific molecular signals. The mutualistic nature of interactions is based on activation of bacterial *nif* genes which encode for the synthesis of the nitrogenase enzyme, the product of which, NH_4^+, is donated to the plant in exchange for products of photosynthesis (i.e., dicarboxylates). However, these mutually beneficial interactions are accompanied by complex regulatory processes which, as in mycorrhiza, involve multiple components common to parasitism (Fig. 2.1), and expressed at the early symbiotic stages when the genes for partners' recognition and signaling operate to elicit nodule morphogenesis. The inducers of this development, Nod factors, are encoded by rhizobial *nod/nol/noe* genes activated by plant flavonoids, which also enhance AM fungal growth and are compounds similar to phytoalexins defending plants from pathogens. Nod factors themselves are modified chitin oligomers, the backbone of which is close in structure to elicitors of defense reactions produced by phytopathogens (Ovtsyna and Staehelin 2005). The subsequent cellular interactions involve many defense-related factors including phenolics, reactive oxygen species, extensions, and lytic enzymes so that nodule development somehow resembles a restricted defense response induced by incompatible pathogens (Gamas et al. 1998; Santos et al. 2001; Morris and Walker 2003). It may be converted into a full-scale defense responses if the bacteria are arrested in synthesis either of surface components (exo- or lipo-polysaccharides), responsible for a dialogue with host defense systems (Niehaus et al. 1993; Kannenberg et al. 1998) or of Nod factors (Martínez-Abarca et al. 1998; Shaw and Long 2003).

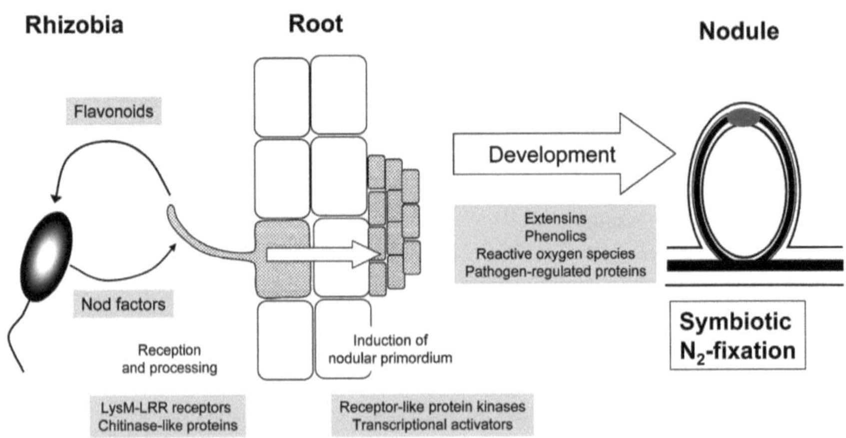

Fig. 2.1 Regulation of legume nodulation by the components common to pathogenic interactions (in *grey boxes*; comments are in Section 2.3)

However, similarities to pathogenic processes are negligible for the late stages of the legume–rhizobia symbiosis. At these stages, partners develop an integrated system of C/N metabolism through which bilateral nutrient exchange is implemented (Fig. 2.2), while for pathogenic systems nutrient flow is usually one-way, from hosts to pathogens (Lewis 1974). Moreover, in the systemic regulation of nodulation, CLAVATA-like receptors and shoot-root signals for sensing whole plant energy and N status are involved (Gresshoff et al. 2004), which are not part of systemic defense processes against pathogens.

Given the pronounced molecular differences between the early and the late symbiotic stages, it is not surprising that the corresponding bacterial genes have different natural histories. The polymorphism of *nod/nol/noe* genes does not correlate with the divergence of the core rhizobial genome (as revealed from 16S rDNA analysis) while it does correlate with the divergence of hosts (Dobert et al. 1994) in interactions where the rhizobia follow the "gene-for-gene" strategy (Tikhonovich et al. 2000). The same applies to genes involved in plant infection by bacterial (Guttman and Sarkar 2004) and fungal (Vanderplank 1982) pathogens. In contrast, the *nif* gene polymorphism

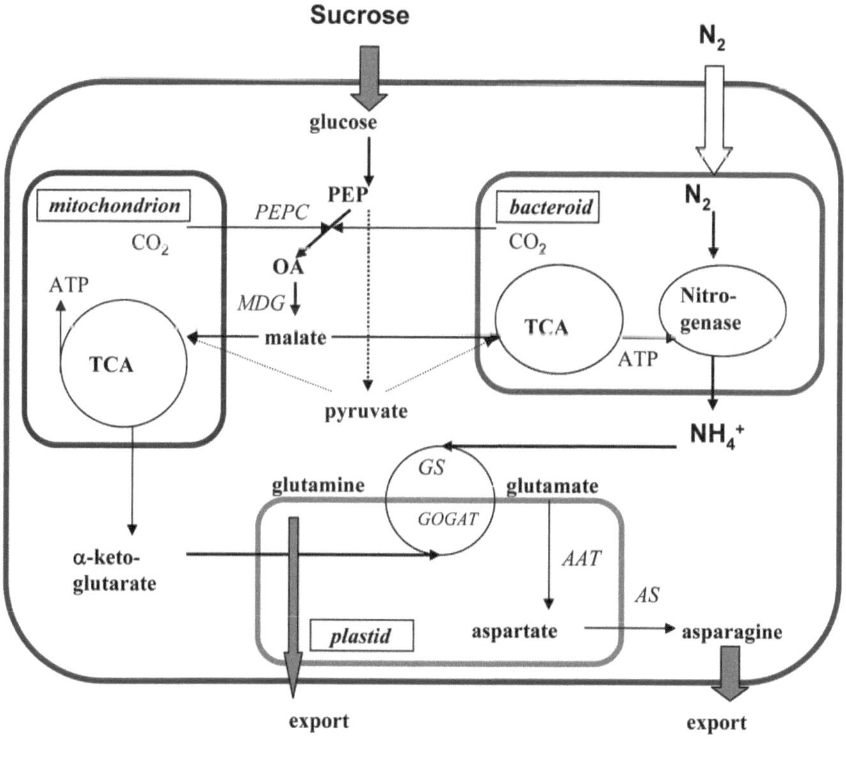

Fig. 2.2 Co-operation of partners at the late stages of legume-rhizobia symbiosis: biochemical basis for the interspecies altruism. *TCA* tricarboxylic acid cycle, *PEPC* phosphoenol pyruvate carboxylase, *MDG* malate dehydrogenase, *GS* glutamine synthetase, *GOGAT* NADP·H-dependent glutamate synthase, *AS* asparagine synthetase, *AAT* aspartate amino transferase, PEP phosphoenol pyruvate, OA oxaloacetate

reflects the divergence of the core bacterial genome, not of the host genome (Dobert et al. 1994). These data suggest that the evolution of pathogenic-like (early) and pathogenic-unlike (late) genes in rhizobia is driven by different mechanisms.

What may be the sources for the phylogenetic discordance between early and late symbiotic genes in microbes? It does not appear logical to suggest some principally different molecular mechanisms for evolution of genes occurring within the same rhizobial genomes. Both groups of genes have been recruited from non-symbiotic systems via the gene duplication–divergence strategy elicited by an increased genomic plasticity (Provorov 1998; Broughton and Perret 1999). It is possible that difference in the mechanisms of evolution for the early and late symbiotic genes is expressed mainly at the population/ecological levels.

A key to understanding the phylogenetic peculiarities of genes involved in mutualism may be found using data from the bacterial side demonstrating that the late symbiotic interactions involve elements of biological altruism. For example, in nearly all rhizobial species, *nif* genes are specifically activated in nodules where combined N is donated to the host (Kaminski et al. 1998; Fig. 2.2). This symbiotic N_2-fixation is associated with the conversion of bacteria into bacteroids (Fig. 2.3)

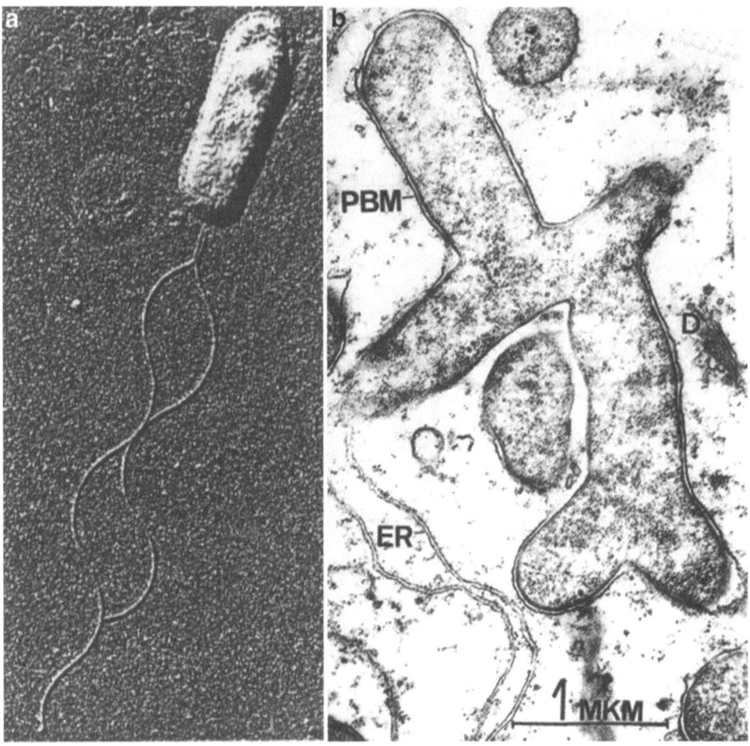

Fig. 2.3 Free-living rhizobia cells (**a**) and bacteroids (**b**): symbiotic differentiation providing a cytological basis for interspecies altruism (**a** is from Jordan (1984), with permission; **b** is provided by V. Lebsky). *PBM* peribacteroid (symbiosome) membrane, *ER* endoplasmic reticulum, *D* dictyosomes

hosted inside a plant intracellular compartment, the symbiosome. In some rhizobial species (*Sinorhizobium meliloti, Rhizobium leguminosarum, R. galegae*) interacting with the evolutionarily advanced "galegoid" legumes (*Galega, Medicago, Trifolium, Vicia*), these bacteroids are non-reproducible and degenerate after a period of functioning (Brewin 1998): the bacteria appear to display a genetically controlled ability to sacrifice themselves to provide the host with nitrogen! However, not all bacteria within a nodule are converted into bacteroids, some microbial cells retain their reproductive activity and after nodule death enter the soil population to enrich it with symbiotically active bacteria (Timmers et al. 2000). These cells do not fix N_2 while they consume host metabolites behaving as the plant parasites or symbiotic cheaters (Kiers et al. 2003). Interestingly, active strains capable of N_2-fixation are often poor soil survivors which may be saved from elimination due to a frequency-dependent selection based on competition for host nodulation (Provorov and Vorobyov 2006). Moreover, it seems possible that host plants can support mutualism by selecting the most effective rhizobia strains from soil and the rhizosphere (Jimenez and Casadesus 1989; Kiers et al. 2003).

May such a strategy of interaction have originated via natural selection operating in plant-associated microbial populations? For a long time this question was enigmatic for those evolutionists who followed Darwin's idea that natural selection cannot support an organism's trait the primary effect of which is to benefit another organism (Person et al. 1962). However, this restriction is valid only whilst natural selection is reduced to its most simple (Darwinian) form aimed at improving fitness in individual organisms. Although mechanisms for this improvement may be diverse (nutrition, consumption of additional energy, adequate behavior under a changing environment), the universal form of benefit was considered as reproductive success of the individual. Clearly, this selective strategy is not compatible with the evolution of mutualism if it is correlated to an altruistically reduced reproduction in microbial cells, as in the case of rhizobia. Many years ago, however, evolutionary ideas were coined which could explain selective mechanisms responsible for symbiotic mutualism.

2.4 Symbiosis and Biological Altruism

Haldane (1932) was among the first who attempted to enlighten on the evolutionary background of altruistic interactions. He considered biological altruism as an intraspecies interaction which improves fitness in one individual at the expense of the other. These interactions were illustrated by parental care which is accompanied by a decrease in parents' fitness while their progenitors save "altruistic genes" for the next generations. Using these instructive ideas, Maynard Smith (1964) and Hamilton (1964) suggested a universal mathematical description of relevant evolutionary processes, speculating that altruism should be supported in populations if the net benefit (*b*) obtained by recipients of altruism, the cost (c) paid by its donors, and genetic relatedness between the donors and recipients (k) satisfy the formulation:

$$k \cdot b - c > 0 \tag{2.1}$$

This simple formula assumes that altruism may be expressed in a group of organisms wherein the total balance of fitness is improved due to its decrease in some individuals (including their programmed death). The involved mechanism was named "kin selection" to stress that it operates within close relatives representing a restricted gene pool of the same family (deme). Developing ideas about how natural selection may operate at the group level, some authors speculated that a possible intergroup (interdeme) selection results in evolution of the altruistic traits (Williams and Williams 1957; Wynne-Edwards 1963). However, these speculations were heavily criticized since under real ecological conditions competition for survival occurs mainly among individuals, not among groups (Mayr 1970; Timofeeff-Ressovsky et al. 1977). Discussions led to a compromise which assumes that even if interdeme selection is involved in adaptive evolution, it is less effective than Darwinian selection due to: (1) limited number of demes; (2) less pronounced interdeme variation (with respect to individual variation); and (3) absence of solid interdeme barriers leading to migration of individuals between demes and restricting interdeme competition (Wade 1977; Wilson 1977).

More problematic was the possibility for interspecies altruism: if non-related organisms follow a mutually beneficial strategy in their symbiotic interactions, which will save the "altruistic genes" after death of their owners (donors of altruism)? In order to extend the notion of altruism to mutualistic symbioses, one can suggest that they represent a type of interaction where the benefit from co-existence for both partners overcomes its cost (Frank 1994; Herre et al. 1999). In a simplified form, this may be expressed as a system of two inequalities:

$$\begin{cases} r \cdot b_1 - c_1 > 0 \\ r \cdot b_2 - c_2 > 0 \end{cases}, \tag{2.2}$$

where $r > 0$ is the coefficient of correlation between the values of benefit obtained by partners. In an ecological meaning, r is close to k from formula (1): both coefficients characterize the maintenance of altruistic traits in the super-organism systems, the stability of which is due to tight and specific interactions of their members. Therefore, mutualistic symbioses like mycorrhiza or nodulation may be represented as a result of reciprocal altruism, which is possible if both partners possess genes with primary effects to improve fitness not of their owners, but of the symbiotic co-inhabitants. In order to pinpoint potential mechanisms responsible for evolution of such genes, it is necessary to identify selective pressures which support interspecies altruism.

2.5 Natural Selection for Interspecies Altruism

Before considering the evolution of mutualistic symbioses, it is useful to recall mechanisms of evolution in pathogenesis. These are based on "negative genetic feedbacks" between the interacting organisms (Pimentel 1968) and assume that those genetic changes improving fitness in one partner usually lead to decreased fitness in the other; increased virulence is beneficial to the pathogen, but deleterious to the host and vice versa. In this case, the co-evolutionary process is based on the interplay of Darwinian selection pressures in both partners creating the "gene-for-gene" systems where, for example, acquisition of new virulence gene(s) in a pathogen leads to decreased host fitness and elicits evolution towards new complementary resistance(s), which in turn stimulate the pathogen to acquire novel virulence, etc. (Frank 1992; Mitchell-Olds and Bergelson 2000).

Models applicable to host–pathogen co-evolution are generally not valid for mutualistic symbioses like mycorrhiza or nodulation where selective processes should support a "positive genetic feedback" because increased fitness in either partner leads to its coordinated increase in the other. Since mutualistic symbiosis may be considered as a product of interspecies altruism, it is logical to suggest its evolution via group selection which supports intraspecies altruism. This appears likely because rhizobial bacteria participate in symbioses as socially organized populations, the structure of which undergoes regular rearrangements during circulation in the "host–environment" system. Analyses of these rearrangements suggest that the intergroup (interdeme) selection may be induced in rhizobia due to preferential propagation of N_2-fixing strains in nodules based on the partners' positive feedbacks (Fig. 2.4a). These feedbacks may be based on allocation of the plant photosynthates specifically into N_2-fixing nodules where excess carbon is partly used for propagation of bacteria to be released into soil after nodule death. Moreover, impairment of N_2 fixation in some nodules may result in "sanctions" from the legume host which restricts nutrition of the bacteria and can suppress them using defense-like reactions (Kiers et al. 2003).

Therefore, rhizobia population dynamics is directed by interdeme selection which results from symbiosis development and is based on competition between nodular clones for host nutrients. In supporting mutualism, interdeme selection from the rhizobial side may be combined with individual selection from the host side. One can suggest that the positive feedback linking partners in mutualistic symbioses is expressed at two levels: in symbiosis metabolism as the exchange of nutrients (Figs. 2.2 and 2.4a), and in co-evolving partner populations as synergistic selective pressures supporting the plant–microbe mutualism (Fig. 2.4b). An additional strategy of natural selection towards increased symbiotic efficiency may be related to differentiation of microsymbiont clones into two subtypes: those which benefit the host through decreased fitness (altruists), and those which consume resources provided by host (egoists) (Fig. 2.5a). Such a separation takes place, for example, after transformation of rhizobia into non-reproducible bacteroids (typical for evolutionarily advanced symbioses formed by "galegoid" legumes) and leads to kin selection represented by inequality (1).

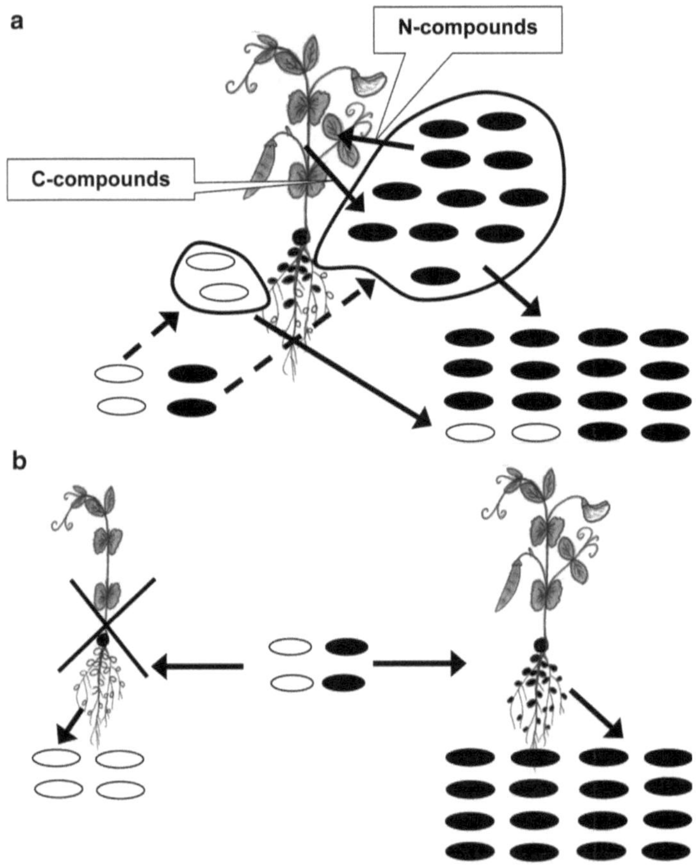

Fig. 2.4 Interdeme selection in the rhizobia population which occurs due to their positive feedback with hosts at the level of the exchange of C and N metabolites (**a**) or of the synergistic selective pressures supporting mutualistic interactions from the plant and bacterial sides (**b**). Nitrogen-fixing bacteria (and nodules infected by them) are represented in *black*, non-fixing bacteria in *white*

It is important to note that in sexually reproducing organisms (e.g., in animals which display parental care), kin selection operates under the restricted relatedness of donors and recipients of altruism (in Mendelian populations, $k \leq 0.5$). In rhizobia, however, kin selection efficiency may be markedly higher since each nodule usually harbors a clone originating from the single bacterial cell which initially infected the root ($k \rightarrow 1$). Surprisingly, "primitive" bacteria when entering into symbiosis with plants acquire more opportunities to implement altruism in their populations than the more "advanced" but free-living animals! In the case of the nodulation symbiosis, the symbiotic altruism may be highly effective because the legume host serves as an effective intermediate for transmitting the altruistic effects within the intranodule clone while the fixed nitrogen may be considered as a pay-off for this service.

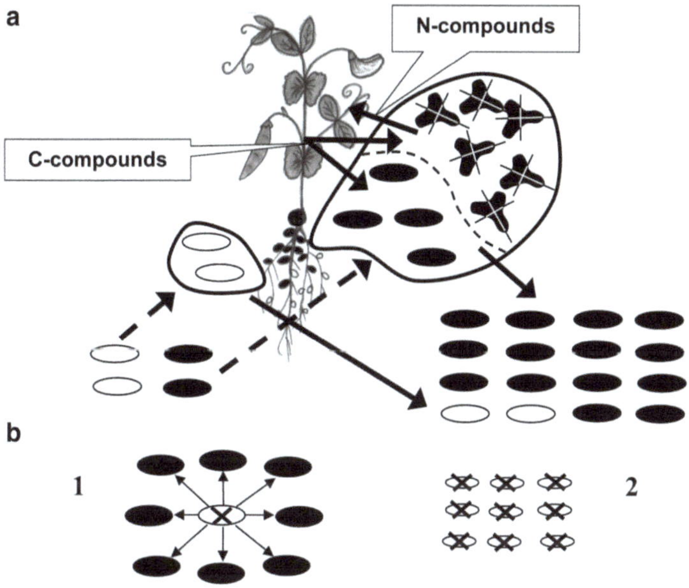

Fig. 2.5 Kin selection in microbial populations resulting from programmed cell death (donors of altruism are *crossed*): (**a**), in endosymbiotic rhizobia clones (non-reproducible bacteroids are represented by Y-like cells); (**b**), in free-living bacterial colonies under stress conditions (*1* the clone surviving due to altruistic interactions between the cellular individuals, *2* the clone eliminated in the absence of altruism; the relevant mechanisms are discussed in Shapiro 1998)

In bacteria, the kin selection is not restricted to symbiotic systems; this selection may also be related to the programmed cell death in the colonies of free-living cells (Fig. 2.5b). However, in mutualistic symbioses, a specific hierarchy of selective pressures is established: individual selection in the host leads to interdeme selection in the microbes (Fig. 2.4) which may be enhanced by the kin selection (Fig. 2.5a). The maximal ecological efficiency of symbiotic mutualism may be reached when all these pressures are operating synergistically. This is evident for the evolutionary advanced "galegoid" legumes where the most pronounced bacteroid differentiation occurs within nodules and where, for example, in *Galega orientalis* and *Medicago sativa* crops, up to 500–600 kg/ha/season of N_2 may be fixed (Provorov and Tikhonovich 2003).

2.6 Conclusions

Symbiotic adaptations are sometimes considered not to be compatible with the involvement of natural selection as a key mechanism for adaptive evolution (Person et al. 1962). In this chapter, we have tried to demonstrate that there is no need to reject the leading role of natural selection in the evolution of mutualism. However, the notion of biological altruism has to be extended from intraspecies interactions towards interspecies ones which assume the operation of group (interdeme, kin) forms of natural selection in the beneficial plant–microbe interactions. Because of the cellular and molecular convergence that can be found between mycorrhizal and rhizobial symbioses, we have taken the legume–rhizobia symbioses as the basis to discuss selective pressures that may be maintaining genes for beneficial traits in soil microbial populations. In rhizobia, for example, these pressures may control the evolution of *nif* genes creating discordance in their phylogenies with respect to *nod/nol/noe* genes (possibly evolving under impacts of individual selection).

The involvement of group-level selection in the evolution of mutualism may also be valid for mycorrhiza. The evolutionary isolation of the major groups of mycorrhizal fungi from plant pathogens (Hibbett et al. 2000; Schußler 2002) suggests the existence of specific (non-Darwinian) mechanisms for evolution of plant-beneficial traits in mycorrhizal fungi. These mechanisms may be based on defined elements of altruism in symbiotic interactions since the major part of nutrients absorbed by the mycobionts from soil is donated to the host. It has been reported that, upon plant inoculation by a mixture of AM fungi that are effective and ineffective with respect to P donation to host, a pronounced allocation of host carbohydrates to the effective strain occurs (Bever 2006). This allocation may provide the background for different forms of group-level selection in mycorrhiza (Kiers and van der Heijden 2006) as has been demonstrated in the legume–rhizobia system. The populations of nuclei harbored by AM fungi may be considered as the evolutionary analogs of internodule rhizobia clones which are the targets for group selection.

In order to progress in evolutionary research in plant–microbe symbioses, much more comparative data about molecular, physiological, morphological, and ecological mechanisms of the partners' interaction are required. A powerful tool for integration of these data into conceived evolutionary models is represented by the mathematical simulation techniques which are used successfully to analyze the natural histories in a wide range of symbiotic and pathogenic organisms (Fleming 1980; Jimenez and Casadesus 1989; Frank 1992; Olivieri and Frank 1994; Simms and Bever 1998; Bever and Simms 2000; Provorov and Vorobyov 2000, 2006).

Acknowledgements The authors are obliged to Sari Timonen, Isabelle Olivieri, and Peter Gresshoff for helpful comments on the manuscript. The presented research is supported by grants from Russian Foundation of Basic Research (06-04-48800, 06-04-89000NWO) and NWO Centre of Excellence (047.018.001). This paper is dedicated to the memory of John Maynard Smith (1920–2004).

References

Abe A, Matsuzawa T, Kuwae A (2005) Type-III effectors: sophisticated bacterial virulence factors. C R Biol 328:413–428

Bever JD (2006) Maintenance of the arbuscular mycorrhizal mutualism. In: Abstracts of 5th Conference on Mycorrhiza "Mycorrhiza for Science and Society". 23–27 July 2006, Granada, Spain, p 72

Bever JD, Simms EL (2000) Evolution of nitrogen fixation in spatially structured populations of *Rhizobium*. Heredity 85:366–372

Bidartondo MI (2005) The evolutionary ecology of myco-heterotrophy. New Phytol 167:335–352

Brewin NJ (1998) Tissue and cell invasion by *Rhizobium*: the structure and development of infection threads and symbiosomes. In: Spaink H, Kondorosi A, Hooykaas PJJ (eds) The Rhizobiaceae. Kluwer, Dordrecht/Boston/London, pp 417–429

Broughton WJ, Perret X (1999) Genealogy of legume-*Rhizobium* symbiosis. Curr Opin Plant Biol 2:305–311

Buttner D, Bonas U (2003) Common infection strategies of plant and animal pathogenic bacteria. Curr Opin Plant Biol 6:312–319

Catara V (2007) *Pseudomonas corrugata*: plant pathogen and/or biological resource? Mol Plant Path 8:233–244

de Bary A (1879) Die Erscheinung der Symbiose. Verlag Von Karl J. Trübner, Strasburg.

Dobert RC, Breil BT, Triplett EW (1994) DNA sequence of the common nodulation genes of *Bradyrhizobium elkanii* and their phylogenetic relationship to those of other nodulating bacteria. Mol Plant-Microbe Interact 7:564–572

Dobzhansky T (1951) Genetics and the origin of species. 3rd edition. Columbia University Press, New York.

Fleming RA (1980) Selection pressures and plant pathogens: robustness of the model. Phytopathology 70:175–180

Frank SA (1992) Models of plant-pathogen co-evolution. Trends Genet 8:213–219

Frank SA (1994) Genetics of mutualism: the evolution of altruism between species. Theor Biol 170:393–400

Gamas P, de Billy F, Truchet G (1998) Symbiosis-specific expression of two *Medicago truncatula* nodulin genes, *MtN1* and *MtN13* encoding products homologous to plant defense proteins. Mol Plant-Microbe Interact 11:393–403

Garmendia I, Aguirreolea J, Giocoechea N (2006) Defence-related enzymes in pepper roots during interactions with arbuscular mycorrhizal fungi and/or *Verticillium dahliae*. BioControl 51:293–310

Gianinazzi-Pearson V, Dumas-Gaudot E, Gollotte A, Tahiri-Alaoni A, Gianinazzi S (1996) Cellular and molecular defense-related root responses to invasion by arbuscular mycorrhizal fungi. New Phytol 133:45–57

Gresshoff PM, Buzas DM, Laniya T, Men A, Jiang O, Schenk PM, Hayward A, Kam J, Li D, Miyahara A, Nontachaiyapoom S, Indrasumunar A, Brcich T, Gualtieri G, Davis P, Carroll B (2004) Systemic regulation of nodulation by a leaf-controlled LRR-receptor kinase. In: Tikhonovich IA, Lugtenberg BJJ, Provorov NA (eds) Biology of plant-microbe interactions, vol 4. IS-MPMI, St. Petersburg, Russia, pp 369–372

Gundel PE, Batista WB, Texeira M, Martinez-Ghersa MA, Omacini M, Ghersa CM (2008) *Neotyphodium* endophyte infection frequency in annual grass populations: relative importance of mutualism and transmission efficiency. Proc Roy Soc B 275:897–905

Guttman DS, Sarkar SF (2004) Evolutionary and functional genomics of host specificity in *Pseudomonas syringae*. In: Tikhonovich IA, Lugtenberg BJJ, Provorov NA (eds) Biology of plant-microbe interactions, vol 4. IS-MPMI, St. Petersburg, Russia, pp 132–137

Hahn M, Mendgen K (2001) Signal and nutrient exchange in biotrophic plant-fungus interfaces. Curr Opin Plant Biol 4:322–327

Haldane JBS (1932) The causes of evolution. Longmans, Green & Co, New York

Hamilton WDJ (1964) The genetical evolution of social behavior. J Theor Biol 7:1–16

Herre EA, Knowlton N, Mueller UG, Rehner SA (1999) The evolution of mutualisms: exploring the paths between conflict and cooperation. Trends Ecol Evol 14:49–53

Hibbett DS, Gilbert LB, Donoghue MJ (2000) Evolutionary instability of ectomycorrhizal symbioses in basidiomycetes. Nature 407:506–508

Jimenez J, Casadesus J (1989) An altruistic model of *Rhizobium*-legume association. J Heredity 80:335–337

Jordan DC (1984) Rhizobiaceae. In: Kreig NR (ed) Bergey's Manual of Systematic Bacteriology, vol 1. Williams and Wilkins, Baltimore/London, pp 234–256
Kaminski PA, Batut J, Boistard P (1998) A survey of symbiotic nitrogen fixation by rhizobia. In: Spaink HP, Kondorosi A, Hooykaas PJJ (eds) The Rhizobiaceae. Molecular biology of model plant-associated bacteria. Kluwer, Dordrecht/Boston/London, pp 431–460
Kannenberg EL, Reuhs BL, Forsberg S, Carlson RW (1998) Lipopolysaccharides and K-antigens: Their structures, biosynthesis and functions. In: Spaink HP, Kondorosi A, Hooykaas PJJ (eds) The Rhizobiaceae. Molecular biology of model plant-associated bacteria. Kluwer, Dordrecht/ Boston/London, pp 119–154
Kapulnik Y, Volpin H, Itzhaki H, Ganon D, Galili S, David R, Shaul O, Elad Y, Chet I, Okon Y (1996) Suppression of defense responses in mycorrhizal alfalfa and tobacco roots. New Phytol 133:59–64
Kiers ET, van der Heijden MGA (2006) Mutualistic stability in the arbuscular mycorrhizal symbiosis: exploring hypotheses of evolutionary cooperation. Ecology 87:1627–1636
Kiers ET, Rousseau RA, West SA, Denison RF (2003) Host sanctions and the legume-*Rhizobium* mutualism. Nature 425:78–81
Kistner C, Parniske M (2002) Evolution of signal transduction in intracellular symbioses. Trends Plant Sci 7:511–518
La Rue TA, Weeden NF (1994) The symbiosis genes of the host. In: Kiss G, Endre G (eds) Proc. 1st Europ. Nitrog. Fixat. Conf. Officina Press, Szeged, Hungary, pp 147–151
Lewis DH (1974) Microorganisms and plants: the evolution of parasitism and mutualism. In: Evolution of Microbial World. Proc. 24th Symp. Soc. General Microbiol. University Press, Cambridge, pp 367–392
Martínez-Abarca F, Herrera-Cervera JA, Bueno P, Sanjuan J, Bisseling T, Olivares J (1998) Involvement of salicylic acid in the establishment of the *Rhizobium meliloti* - alfalfa symbiosis. Mol Plant Microbe Interact 11:153–155
Maynard Smith J (1964) Group selection and kin selection. Nature 201:1145–1147
Maynard Smith J (1989) Generating novelty by symbiosis. Nature 341:284–285
Mayr E (1970) Populations, species, and evolution. The Belknap Press of Harvard University Press, Cambridge, Massachusetts
Metzler MC, Laine KJ, De Boer SH (1997) The status of molecular biological research on the plant pathogenic genus *Clavibacter*. FEMS Microbiol Lett 150:1–8
Mitchell-Olds T, Bergelson J (2000) Biotic interactions: genomics and coevolution. Curr Opin Plant Biol 3:273–277
Morris ER, Walker JC (2003) Receptor-like protein kinases: the keys to response. Curr Opin Plant Biol 6:339–342
Niehaus K, Kapp D, Pühler A (1993) Plant defense and delayed infection of alfalfa pseudonodules induced by an exopolysaccharide (EPS-1)-deficient *Rhizobium meliloti* mutant. Planta 190:415–425
Olivieri I, Frank SA (1994) The evolution of nodulation in *Rhizobium*: altruism in the rhizosphere. J Heredity 85:46–47
Ovtsyna AO, Staehelin C (2005) Bacterial signals required for the *Rhizobium*-legume symbiosis. Recent Res Dev Microbiol 7:631–648
Parniske M (2004) Molecular genetics of the arbuscular mycorrhizal simbiosis. Curr Opin Plant Biol 7:414–421
Person C, Samborski DJ, Rohringer R (1962) The gene-for-gene concept. Nature 194:561–562
Pimentel D (1968) Population regulation and genetic feedback. Science 159:1432–1437
Provorov NA (1998) Coevolution of rhizobia with legumes: facts and hypotheses. Symbiosis 24:337–367
Provorov NA, Tikhonovich IA (2003) Genetic resources for improving nitrogen fixation in legume-rhizobia symbiosis. Genetic Res Crop Evol 50:89–99
Provorov NA, Vorobyov NI (2000) Population genetics of rhizobia: construction and analysis of an "infection and release" model. J Theor Biol 205:105–119

Provorov NA, Vorobyov NI (2006) Interplay of Darwinian and frequency-dependent selection in the host-associated microbial populations. Theor Popul Biol 70:262–272

Provorov NA, Borisov AY, Tikhonovich IA (2002) Developmental genetics and evolution of symbiotic structures in nitrogen-fixing nodules and arbuscular mycorrhiza. J Theor Biol 214:215–232

Santos R, Herouart D, Sigaud S, Touati D, Puppo A (2001) Oxidative burst in alfalfa-*Sinorhizobium meliloti* symbiotic interaction. Mol Plant Microbe Interact 14:86–89

Schardl CL, Leuchtmann A, Chung KR, Penny D, Siegel MR (1997). Coevolution by common descent of fungal symbionts (*Epichloe* spp.) and grass hosts. Mol Biol Evol 14:133–143

Schußler A (2002) Molecular phylogeny, taxonomy and evolution of *Geosiphon pyriformis* and arbuscular mycorrhizal fungi. Plant Soil 244:75–83

Shapiro JA (1998) Thinking about bacteria as multicellular organisms. Ann Rev Microbiol 52:81–104

Shaw SL, Long S (2003) Nod factor inhibition of reactive oxygen efflux in a host legume. Plant Physiol 132:2196–2204

Simms EL, Bever JD (1998) Evolutionary dynamics of rhizopine within spatially structured *Rhizobium* populations. Proc R Soc Lond B 265:1713–1719

Soto MJ, Sanjuan J, Olivares J (2006) Rhizobia and plant pathogenic bacteria: Common infection weapons. Microbiology 152:3167–3174

Spaink HP, Kondorosi A, Hooykaas PJJ (eds) (1998) The Rhizobiaceae. Molecular Biology of Model Plant-Associated Bacteria. Kluwer, Dordrecht/Boston/London.

Tikhonovich IA, Provorov NA (2007) Beneficial plant-microbe interactions. In: Dyakov YT, Dzhavakhiya V, Korpela T (eds) Comprehensive and molecular phytopathology. Elsevier, Amsterdam, pp 365–420

Tikhonovich IA, Kozhemyakov AP, Ovtsyna AO, Provorov NA (2000) Construction of highly efficient symbiotic systems. In: Provorov NA, Tikhonovich IA, Veronesi F (eds) New Approaches and Techniques in Breeding Sustainable Fodder Crops and Amenity Grasses. Biont, St. Petersburg, pp 131–135

Timmers ACS, Soupene E, Auriac MC, de Billy F, Vasse J, Boistard P, Truchet G (2000) Saprophytic intracellular rhizobia in alfalfa nodules. Mol Plant Microbe Interact 13:1204–1213

Timofeeff-Ressovsky NW, Jablokov AV, Glotov NV (1977) Grundriss der Populationslehre. Gustav Fisher Verlag, Jena

Vanderplank JE (1982) Host-pathogen interaction in plant disease. Academic Press, New York/London

Wade MJ (1977) An experimental study of group selection. Evolution 31:134–153

White JF, Sullivan RF, Moy M, Meyer W, Cabral D (2002) Evolution of *Epichloë/Neotyphodium* endophytes and other Clavicipitalean biotrophs. In: Seckbach J (ed) Symbiosis: Mechanisms and Model Systems. Kluwer, Dordrecht/Boston/London, pp 413–424

Williams GC, Williams DC (1957) Natural selection of individually harmful social adaptations among sibs with special reference to social insects. Evolution 11:32–39

Wilson DS (1977) Structured demes and the evolution of group-advantageous traits. Am Nat 111:157–185

Wynne-Edwards VC (1963) Intergroup selection in the evolution of social systems. Nature 4907:623–626

Chapter 3
Dissection of Genetic Cell Programmes Driving Early Arbuscular Mycorrhiza Interactions

Vivienne Gianinazzi-Pearson, Marie Tollot, and Pascale M. A. Seddas

Abstract The persistence through evolution of the arbuscular mycorrhiza (AM) symbiosis between Glomeromycota and plants is probably due to a widespread molecular dialogue between the two partners. Most studies have focussed on established mycorrhizal systems whilst evidence for cellular commitment of the symbiotic partners during early developmental phases is recent. Whereas spore germination by AM fungi can occur spontaneously, subsequent hyphal branching, appressoria differentiation, root penetration and intraradical development leading to symbiosis establishment are under the control of molecular interactions between the two partners. In this chapter, recent work on AM fungus–plant interactions is presented in such a way as to highlight the subtlety of this ancestral molecular complicity between the two partners.

3.1 Introduction

Arbuscular mycorrhizas have existed since Ordovician times (Redecker et al. 2000), and the presence of this root symbiosis in most families of extant land plants can only be explained by the maintenance of compatibility systems as new species appeared during evolution. Interest in this ancient and widespread plant symbiosis has expanded exponentially over the last few years, both from a fundamental viewpoint of inter-organism interactions and for their potential exploitation in plant production systems. Identification of molecular codes driving the coordinated development of plant and fungal partners provides not only the basis for deciphering implied cell programmes but also a starting block for optimising manipulation and management of the symbiosis for sustainable crop production.

Establishment of the arbuscular mycorrhizal symbiosis follows a conserved sequence of developmental steps which are independent of the fungal and plant

V. Gianinazzi-Pearson, M. Tollot, and P.M.A. Seddas (✉)
UMR INRA 1088/CNRS 5184/Université de Bourgogne Plante-Microbe Environnement, BP 86510, 21065 Dijon Cedex, France
e-mail: pseddas@dijon.inra.fr

C. Azcón-Aguilar et al. (eds.), *Mycorrhizas - Functional Processes and Ecological Impact*,
DOI: 10/1007-978-3-540-87978-7_3,

species involved. In events prior to root penetration, hyphal branching from germinating spores and appressorium differentiation at the root surface are activated by host plants. Once this first interface of contact is established, plant accommodation of the fungal symbiont initiates and drives root penetration. Most investigations of cell programme modulation during arbuscular mycorrhiza interactions have documented the mycorrhizal transcriptome or proteome in the established symbiosis, providing evidence for the existence of multiple regulatory pathways (see Liu et al. 2007 for references). Data about cellular and molecular interactions before and during initial contact between symbionts are far more limited due to the obligate biotrophic lifestyle of arbuscular mycorrhizal (AM) fungi (Glomeromycota) and their asynchronous pattern of root colonisation (Gianinazzi-Pearson et al. 2007). In this chapter. we review recent evidence for cellular commitment of the symbiotic partners during these early developmental phases and discuss their relationships to compatibility events driving the mycorrhizal route.

3.2 Signal Perception and Recognition Events Prior to Cell Contact

Cell-to-cell interactions leading to the finely-tuned readjustments in plant and fungal tissues during mycorrhiza development must be the outcome of sophisticated signalling events between the symbiotic partners. The first events occur in the rhizosphere where plant exudates not only stimulate or inhibit microbial activities through the production of mucilage and metabolites, but also act as messengers which initiate molecular dialogues between roots and microorganisms. Although spores of the Glomeromycota can germinate asymbiotically, stimulation of hyphal growth and/or branching requires root exudates from host plants (Gianinazzi-Pearson et al. 1989; Giovannetti et al. 1996; Bécard et al. 2004). The inability of non-host plants to induce such fungal responses has been put down to their lack of stimulatory signal compounds (Gianinazzi-Pearson et al. 1989; Buée et al. 2000) or their release of inhibitory compounds (Nagahashi and Douds 2000; Gadkar et al. 2003). Enhanced hyphal branching by host root exudates is accompanied by increased fungal mitotic activity, respiration, biogenesis of mitochondria and effects on mitochondrial shape and orientation (Buée et al. 2000; Tamasloukht et al. 2003; Besserer et al. 2006). Little is known about the fungal genes responding to plant signals prior to root contact, although rapid activation of a number of genes associated with increased cell activity and with predicted functions, for example, in energy metabolism, transcription or protein synthesis has been detected in *Gigaspora margarita* or *Glomus intraradices* during presymbiotic responses to carrot or *Medicago truncatula* root exudates, respectively (Tamasloukht et al. 2003; P. Seddas et al., unpublished results). In the absence of root exudates, spore germination is followed by premature hyphal growth arrest and cytoplasm retraction long before spore reserves are exhausted. Requena et al. (2002) suggested that, in *Glomus mosseae*, a gene coding a putative hedgehog protein with GTPase activity

(*GmGIN1*) could be involved in programmed cell death of hyphae in the absence of a signal from the host plant. To date, there exists no evidence for the involvement of host root exudates in the regulation of genes involved in the obligate biotrophy of the Glomeromycota (Trépanier et al. 2005).

3.2.1 Fungal Perception of Plant Signals

Several classes of plant molecules have been proposed as potential signals involved in enhancing hyphal growth and/or branching in the Glomeromycota. For example, root exuded flavonoids derived from the phenylpropanoid pathway can favour spore germination, activate fungal transcription (Fig. 3.1) and enhance hyphal growth, branching and/or root colonisation by arbuscular mycorrhizal fungi. However, their effects are not uniform and can vary from positive to neutral or negative between the fungal species or genus, the type of molecule and/or its concentration (Vierheilig et al. 1998). This has led to the suggestion that a possible structure–activity relationship exists (Bécard et al. 1992; Scervino et al. 2006), and that the Glomeromycota may have genus- or species-specific flavonoid requirements during symbiotic interactions (Vierheilig et al. 1998; Scervino et al. 2005). Since root exudates from a maize mutant deficient for chalcone synthase in the

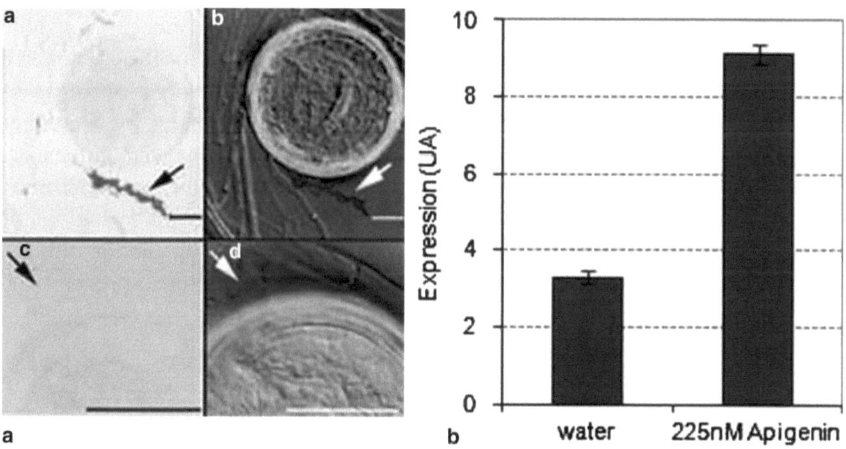

Fig. 3.1 Example of transcriptional activation, based on a stearoyl-CoA-desaturase gene, during spore germination of *Glomus intraradices*. (**a**) Localisation of gene transcripts (*arrows*) by direct fluorescent in situ RT-PCR, using specific 5' Texas Red labelling primers (Seddas et al. 2008), in hyphae growing from spores germinated in 225 nM apigenin (a, b) or water (c, d), at 25° C in 2.2% CO_2. (a, c): Nomarski image; (b, d): fluorescent confocal image. All samples were fixed with DMSO and permeabilised by a chitinase/pectinase treatment. *Bar* = 50 μm. (**b**) Quantification of transcript accumulation in germinating spores by qRT-PCR. PCR products were analysed using SDS 2.2 software. Gene expression data were plotted as $2^{(40-CT)}/10$ and normalised against the housekeeping *TEF* gene (Gianinazzi-Pearson et al. 2006)

flavonoid biosynthetic pathway still induce the hyphal branching response (Buée et al. 2000), it has been concluded that other signal molecules should be perceived by mycorrhizal fungi. Another group of plant signal molecules which has been proposed to be involved in this early fungal perception of host roots belongs to the strigolactone family, derivatives of the apocarotenoid biosynthetic pathway (Bouwmeester et al. 2007). Strigolactones have been shown to be among components of root exudates from *Sorghum bicolor* and *Lotus japonicus* which activate spore germination and/or hyphal branching of different species of Glomeromycota (Akiyama et al. 2005; Besserer et al. 2006). Synthetic strigolactones mimic this effect of root exudates as well as those on respiration rate and on mitochondria biogenesis, morphology, and motility in branching hyphae (Besserer et al. 2006). However, the central role of strigolactones in plant–fungal signalling events during presymbiotic mycorrhizal interactions has yet to be confirmed. Strigolactones are produced by the non-mycorrhizal plant *Arabidopsis thaliana* (Westwood 2000; Goldwasser and Yoder 2001), and plant mutants defective for strigolactone production have not yet been obtained. Their effects on non-target saprophytic or pathogenic fungi have been studied recently. No effect was observed for hyphal branching of soil filamentous fungi or for microconidia germination of *Fusarium oxysporum* (Steinkellner et al. 2007), but a synthetic strigolactone rapidly stimulated cell respiration of the biotrophic pathogens *Ustilago maydis* and *Sporisorium reilianum* at concentrations active on arbuscular mycorrhizal fungi (Sabbagh 2008). These observations suggest that strigolactones may not be specific signals for AM fungi but could have an incidence on other plant interacting fungi.

In summary, both flavonoids and strigolactones are widely occurring molecules in the plant kingdom and, to date, there is no evidence for the existence of a metabolite acting specifically as a universal signal for recognition of a host plant by arbuscular mycorrhizal fungi. Consequently, mechanisms by which host root exudates enhance hyphal growth or branching in these fungi remain obscure. Bécard et al. (2004) have proposed a more complex scenario in which there will be an ensemble of metabolic changes induced by root factors which will enable a symbiotic fungus to exploit its own growth potential and so ensure the developmental switch from the asymbiotic stage of spore germination to presymbiotic hyphal branching.

3.2.2 Plant Perception of Fungal Signals

Arbuscular mycorrhizal fungi also need to communicate to their host plants. Initial indications that fungal signals are perceived by a host plant before physical contact between the symbionts came from observations that root flavonoid contents in alfalfa increased in the presence of mycelium and spores of *G. intraradices* (Larose et al. 2002). More direct evidence has been provided in studies of plant gene responses to hyphae growing from germinated spores of *Gigaspora* or *Glomus* species in the vicinity of host roots but separated by a permeable membrane. Diffusible fungal molecules (so-called Myc factors) can induce transcription by *M. truncatula* roots of an *MtENOD11* gene (Kosuta et al. 2003), the expression of which is synchronous with the induction of hyphal branching and does not occur with dead spores or fungal pathogens. Other plant genes which respond to the diffusible Myc factors encode proteins related to calcium-dependent signal transduction pathways, such as calcium lipid-binding protein, casein kinase, 14.3.3 protein which can interact with calcium-dependent kinases, and annexin which belongs to a family of calcium-binding proteins (Weidmann et al. 2004). Monitoring of gene responses in *M. truncatula* plants carrying a single mutation for the gene *MtDMI3/MtSYM13*, which is necessary for root colonisation by symbiotic microorganisms (Catoira et al. 2000), has provided evidence that signal perception in presymbiotic mycorrhizal interactions is independent of the Nod factor transduction pathway leading to nodulation. The signal transduction pathway elicited by Myc factors is not activated by *Sinorhizobium meliloti* (Weidmann et al. 2004), and mycorrhiza-defective mutants of *M. truncatula* still respond to the fungal signals with induction of *MtENOD11*, but not to Nod factors (Kosuta et al. 2003). *MtDMI3/MtSYM13* codes for a nuclear localised calcium and calmodulin-binding protein kinase which is assumed to be involved in deciphering intracellular calcium oscillations (Oldroyd and Downie 2006). Calcium is an intracellular messenger coordinating responses to numerous developmental cues and environmental challenges, and diffusible molecules released by Glomeromycota spores can induce a transient and rapid calcium response in soybean cell cultures (Navazio et al. 2006). All these observations are concomitant with the role of a calcium-mediated signalling cascade pathway downstream of perception by host plant cells of molecules released by arbuscular mycorrhizal fungi.

3.3 Symbiotic Partnership Through Early Morphological Integration

Cell-to-cell contact between an arbuscular mycorrhizal fungus and a host root provokes morphological rearrangements in both partners, characterised by differentiation of hyphae into appressoria and biogenesis in epidermal cells of an intracellular compartment to accommodate the penetrating fungus. Studies of cell processes related to this developmental stage tend to confirm the implication of calcium in the early interactions between mycorrhizal symbionts.

3.3.1 Appressoria Stage of Interactions

On the fungal side, appressorium formation is a host-specific event which indicates recognition of the host root surface by the fungal hyphae (Giovannetti et al. 1994). There is a paucity of information concerning regulatory processes involved in this morphogenetic phenomenon in the Glomeromycota. Triggering of transcriptome modifications in germinated sporocarps of *G. mosseae* has been reported in synchrony with appressorium formation on a root of parsley (Breuninger and Requena 2004). A number of up-regulated fungal genes were identified which encode proteins with functions in signalling, transduction, cytoskeleton, general cell metabolism or defence/stress responses, of which several have a potential role in calcium-based signalling pathways. More recently, we have focussed on transcripts of a subset of genes, with predicted functions in transcription, protein synthesis, primary/secondary metabolism or of unknown function, to monitor *G. intraradices* gene expression during appressorium formation on host and non-host roots. Whilst most of the genes are up-regulated in compatible interactions with *M. truncatula*, several showed no significant activation or down-regulation during incompatible interactions with plants mutated for the symbiosis-related gene *MtDMI3/MtSYM13* (P. Seddas et al., unpublished results). Alkaline phosphatase activity in appressoria developing on roots of *dmi3/Mtsym13* mutants shows that these fungal structures remain physiologically active even though penetration of plant tissues does not occur (Fig. 3.2a). It would appear therefore that the lack of calcium and calmodulin-dependent protein kinase synthesis alters signalling and/or recognition processes downstream of appressorium formation so that fungal gene activation and root colonisation by the symbiont are impeded.

On the plant side, cell responses are activated by appressorium formation by an arbuscular mycorrhizal fungus and these can be detected at both molecular and structural levels. Expression profiling of defence or nodulation-related genes in *Pisum sativum* and *M. truncatula* (Albrecht et al. 1998; Ruiz-Lozano et al. 1999; Chabaud et al. 2002) gave first indications of transcriptional activation by fungal contact with root epidermal cells prior to hyphal penetration. In more recent investigations based on non-target transcriptome analyses, a number of plant genes have been identified that are up-regulated by appressoria and which reflect activation of signal perception and transduction pathways in *M. truncatula* roots (Weidmann et al.

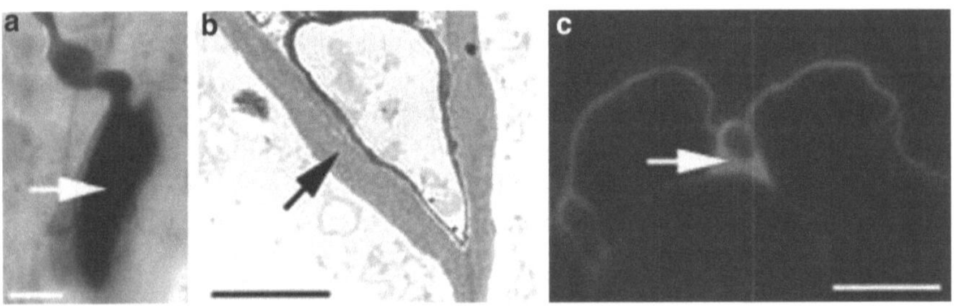

Fig. 3.2 Interactions between the *Medicago truncatula* mutant *dmi3/Mtsym13* and *G. intraradices*. (**a**) An appressorium (*arrow*) stained for alkaline phosphatase activity in contact with root epidermal cells, 7 days after inoculation (dai). *Bar* = 10 μm. (**b**) Micrograph of wall thickenings (*arrow*) in root epidermal cells below an appressorium. *Bar* = 2 μm. (**c**) UV-fluorescent cell wall thickenings (*arrow*), indicating phenolics, below an appressorium in contact with epidermal root cells, 10 dai. *Bar* = 10 μm

2004; Sanchez et al. 2005). In addition to those which respond to diffusible Myc factors (see above), several code for transcription factors and signal transduction-related proteins which are also transiently activated by the growth-promoting bacterium *Pseudomonas fluorescens* (Sanchez et al. 2005; Gianinazzi-Pearson et al. 2006). Again, these signal perception/transduction-related genes are not activated when appressoria develop at the root surface of the *dmi3/Mtsym13* mutant lacking the calcium/calmodulin-dependent protein kinase. However, calcium may not be the only intracellular messenger to play a role in early root responses to arbuscular mycorrhizal fungi. Nitric oxide (NO) is another recently identified key signalling molecule in plants which has been hypothesised to be involved in arbuscular mycorrhiza interactions (Weidmann et al. 2004; Vieweg et al. 2005). The mobile nature of NO, and its chemical reactivity with various cell targets, makes it a potentially important molecule in cell responses. Nitrate reductase and nitrite reductase are enzyme systems which are suspected to participate in plant NO production (Yamasaki and Sakihama 2000; Stöhr and Stremlau 2006). Transcripts encoding these enzymes accumulate in *M. truncatula* roots in response to appressorium formation by *G. intraradices* whereas their transcription is not affected when appressoria develop on roots of the mycorrhiza-deficient *dmi3/Mtsym13* mutant (Gianinazzi-Pearson et al. 2007).

3.3.2 Root Penetration by AM Fungi

Appressoria give rise to penetration hyphae which can enter root tissues either intercellularly or across epidermal cells, and the genetic programmes involved in fungal accommodation by the host plant will no doubt differ according to the mode of penetration. In hemibiotrophic fungal pathogens, infection-related morphogenesis is promoted by the perception of host plant signals which initiate signal transduction pathways (Xu and Hamer 1996; Xu et al. 1998). In line with the hypothesis that similar genetic programmes may exist in arbuscular mycorrhizal fungi, we have found transcription factor genes that are active in appressoria and up-regulated in *G. intraradices* during intercellular root penetration of wild-type *M. truncatula* but not during interactions with the *dmi3/Mtsym13* mutant resistant to arbuscular mycorrhizal fungi (Fig. 3.3). The *GintTFA/R I* gene codes a homologue of a transcription activator/repressor (PacC) which regulates morphogenesis in *U. maydis,* whilst *GintTFA II* is homologue to a transcription activator gene (*RCD*) implicated in sexual development of yeast (Okazaki et al. 1998; Aréchiga-Carvajal and Ruiz-Herrera 2005). In the case of *Gigaspora* species which colonise *M. truncatula* roots by crossing epidermal cells, the appressorium stage of mycorrhiza interactions is accompanied by induction in the host cells of an intracellular structure, a pre-penetration apparatus, considered to be responsible for initiating trans-cellular accommodation of the invading fungus (Genre et al. 2005). This epidermal cell response is triggered by repositioning of the plant nucleus in the vicinity of the contact site followed by reorganisation of the endoplasmic reticulum and cytoskeleton,

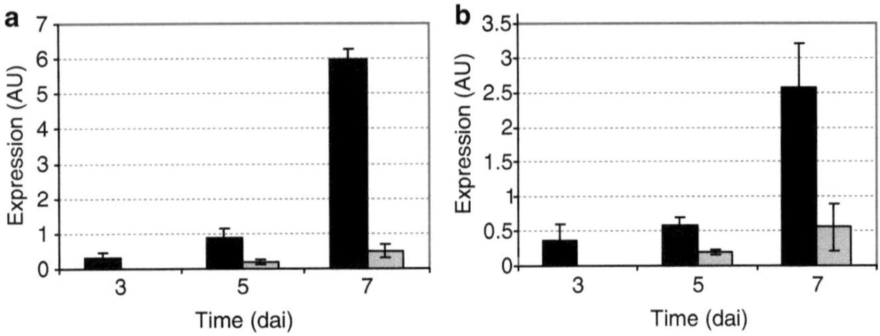

Fig. 3.3 Expression (arbitrary units) of the putative transcription factor genes *GintTFAII* (**a**) and *GintTFA/R1* (**b**) in *G. intraradices* (BEG 141) interacting with roots of *M. truncatula* J5 (wild-type, *black columns*) or TRV25 (*dmi3/Mtsym13* mutant, *grey columns*). On wild-type roots, appressoria begin to form at 3 dai, hyphae penetrate roots 5 dai and form first arbuscules 7 dai, whilst appressoria first develop on mutant roots 5 dai

and polarisation of microfilaments. Migration of the plant nucleus to the opposite cell wall (towards the cortex) then induces the formation of an apoplastic tunnel within which fungal penetration occurs. Observations on interactions with mycorrhiza-defective *M. truncatula* mutants indicate that nuclear movement towards sites of fungal contact is not dependent on the symbiosis-related genes, such as *MtDMI2/MtSYM2* which codes a receptor-like kinase or *MtDMI3/MtSYM13*, whilst assembly of the intracellular apoplastic tunnel is, providing further evidence for the active involvement of symbiosis-related plant genes in the fine control of root colonisation. Monitoring of gene expression during this process has revealed that an expansin-like gene is strongly up-regulated in the epidermis of wild-type *M. truncatula,* which may reflect cell wall loosening necessary to accommodate the fungal symbiont during this mode of root penetration (Siciliano et al. 2007).

3.4 Intersymbiont Compatibility: Getting Around Host Defence Systems?

Several plant genes that are induced in response to appressoria and before root penetration belong to protein-encoding families associated with defence strategies against pests or pathogens (García-Garrido and Ocampo 2002; Brechenmacher et al. 2004; Massoumou et al. 2007; Siciliano et al. 2007). However, the weak and/or transient priming of plant defence responses during symbiotic interactions contrasts markedly from what is typical of plant invasion by pathogenic micro-organisms (see Chapter 9 by Pozo et al.). In addition, plant mutant genotypes resistant to mycorrhizal fungi remain susceptible to pathogenic organisms (Gianinazzi-Pearson et al.

1994; Catoira et al. 2000; Gao et al. 2006; Mellersh and Parniske 2006), suggesting the existence of specific compatibility systems towards the fungal symbionts. Since physiologically active appressoria are formed at the surface of mycorrhiza-resistant plant mutants, but the pre-penetration apparatus is not formed, inactivation of symbiosis-related plant genes must affect fungal penetration processes rather than general cell activity. A possible way is through inhibitory root responses to the mycorrhizal fungi. This was first evoked following observations that *G. mosseae* elicited wall appositions, containing phenolics and callose, usually associated with the activation of defence reactions to pathogens in epidermal and hypodermal cells of roots of a "locus a" pea mutant (=*Pssym9/30*) (Gollotte et al. 1993). Not surprisingly, such wall reactions are also induced beneath appressoria by roots of *M. truncatula* mutated for *DMI3/MtSYM13* (orthologue of *PsSYM9/30*) (Fig. 3.2b, c). Subsequent reports that resistance to mycorrhizal fungi in *Pssym9/30* plants is accompanied by increased defence-related gene expression (Ruiz-Lozano et al. 1999) and enhanced salicylic acid accumulation in roots (Blilou et al. 1999) have led to the speculation that a defence mechanism associated with cell wall apposition and defence gene expression, mediated by salicylic acid, could result from inactivation of symbiosis-related plant genes (Gollotte et al. 1993; García-Garrido and Ocampo 2002). This is corroborated by findings that salicylic acid applied to roots has no effect on appressorium formation but delays root penetration and colonisation (Blilou et al. 2000). Pozo and Azcón-Aguilar (2007) have proposed that establishment of the arbuscular mycorrhiza symbiosis implies partial suppression of salicylic acid-dependent plant responses in initial stages which are compensated by enhancement of jasmonic acid-regulated responses during later stages (see Chapter 9 by Pozo et al. for further information).

First analyses of microarray transcriptome profiles of *G. intraradices*-inoculated versus non-inoculated *M. truncatula* wild-type and *dmi3/Mtsym13* mutant roots has revealed a high number of differentially regulated genes in the incompatible interactions (Seddas, Kuester, Becker & Gianinazzi-Pearson, unpublished results). As may be expected from the wall reactions to appressoria in cells of *dmi3/Mtsym13* roots, genes implicated in wall constituent synthesis are activated. In addition, there is over-expression of disease resistance-related genes in mutant roots but this does not appear to be responsible for the mycorrhiza-resistant phenotype of *dmi3/Mtsym13* roots since these genes are down-regulated by interactions with the symbiotic fungus. Likewise, expression of genes involved in secondary metabolite synthesis which could relate to a rise in jasmonate levels (Hause et al. 2007) is reduced with appressorium formation, indicating that the molecular bases of regulatory mechanisms reside elsewhere. In this context, cell processes modified by symbiosis-related plant genes may impact on root interactions by directly modulating AM fungal activity not only via the production of stimulatory plant molecules but also by controlling the production of inhibitory plant factors. This hypothesis is reinforced by the observation that several genes of *G. intraradices* are down-regulated when root penetration is prevented on *dmi3/Mtsym13* roots as compared to an interaction leading to symbiosis establishment (see above).

All these considerations raise the question of the role of different symbiosis-related plant genes in recognition and signalling events between host plants and the Glomeromycota, as well as their implication in decisive processes leading to arbuscular mycorrhiza establishment. In addition, whilst appressorium formation at the host root surface is a common event, patterns of root penetration (inter-cellular, trans-cellular) and colonisation (Paris/Arum) can vary between fungal–plant combinations so that generalisations about the cascade of cell programmes which drive compatibility at this early stage will need to be modulated.

Acknowledgements The authors are grateful to INRA and the Conseil Régional de Bourgogne for financial support for part of the work reported here.

References

Akiyama K, Matsuzaki K, Hayashi H (2005) Plant sesquiterpenes induce hyphal branching in arbuscular mycorrhizal fungi. Nature 435:824–827

Albrecht C, Geurts R, Lapeyrie F, Bisseling T (1998) Endomycorrhizae and rhizobial Nod factors both require*SYM8* to induce the expression of the early nodulin genes*PsENOD5* and*PsENOD12A*. Plant J 15:605–614

Aréchiga-Carvajal ET, Ruiz-Herrera J (2005) The RIM101/pacC homologue from the basidiomycete*Ustilago maydis* is functional in multiple pH-sensitive phenomena. Euk Cell 4:999–1008

Bécard G, Douds DD, Pfeffer PE (1992) Extensive hyphal growth of vesicular-arbuscular mycorrhizal fungi in the presence of CO2 and flavonols. Appl Env Microb 58:821–825

Bécard G, Kosuta S, Tamasloukht M, Séjalon-Delmas N, Roux C (2004) Partner communication in the arbuscular mycorrhizal interaction. Can J Bot 82:1186–1197

Besserer A, Puech-Pagès V, Kiefer P, Gomez-Roldan V, Jauneau A, Roy S, Portais JC, Roux C, Bécard G, Séjalon-Delmas N (2006) Strigolactones stimulate arbuscular mycorrhizal fungi by activating mitochondria. PLoS Biol 4:e226

Blilou I, Ocampo JA, Garcia-Garrido JM (1999) Resistance of pea roots to an endomycorrhizal fungus or*Rhizobium* correlates with enhanced levels of endogenous salicylic acid. J Exp Bot 50:1663–1668

Blilou I, Ocampo JA, Garcia-Garrido JM (2000) Induction of*Ltp* (Lipid transfer protein) and*Pal* (Phenylalanine ammonia-lyase) gene expression in rice roots colonized by the arbuscular mycorrhizal fungus*Glomus mosseae*. J Exp Bot 51:1969–1977

Bouwmeester HJ, Toux C, Lopez-Raez JA, Bécard G (2007) Rhizosphere communication of plants, parasitic plants and AM fungi. Trends Plant Sci 12:224–230

Brechenmacher L, Weidmann S, van Tuinen D, Chatagnier O, Gianinazzi S, Franken P, Gianinazzi-Pearson V (2004) Expression profiling of up-regulated plant and fungal genes in early and late stages of*Medicago truncatula-Glomus mosseae* interactions. Mycorrhiza 14:253–262

Breuninger M, Requena N (2004) Recognition events in AM symbiosis: analysis of fungal gene expression at the early appressorium stage. Fungal Gen Biol 41:794–804

Buée M, Rossignol M, Jauneau A, Ranjeva R, Bécard G (2000) The pre-symbiotic growth of arbuscular mycorrhizal fungi is induced by a branching factor partially purified from plant root exudates. Mol Plant-Microbe Interact 13:693–698

Catoira R, Galera R, de Billy F, Penmetsa V, Journet E-P, Mailler F, Rosenberg C, Cook D, Gough C, Dénarié J (2000) Four genes of*Medicago truncatula* controlling components of a Nod factor transduction pathway. Plant Cell 12:1647–1665

Chabaud M, Venard C, Defaux-Petras A, Bécard G, Barker DG (2002) Targeted inoculation of *Medicago truncatula* in vitro root cultures reveals *MtENOD11* expression during early stages of infection by arbuscular mycorrhizal fungi. New Phytol 156:265–273

Gadkar V, David-Schwartz R, Nagahashi G, Douds DD, Wininger S, Kapulnik Y (2003) Root exudate of *pmi* tomato mutant M161 reduces AM fungal proliferation in vitro. FEMS Microbiol Lett 223:193–198

Gao LL, Smith FA, Smith SE (2006) The rmc locus does not affect plant interactions or defence-related gene expression when tomato (*Solanum lycopersicon*) is infected with the root fungal parasite *Rhizoctonia*. Funct Plant Biol 33:289–296

García-Garrido JM, Ocampo JA (2002) Regulation of the plant defence response in arbuscular mycorrhizal symbiosis. J Exp Bot 53:1377–1386

Genre A, Chabaud M, Timmers T, Bonfante P, Barker DG (2005) Arbuscular mycorrhizal fungi elicit a novel intracellular apparatus in *Medicago truncatula* root epidermal cells before infection. Plant Cell 17:3489–3499

Gianinazzi-Pearson V, Branzanti B, Gianinazzi S (1989) In vitro enhancement of spore germination and hyphal growth of a vesicular-arbuscular mycorrhizal fungus by host root exudates and plant flavonoids. Symbiosis 7:243–255

Gianinazzi-Pearson V, Gollotte A, Dumas-Gaudot E, Franken P, Gianinazzi S (1994) Gene expression and molecular modifications associated with plant responses to infection by arbuscular mycorrhizal fungi. In: Downie DM, Osbourn AE (eds) Advances in molecular genetics of plant-microbe interactions, vol 3. Kluwer, Boston, pp. 179–186

Gianinazzi-Pearson V, Weidmann S, Seddas P, Massoumo M, van Tuinen D, Gianinazzi S (2006) Signal-related gene responses in beneficial root interactions with non-rhizobial microorganisms. In: Sanchez F, Quinto C, Lopez-Lara IM, Geiger O (eds) Biology of plant-microbe interactions vol 5. IS-MPMI, St Paul, Minnesota, pp. 516–524

Gianinazzi-Pearson V, Séjalon-Delmas N, Genre A, Jeandroz S, Bonfante P (2007) Plants and arbuscular mycorrhizal fungi: cues and communication in the early steps of symbiotic interactions. Adv Bot Res 46:181–219

Giovannetti M, Sbrana C, Logi C (1994) Early processes involved in host recognition by arbuscular mycorrhizal fungi. New Phytol 127:703–709

Giovannetti M, Sbrana C, Citernesi AS, Avio L (1996) Analysis of factors involved in fungal recognition responses to host-derived signals by arbuscular mycorrhizal fungi. New Phytol 133:65–71

Goldwasser Y, Yoder JI (2001) Differential induction of *Orobanche* seed germination by *Arabidopsis thaliana*. Plant Sci 160:951–959

Gollotte A, Gianinazzi-Pearson V, Giovannetti M, Sbrana C, Avio L, Gianinazzi S (1993) Cellular localization and cytochemical probing of resistance reactions to arbuscular mycorrhizal fungi in a 'locus A' myc- mutant of *Pisum sativum* (L.). Planta 191:112–122

Hause B, Mrosk C, Isayenkov S, Strack D (2007) Jasmonates in arbuscular mycorrhizal interactions. Phytochemistry 68:101–110

Kosuta S, Chabaud M, Lougnon G, Gough C, Dénarie J, Barker DG, Bécard G (2003) A diffusible factor from arbuscular mycorrhizal fungi induces symbiosis-specific *MtENOD11* expression in roots of *Medicago truncatula*. Plant Physiol 131:952–962

Larose G, Chênevert R, Moutoglis P, Gagne S, Piché Y, Vierheilig H (2002) Flavonoid levels in roots of *Medicago sativa* are modulated by the development stage of the symbiosis and the root colonizing arbuscular-mycorrhizal fungus. J Plant Phys 159:1329–1339

Liu J, Maldonado-Mendoza I, Lopez-Meyer M, Cheung F, Tow CD, Harrison MJ (2007) Arbuscular mycorrhizal symbiosis is accompanied by local and systemic alterations in gene expression and an increase in disease resistance in the shoots. Plant J 50:529–544

Massoumou M, van Tuinen D, Chatagnier O, Arnould C, Brechenmacher L, Sanchez L, Selim S, Gianinazzi S, Gianinazzi-Pearson V (2007) *Medicago truncatula* gene responses specific to arbuscular mycorrhiza interactions with different species and genera of Glomeromycota. Mycorrhiza 17:223–234

Mellersh D, Parniske M (2006). Common symbiosis genes of *Lotus japonicus* are not required for intracellular accommodation of the rust fungus *Uromyces loti*. New Phytol 170:641–644

Nagahashi G, Douds DD (2000) Partial separation of root exudate components and their effects upon the growth of germinated spores of AM fungi. Mycol Res 104:1453–1464
Navazio N, Moscatiello R, Genre A, Novero M, Baldan B, Bonfante P, Mariani P (2006) A diffusible signal from arbuscular mycorrhizal fungi elicits a transient cytosolic calcium elevation in host plant cells. Plant Physiol 144:673–681
Okazaki N, Okasaki K, Watanabe Y, Kato-Hayashi M, Yamamoto M, Okayama H (1998) Novel factor highly conserved among eukaryotes controls sexual development in fission yeast. Mol Cell Biol 18:887–895
Oldroyd GED, Downie JA (2006) Nuclear calcium changes at the core of symbiosis signalling. Curr Op Plant Biol 9:351–357
Pozo MJ, Azcon-Aguilar C (2007) Unraveling mycorrhiza-induced resistance. Curr Op Plant Biol 10:393–398
Redecker D, Kodner R, Graham LE (2000) Glomalean fungi from the Ordovician. Science 289:1920–1921
Requena N, Mann P, Hampp R, Franken P (2002) Early developmentally regulated genes in the arbuscular mycorrhizal fungus*Glomus mosseae*: identification of*GmGIN1*, a novel gene with homology to the C-terminus of metazoan hedgehog proteins. Plant Soil 244:129–139
Ruiz-Lozano JM, Roussel H, Gianinazzi S, Gianinazzi-Pearson V (1999) Defence genes are differentially induced by a mycorrhizal fungus and*Rhizobium* in wild type and symbiosis-defective pea genotypes. Mol Plant-Microbe Interact 12:976–984
Sabbagh SK (2008) Adaptation à la pénétration racinaire de deux Ustilaginaceae parasites du maïs*Ustilago maydis* et*Sporisorium reilianum* – Analyse microscopique et transcriptomique. Doctoral thesis, University of Toulouse III, F, 97 pages
Sanchez L, Weidmann S, Arnould C, Bernard AR, Gianinazzi S, Gianinazzi-Pearson V (2005) *P. fluorescens* and*G. mosseae* trigger*DMI3*-dependent activation of genes related to a signal transduction pathway in roots of*Medicago truncatula*. Plant Physiol 139:1065–1077
Scervino JM, Ponce MA, Erra-Bassells R, Bompadre J, Vierheilig H, Ocampo JA, Godeas A (2005) Flavonoids exhibit fungal species and genus effects on the presymbiotic growth of*Gigaspora* and*Glomus*. Mycol Res 109:789–794
Scervino JM, Ponce MA, Erra-Bassells R, Bompadre MJ, Vierheilig H, Ocampo JA, Godeas A (2006) Glycosidation of apigenin results in a loss of its activity on different growth parameters of arbuscular mycorrhizal fungi from the genus*Glomus* and*Gigaspora*. Soil Biol Biochem 38:2919–2922
Seddas PMA, Arnould C, Tollot M, Arias CM, Gianinazzi-Pearson V (2008) Spatial monitoring of gene activity in extraradical and intraradical developmental stages of arbuscular mycorrhizal fungi by direct fluorescent in situ RT-PCR. Fungal Gen Biol 45:1155–1165
Siciliano V, Genre A, Balestrini R, Cappellazzo G, deWit PJGM, Bonfante P (2007) transcriptome analysis of arbuscular mycorrhizal roots during development of the prepenetration apparatus. Plant Physiol 144:1455–1466
Steinkellner S, Lendzemo V, Langer I, Schweiger P, Khaosaad T, Toussaint JP, Vierheilig H (2007) Flavonoids and strigolactones in root exudates as signals in symbiotic and pathogenic plant-fungus interactions. Molecules 12:1290–1306
Stöhr C, Stremlau S (2006) Formation and possible roles of nitric oxide in plant roots. J Exp Bot 57:463–470
Tamasloukht MB, Sejalon-Delmas N, Kluver A, Roux C, Bécard G, Franken P (2003) Root factors induce mitochondrial-related gene expression and fungal respiration during the developmental switch from asymbiosis to presymbiosis in the arbuscular mycorrhizal fungus*Gigaspora rosea*. Plant Physiol 131:1468–1478
Trépanier M, Bécard G, Moutoglis P, Willemot C, Gagné S, Avis TJ, Rioux JA (2005) Dependence of arbuscular mycorrhizal fungi on their plant host for palmitic acid synthesis. Appl Env Microb 71:5341–5347
Vierheilig H, Bago B, Albrecht C, Poulin M-P, Piché Y (1998) Flavonoids and arbuscular-mycorrhizal fungi. In: Manthey JA, Buslig BS (eds) Flavonoids in the living system, Plenum Press, New York, pp. 9–33

Vieweg MF, Hohnjec N, Küster H (2005) Two genes encoding truncated hemoglobins are regulated during root nodule and arbuscular mycorrhiza symbioses of *Medicago truncatula* Planta 220:757–766

Weidmann S, Sanchez L, Descombin J, Chatagnier O, Gianinazzi S, Gianinazzi-Pearson V (2004) Fungal elicitation of signal transduction-related plant genes precedes mycorrhiza establishment and requires the dmi3 gene in *Medicago truncatula*. Mol Plant-Microbe Interact 17:1385–1393

Westwood JH (2000) Characterization of the *Orobanche-Arabidopsis* system for studying parasite-host interactions. Weed Sci 48:742–748

Xu JR, Hamer JE (1996) MAP kinase and cAMP signaling regulate infection structure formation and pathogenic growth in the rice blast fungus *Magnaporthe grisae*. Genes Devel 10:2696–2706

Xu JR, Staiger CJ, Hamer JE (1998) Inactivation of the mitogen-activated protein kinase Mps1 from the rice blast fungus prevents penetration of the host cells but allows activation of plant defence responses. Proc Natl Acad Sci USA 95:12713–12718

Yamasaki H, Sakihama Y (2000) Simultaneous production of nitric oxide and peroxynitrite by plant nitrate reductase: in vitro evidence for the NR-dependent formation of active nitrogen species. FEBS Lett 468:89–92

Chapter 4
Analysis of Mycorrhizal Functioning Using Transcriptomics

Matthias Arlt, Dietmar Schwarz, and Philipp Franken

Abstract Non-targeted approaches are not based on hypotheses, but screen for molecule pattern differences in organisms under particular conditions or at certain developmental stages. Patterns of RNA accumulation have been intensively analysed during the last years in order to get insights into mycorrhizal symbioses. This chapter will give an overview about these attempts and the results. As an example about a non-targeted strategy, the establishment and analysis of a subtractive cDNA library of tomato mycorrhiza will be presented. A symbiosis-induced gene encoding a copper-binding protein was studied in more detail.

4.1 Introduction

The functions of mycorrhizal interactions are traditionally divided into three groups based on their impact on the plant: mycorrhizal fungi supply mineral nutrients to roots (biofertilisers), they influence plant development (bioregulators) and they enable mycorrhizal plants to overcome biotic and abiotic stress (bioprotectors) (Smith and Read 1997). One strategy to elucidate the physiological basis of the symbiosis is to analyse particular processes or factors which are hypothesised to be involved in these three functions. Hence, on the one hand, the content and distribution of, e.g., mineral nutrients, carbohydrates, phytohormones or defence-related compounds are monitored. On the other hand, plant and fungal genes encoding proteins which regulate the transport and the metabolism of these factors are cloned, their expression is investigated and their role for symbiotic functioning proved in transgenic organisms. Such targeted approaches are included in this book (see Chapter 6 by Ferrol and Perez-Tienda and Chapter 8 by González-Guerrero et al.; see also Gianinazzi-Pearson et al. 1996). In contrast, non-targeted approaches

M. Arlt, D. Schwarz, and P. Franken (✉)
Institute of Vegetable and Ornamental Crops, Theodor-Echtermeyer-Weg, 114979 Grossbeeren, Germany
e-mail: franken@igzev.de

C. Azcón-Aguilar et al. (eds.), *Mycorrhizas - Functional Processes and Ecological Impact*, 47
DOI: 10/1007-978-3-540-87978-7_4, © Springer-Verlag Berlin Heidelberg 2009

are not based on a particular hypothesis. In such approaches, plants and fungi living asymbiotically are compared with specific stages of their symbiotic interaction. The resulting differences are subsequently characterised in more detail. Subject of screening could be metabolites, proteins or transcripts; the latter are discussed in this chapter.

The methods which are being used to monitor RNA accumulation and the results which were obtained have been comprehensively overviewed for ecto- and endomycorrhizal symbioses (Franken and Krajinski 2006). Therefore, this chapter will only briefly describe steps in mycorrhizal research leading to transcriptomics and summarise all the results in tables, before giving an example of a typical approach to gain insight into RNA accumulation patterns during an arbuscular mycorrhizal (AM) symbiosis.

4.2 From Identification of Single Genes to Transcriptomics

The non-targeted identification of genes regulated in ectomycorrhizal (ECM) symbioses started with the establishment and screening of a cDNA library of the interaction between *Eucalyptus globulus* and *Pisolithus tinctorius* (Tagu et al. 1993). Using this model system, expressed sequence tag (EST) collections were established and used for the identification and the characterisation of numerous plant and fungal genes potentially involved in mycorrhizal functioning. In a broader attempt, all ESTs were finally clustered, annotated and screened with cDNA derived from different stages of the interaction (Duplessis et al. 2005). This resulted in the identification of five classes of plant and fungal genes with distinct expression patterns at particular steps of the interaction. Other fungi used for transcriptional analyses of ectomycorrhizal interactions were *Tuber borchii* and *Piloderma croceum* (Table 4.1). Particular analyses were aimed in these and other systems (*Amanita muscaria*) at the interaction of the ectomycorrhizal fungi with mycorrhiza helper bacteria, to their fruit body development or to the influence of nitrogen starvation (Table 4.2). In all these cases, however, it turned out that confirmation of the putative role of the identified genes in symbiotic functioning was difficult because transformation systems were not available. Based on this experience, other researchers selected ECM fungi which could be transformed. This resulted in the model systems *Hebeloma cylindrosporum*, *Paxillus involutus* and *Laccaria bicolor*, where RNA accumulation patterns were screened (Tables 4.1 and 4.2) and the function of selected genes has been analysed (reviewed in Franken and Krajinski 2006). The international community finally chose poplar (*Populus*) and *Laccaria bicolor* as the first plant and fungal organisms forming ECM to sequence the whole symbiotic genome (Martin et al. 2004). The poplar sequencing project was finished in 2004 (www.ornl.gov/sci/ipgc/home.htm) and the genome sequence of the fungus was recently published (Martin et al. 2008). This now opens the possibility to simultaneously analyse the expression of the genomes of two ectomycorrhizal partners during their interaction.

Table 4.1 Identification of ectomycorrhiza-regulated genes by transcriptomic approaches

Plant	Fungus	Experimental approach[a]	No of regulated genes[b]	References
Eucalyptus globulus	*Pisolithus tinctorius*	Differential screening	6	Tagu et al. (1993)
Pinus resinosa	*Laccaria bicolor*	Differential display	5	Kim et al. (1999)
E. globulus	*P. tinctorius*	Array	65	Voiblet et al. (2001)
P. resinosa	*L. bicolor*	Differential display	107	Podila et al. (2002)
Tilia platyphyllos	*Tuber borchii*	Differential screening	~30	Polidori et al. (2002)
Tilia americana	*T. borchii*	Subtractive hybridisation	58	Menotta et al. (2004)
Quercus robur	*Piloderma croceum*	Subtractive hybridisation	50	Krüger et al. (2004)
Betula pendula	*Paxillus involutus*	Array	~230	Johansson et al. (2004)
B. pendula	*P. involutus* (diff. strains)	Array	66	Le Quere et al. (2004)
E. globulus	*P. tinctorius*	Array	395	Duplessis et al. (2005)
B. pendula	*P. involutus*	Array	389	Le Quere et al. (2005)
Pinus sylvestris	*Tricholoma vaccinum*	Differential display	76	Krause and Kothe (2006)
Q. robur	*P. croceum*	Array	75	Frettinger et al. (2007)
P. sylvestris	*L. bicolor*	Array	236	Heller et al. (2008)

[a]Only non-targeted approaches directed to differentially expressed genes are mentioned.
[b]Numbers cannot be compared, because different methods for verification and different cut off levels were used.

Table 4.2 Identification of regulated genes from ectomycorrhizal fungi by transcriptomic approaches

Fungal species	Subject of investigation	Experimental approach[a]	No of regulated genes[b]	References
T. borchii	Vegetative mycelium versus fruit bodies	Differential RNA display	5	Zeppa et al. (2000)
T. borchii	Vegetative mycelium versus fruit bodies	Array	43	Lacourt et al. (2002)
P. involutus	Influence of heavy metal	Array	94	Jacob et al. (2004)
Amanita muscaria	Influence of helper bacterium	Subtractive hybridisation	58	Schrey et al. (2005)
T. borchii	Nitrogen starvation	Array	51	Montanini et al. (2006)

[a]Only non-targeted approaches directed to differentially expressed genes are mentioned.
[b]Numbers cannot be compared, because different methods for verification and different cut off levels were used.

Similar to ectomycorrhiza, AM transcriptomics started with the screening of cDNA libraries (Tahiri-Alaoui and Antoniw 1996; Burleigh and Harrison 1997; Murphy et al. 1997). Other techniques, such as differential RNA display and subtractive hybridisation, followed for the identification of small numbers of AM regulated genes (Martin-Laurent et al. 1997; van Buuren et al. 1999). The availability of particular pea mutants opened the possibility to target such analyses to certain stages of the interaction (Table 4.3). This resulted in the identification of genes specifically regulated during early or late stages of the AM symbiosis (Lapopin et al. 1999; Roussel et al. 2001; Grunwald et al. 2004). Non-targeted transcriptional analyses of AM were, however, always confronted with a general problem: while nearly 50% of the transcripts in RNA populations of ECM are of fungal origin, in the case of AM the amount is lower than 5%. It is therefore not surprising that the majority of transcripts detected in RNA populations of AM interactions were derived from the plant. In order to be able to identify more AM fungal genes, RNA was extracted from pure fungal material and used for the identification of single genes, the establishment of EST collections and the analysis of development-regulated gene expression in AM fungi (Table 4.4). New investigations were focussed on the regulation of fungal gene expression in extraradical hyphae by heavy metal contamination (Waschke et al. 2006) or by nutrient supply (Cappellazzo et al. 2007). In a recent analysis, the method of laser dissection microscopy was used to also analyse fungal RNA accumulation in intraradical structures (Balestrini et al. 2007).

Two legumes were chosen, *Lotus japonicum* and *Medicago truncatula*, as model plants for more comprehensive analyses of AM interactions (Cook et al. 1997). While the former plant was used more for mutant analyses, large collections of ESTs of the second plant were established and used for the production of microarrays and for in silico analyses of gene expression (Table 4.3). One microarray carrying about 16,000 *M. truncatula* genes has been used to compare expression patterns in roots colonised by *Glomus mosseae* and *Glomus intraradices* (Hohnjec et al. 2005) or by *G. intraradices* and *Gigaspora gigantea* (Liu et al. 2007). The expression of individual genes was also used to compare the interaction of *M. truncatula* with different AM fungal isolates (Massoumou et al. 2007; Feddermann et al. 2008). Further analyses showed that many of these mycorrhiza-regulated genes were specifically expressed in arbusculated cells of *M. truncatula* (for review see Hohnjec et al. 2006) or during the formation of the prepenetration apparatus (Siciliano et al. 2007), a structure which is formed upon physical contact of hyphae with the root epidermis (Genre et al. 2005; see also Chapter 3 by Gianinazzi-Pearson et al.). In the mean time, sequencing of the rice genome was completed. Being a mycorrhizal plant, a whole genome chip could be screened, and numerous AM-regulated genes were identified in this important model crop (Güimil et al. 2005).

Although mycorrhizal structures and the general outcome of the symbiosis are similar, it is obvious that interactions between different plants and AM fungal isolates are functionally diverse (for review see Franken and George 2006). It is therefore necessary to integrate, in addition to the two legume models and the monocotyledon rice, species from other plant families into non-targeted transcriptional analyses. The Solanaceae would be one of the first choices because the family

Table 4.3 Identification of arbuscular mycorrhiza-regulated genes by transcriptomic approaches

Plant	Fungus	Experimental approach[a]	No of regulated genes[b,c]	References
Tomato	*Glomus mosseae*	Differential screening	4	Tahiri-Alaoui and Antoniw (1996)
Barley	*Glomus intraradices*	Differential screening	2	Murphy et al. (1997)
Medicago truncatula	*Glomus versiforme*	Differential screening	1	Burleigh and Harrison (1997)
Pisum sativum	*G. mosseae*	Differential RNA display	7	Martin-Laurent et al. (1997)
M. truncatula	*G. versiforme*	Subtractive hybridisation	3	van Buuren et al. (1999)
P. sativum (late mutant)	*G. mosseae*	Differential RNA display	4	Lapopin et al. (1999)
P. sativum (early mutant)	*G. mosseae*	Differential RNA display	6	Roussel et al. (2001)
M. truncatula	*G. intraradices*	Electronic Northern	260	Journet et al. (2002)
M. truncatula	*G. intraradices*	Subtractive hybridisation	20	Wulf et al. (2003)
M. truncatula	*G. versiforme*	2.3 K array	92	Liu et al. (2003)
P. sativum (late mutant)	*G. mosseae*	Subtractive hybridisation	25	Grunwald et al. (2004)
M. truncatula	*G. mosseae*	Subtractive hybridisation	18	Brechenmacher et al. (2004)
M. truncatula	*G. mosseae*	Subtractive hybridisation	29	Weidmann et al. (2004)
M. truncatula	*G. intraradices*	6 K array	~1100	Manthey et al. (2004)
M. truncatula	*G. intraradices*	Electronic Northern	115	Frenzel et al. (2005)
M. truncatula	*G. intraradices* and G. mosseae	16 K array	1378 and 1087 (~400 co-regulated)	Hohnjec et al. (2005)
Lotus japonicum	*G. intraradices*	AFLP	9	Kistner et al. (2005)
Oryza sativa	*G. intraradices*	50 K array	224	Güimil et al. (2005)
M. truncatula	*Gigaspora margarita*	Subtractive hybridisation	15	Siciliano et al. (2007)
M. truncatula	*Gigaspora gigantea* and *G. intraradices*	16 K array	647 and 157 (56 co-induced)	Liu et al. (2007)

[a]Only non-targeted approaches directed to differentially expressed genes are mentioned.
[b]Numbers cannot be compared, because different methods for verification and different cut off levels were used.
[c]Mainly plant genes.

contains numerous important crops such as potato, tomato and petunia. Tomato is moreover an excellent model plant since it has been extensively used in mycorrhizal research in order to analyse the exchange of phosphate and carbohydrates

Table 4.4 Identification of regulated genes from arbuscular mycorrhizal fungi by transcriptomic approaches

Fungal species	Subject of investigation	Experimental approach[a]	No of regulated genes[b]	References
G. mosseae	Bacteria on asymbiosis	Differential RNA display	1	Requena et al. (1999)
Gigaspora rosea	Presymbiosis	Differential RNA display	7	Tamasloukht et al. (2003)
G. mosseae	Asymbiosis versus EM[c]	Subtractive hybridisation	9	Requena et al. (2002)
G. rosea	Presymbiosis	Subtractive hybridisation	38	Tamasloukht et al. (2007)
G. mosseae	Appressorium development	Subtractive hybridisation	31	Breuninger and Requena (2004)
G. intraradices	Influence of heavy metal on EM	Subtractive hybridisation	31	Waschke et al. (2006)
G. intraradices	Influence of bacteria on EM	Subtractive hybridisation	43	Hildebrandt et al. (2006)
G. intraradices	Influence of nitrogen on EM	Subtractive hybridisation	34	Cappellazzo et al. (2007)

[a]Only non-targeted approaches directed to differentially expressed genes are mentioned.
[b]Numbers cannot be compared, because different methods for verification and different cut off levels were used.
[c]Extraradical mycelium.

(Nagy et al. 2005; Schaarschmidt et al. 2007), and for mutant analyses (Barker et al. 1998; David-Schwartz et al. 2003). Although the first non-targeted analysis of AM was carried out with tomato (Tahiri-Alaoui and Antoniw 1996), it was thereafter not further used in transcriptomics. In the following sections, the establishment of tomato as an experimental system and the construction and analysis of a subtractive cDNA library will be described as an example for a non-targeted approach aimed at the identification of mycorrhiza-regulated genes in this model crop.

4.3 Establishment of the Experimental System

In order to identify genes which are AM-specifically regulated, it is useful to avoid experimental situations where secondary effects of the colonisation by the AM fungus, like changes in plant nutrition or developmental patterns, could dominate. Different tomato cultivars were inoculated with mycorrhizal fungi and cultivated under low phosphate conditions. Estimation of mycorrhization parameters 6 ;weeks after inoculation showed relatively high values for the cultivar ‘Hildares’. Growth responses of the plants were negligible minimising secondary effects on gene expression. In contrast, phosphate concentrations increased by approximately 60% in mycorrhizal plants of the cultivar ‘Hildares’ compared to the corresponding controls. However, transcriptional analysis of *M. truncatula* mycorrhiza indicated a relatively small overlap between symbiosis- and phosphate-regulated genes (Liu et al. 2003; Hohnjec et al. 2005). Furthermore, phosphate regulation of putative AM-induced genes can be easily verified in subsequent analyses, while it is difficult to simulate growth responses. In addition to nutritional effects and changes in growth parameters, mycorrhiza-responsive genes could also be regulated by the biotic stress due to intensive fungal colonisation of the root tissues. It is therefore not surprising that mycorrhizal induction of pathogenesis- and defence-related genes has been shown in numerous targeted analyses (for reviews see Gianinazzi-Pearson et al. 1996; Garcia-Garrido and Ocampo 2002) and in transcriptomic approaches (Güimil et al. 2005). A tomato–*Pythium aphanidermatum* pathosystem, which had been established for soilless cultures (Schwarz and Grosch 2003), was therefore adopted in the experimental treatments.

4.4 Construction and Analysis of a Subtractive cDNA Library

Based on preliminary experiments, tomato plants cv. 'Hildares' were transferred 2 weeks after germination to a substrate containing the AM fungus *Glomus mosseae* BEG 12 (Biorize, Dijon). Plants were grown together with mock-inoculated control plants and fertilised daily using a half-strength Hoagland nutrient solution with 10% of the usual phosphate concentration. Four weeks later, all plants were inoculated with the phytopathogenic fungus *P. aphanidermatum*. Upon harvest, 42% of the root cortex was colonised by the AM fungus and 33% harboured fungal arbuscules. Carrot-agar tests indicated 50% root colonisation by the pathogen. RNA was extracted and cDNA synthesized. Subsequently cDNA from roots co-inoculated with *G. mosseae* and *P. aphanidermatum* was subtracted by cDNA from roots colonised only by the pathogen following a protocol described by Wulf et al. (2003) and Brechenmacher et al. (2004). The remaining cDNA was cloned into a plasmid vector. This cDNA library was supposed to be enriched for clones harbouring fragments of genes being induced in mycorrhizal roots, but not regulated by the pathogen.

The inserts of 192 clones were hybridized in a dot blot analysis to labelled cDNA of mycorrhizal roots and controls from two biological repetitions. This revealed 20 fragments putatively belonging to mycorrhiza-induced genes. These fragments were analysed and their sequences were screened for similarities to known genes by BlastX (Table 4.5). Ten fragments gave no significant results and another five were similar to genes encoding proteins with unknown function. Two of the fragments (2D10, 2G12) are putatively derived from RNA expressed in plastids. This does not necessarily indicate a specific function of the encoded proteins, but could be a secondary effect of the proliferation of plastids during arbuscule development (Fester et al. 2007). A specific function can also not be assigned to the gene encoding a mediator of RNA polymerase activity. Such transcription factors could be general regulators of a number of genes and therefore be involved in different functions of the AM symbiosis. One of the fragments codes for a protein with 83% identical amino acids to a type 2 metallothionein. The corresponding protein in water melon shows drought-induced expression and is able to detoxify hydroxyl radicals (Akashi et al. 2004). RNA

Table 4.5 Mycorrhiza-induced genes from tomato

Clone	Length (bp)	Similarity[a]	Origin	e-Value[b]
1	513	Type-2 metal-lothionein	*Citrullus lanatus*	$5e^{-22}$
2	381	No homology	–	–
3	407	Unnamed protein product	*Vitis vinifera*	$1e^{-28}$
5	652	No homology	–	–
7	723	No homology	–	–
8	400	No homology	–	–
9	640	RNA polymerase mediator	*Xenopus tropicalis*	$1e^{-12}$
10	550	No homology	–	–
12	538	Senescence-associated	*Pisum sativum*	$5e^{-29}$
14	359	Senescence-associated	*Cryptosporidium hominis*	$4e^{-29}$
16	454	No homology	–	–
18	542	No homology	–	–
19	563	Unnamed protein product	*V. vinifera*	$2e^{-43}$
24	557	No homology	–	–
25	736	No homology	–	–
11D7	406	No homology	–	–
2D10	546	RNA polymerase β subunit	*Lactuca sativa*	$8e^{-94}$
2G12	606	Rubisco	*L. sativa*	$9e^{-103}$
11H7	498	Dicyanin	*Lycopersicon esculentum*	$3e^{-20}$
11C3	547	Hypothetical protein	*Ustilago maydis*	$5e^{-17}$

[a]According to BlastX analysis on 16 October 2007.
[b]Results are shown for *e*-value < e^{-10}.

accumulation of metallothionein genes were also detected in transcriptomic analyses of mycorrhiza in *P. sativum* (Grunwald et al. 2004). Their activity could be associated with arbuscule degradation since H_2O_2 is generated during this process (Fester and Hause 2005). The induction of the gene encoding senescence-associated proteins (represented by the two ESTs 12 and 14) also fits into this picture. A general higher expression of genes in mycorrhiza usually being induced by oxidative stress could be one of the reasons why mycorrhizal plants show increased stability during abiotic stress situations compared to non-mycorrhizal controls.

4.5 Expression of Copper-Binding Proteins

Most interestingly, one of the fragments encoded a peptide with a copper-binding domain and showed homology to a tomato protein called dicyanin. Based on the sequences, specific oligonucleotides were designed for the known tomato gene

(*LeCbp1*) and the newly identified gene (*LeCbp2*). RNA was extracted from tomato roots (10% P) inoculated with the AM fungus or with the pathogen and from non-inoculated control roots (fertilised with 10% or 100% P). The corresponding cDNAs were used for real-time PCR expression analysis of the two genes, using the gene *LeTef1* encoding the translation factor EF-1α for calibration. These analyses confirmed that *LeCbp2* is strongly induced (20-fold) in roots colonised with the AM fungus (Fig. 4.1). No significant differences were observed after inoculation with the pathogen or in roots fertilised with optimal amounts of phosphate. In contrast, *LeCbp1* was not regulated by the two root-colonising fungi, but more than three times induced by phosphate (Fig. 4.1).

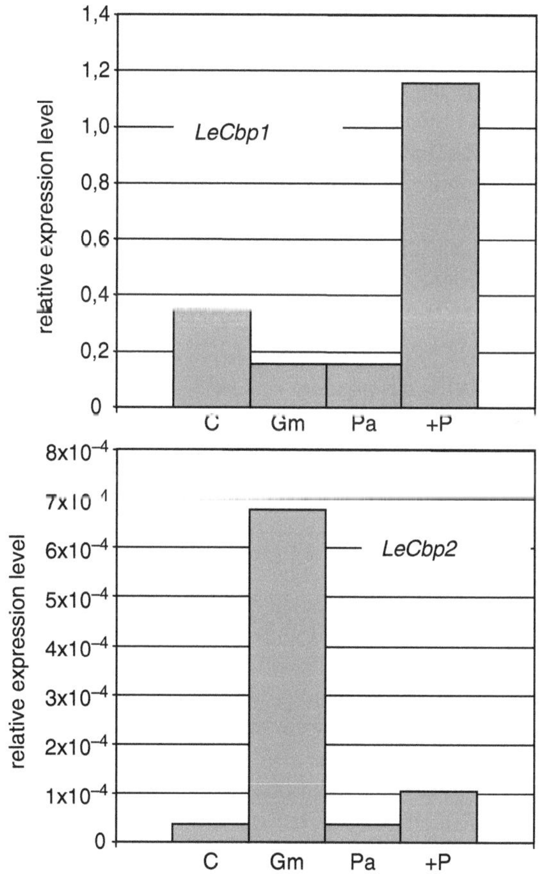

Fig. 4.1 RNA accumulation of genes encoding copper-binding proteins. RNA was extracted from tomato roots fertilised with 10% phosphate and mock-inoculated (*C*), inoculated with the arbuscular mycorrhizal fungus *Glomus mosseae* (*Gm*) or the root pathogen *Pythium aphanidermatum* (*Pa*) or from roots fertilised with 100% phosphate (*+P*). cDNA was synthesised and used for real-time PCR with primers corresponding to the genes *LeCbp1* and *LeCbp2*. The figure shows the RNA accumulation level relative to the values obtained for the constitutive expressed gene *LeTef1*

Proteins with copper-binding domains belong to a large gene family. The best characterised members are the plastocyanins which are involved in electron transfer in chloroplasts (Burkey et al. 1996). Hence, the putative CBP2 could play a role in the proliferating plastid network around the fungal arbuscules (Fester et al. 2007). Interestingly, among the gene family encoding plastocyanins at least one gene is specifically expressed in arbusculated cells (Hohnjec et al. 2005; see also Chapter 5 by Recorbet et al.). Copper-binding proteins are also involved in copper-homeostasis, as discussed for the tomato gene Cbp1 upon infection of leaves with *Botrytis cinerea* (Company and Gonzalez-Bosch 2003). In *Arabidopsis*, such copper chaperons transport the metal ion to proteins like cytochrome C oxidase which are needed during biotic and abiotic stress (Attallah et al. 2007). It could be that the higher expression of copper-containing proteins in arbusculated cells results in increased inner cell transport and the induction of such small copper-binding proteins as CBP2. The induction of Cbp1 in roots fertilised with high amounts of phosphate is more difficult to explain. One possibility is the role of ethylene signalling in the response of P-rich plants, where it prevents root hair elongation (Franco-Zorrilla et al. 2004). This response requires correct ethylene-sensing and the functioning of the respective receptors is dependent on copper (Hirayama et al. 1999). High amounts of phosphate could therefore induce the expression of copper-binding proteins to ensure high ethylene sensitivity and an appropriate plant response.

4.6 Concluding Remarks

Transcriptomic approaches have broadened our understanding of mycorrhizal functioning over the past years. Numerous plant genes which have been identified to be mycorrhizal-regulated are involved in the mutual exchange of nutrients and carbohydrates, in plant defence responses and in phytohormone metabolism. In order to exploit these mycorrhizal functions for applications in plant production, however, it is necessary to work with crops. Tomato is an ideal plant since it combines the fact that it is the most consumed vegetable with the attribute of being a model plant. The usefulness of tomato has been shown for targeted approaches concerning phosphate and carbohydrate metabolism in the symbiotic interaction. The present study shows that it is also an amenable model plant to identify mycorrhiza-regulated genes by a non-targeted approach. The role of such genes should be easily confirmed since transformation of tomato is a routine process. Finally, the breeding of new cultivars with improved mycorrhizal functioning traits, based on the molecular data, will be facilitated by the large genetic resources that exist for this model crop.

Acknowledgement We thank Katja Boldt for measuring P content in tomato plants.

References

Akashi K, Nishimura N, Ishida Y, Yokota A (2004) Potent hydroxyl radical-scavenging activity of drought-induced type-2 metallothionein in wild watermelon. Biochem Biophys Res Commun 323:72–78

Attallah CV, Welchen E, Pujol C, Bonnard G, Gonzalez DH (2007) Characterization of *Arabidopsis thaliana* genes encoding functional homologues of the yeast metal chaperone Cox19p, involved in cytochrome c oxidase biogenesis. Plant Mol Biol 65:343–355

Balestrini R, Gomez-Ariza J, Lanfranco L, Bonfante P (2007) Laser microdissection reveals that transcripts for five plant and one fungal phosphate transporter genes are contemporaneously present in arbusculated cells. Mol Plant-Microbe Interact 20:1055–1062

Barker SJ, Stummer B, Gao L, Dispain I, Oconnor PJ, Smith SE (1998) A mutant in *Lycopersicon esculentum* Mill. with highly reduced VA mycorrhizal colonization - isolation and preliminary characterisation. Plant J 15:791–797

Brechenmacher L, Weidmann S, Van Tuinen D, Chatagnier O, Gianinazzi S, Franken P, Gianinazzi-Pearson V (2004) Expression profiling of up-regulated plant and fungal genes in early and late stages of *Medicago truncatula-Glomus mosseae* interactions. Mycorrhiza 14:253–262

Breuninger M, Requena N (2004) Recognition events in AM symbiosis: analysis of fungal gene expression at the early appressorium stage. Fung Gen Biol 41:794–804

Burkey KO, Gizlice Z, Carter TE (1996) Genetic variation in soybean photosynthetic electron transport capacity is related to plastocyanin concentration in the chloroplast. Photosynth Res 49:141–149

Burleigh SH, Harrison MJ (1997) A novel gene whose expression in *Medicago truncatula* is suppressed in response to colonization by vesicular-arbuscular mycorrhizal fungi and to phosphate nutrition. Plant Mol Biol 34:199–208

Cappellazzo G, Lanfranco L, Bonfante P (2007) A limiting source of organic nitrogen induces specific transcriptional responses in the extraradical structures of the endomycorrhizal fungus *Glomus intraradices*. Curr Genet 51:59–70

Company P, Gonzalez-Bosch C (2003) Identification of a copper chaperone from tomato fruits infected with *Botrytis cinerea* by differential display. Biochem Biophys Res Commun 304:825–830

Cook DR, VandenBosch K, de Bruijn FJ, Huguet T (1997) Model legumes get the nod. Plant Cell 9:275–281

David-Schwartz R, Gadkar V, Wininger S, Bendov R, Galili G, Levy AA, Kapulnik Y (2003) Isolation of a premycorrhizal infection (pmi2) mutant of tomato, resistant to arbuscular mycorrhizal fungal colonization. Mol Plant-Microbe Interact 16:382–388

Duplessis S, Courty PE, Tagu D, Martin F (2005) Transcript patterns associated with ectomycorrhiza development in *Eucalyptus globulus* and *Pisolithus microcarpus*. New Phytol 165:599–611

Feddermann N, Boller T, Salzer P, Elfstrand S, Wiemken A, Elfstrand M (2008) *Medicago truncatula* shows distinct patterns of mycorrhiza-related gene expression after inoculation with three different arbuscular mycorrhizal fungi. Planta 227:671–680

Fester T, Hause G (2005) Accumulation of reactive oxygen species in arbuscular mycorrhizal roots. Mycorrhiza 15:373–379

Fester T, Lohse S, Halfmann K (2007) "Chromoplast" development in arbuscular mycorrhizal roots. Phytochemistry 68:92–100

Franco-Zorrilla JM, Gonzalez E, Bustos R, Linhares F, Leyva A, Paz-Ares J (2004) The transcriptional control of plant responses to phosphate limitation. J Exp Bot 55:285–293

Franken P, George E (2006) Diversity of arbuscular mycorrhizal fungi. In: Benckiser G, Schnell S (eds) Biodiversity in agricultural production systems. CRC Press, Boca Raton, pp 189–203

Franken P, Krajinski F (2006) Transcriptomics for determining gene expression in symbiotic root-fungus interactions. In: Cooper J, Rao JR (eds) Molecular techniques for soil, rhizosphere and plant microorganisms. CAB International, Oxfordshire, pp 213–235

Frenzel A, Manthey K, Perlick AM, Meyer F, Puhler A, Kuster H, Krajinski F (2005) Combined transcriptome profiling reveals a novel family of arbuscular mycorrhizal-specific *Medicago truncatula* lectin genes. Mol Plant-Microbe Interact 18:771–782

Frettinger P, Derory J, Herrmann S, Plomion C, Lapeyrie F, Oelmuller R, Martin F, Buscot F (2007) Transcriptional changes in two types of pre-mycorrhizal roots and in ectomycorrhizas of oak microcuttings inoculated with *Piloderma croceum*. Planta 225:331–340

Garcia-Garrido JM, Ocampo JA (2002) Regulation of the plant defence response in arbuscular mycorrhizal symbiosis. J Exp Bot 53:1377–1386

Genre A, Chabaud M, Timmers T, Bonfante P, Barker DG (2005) Arbuscular mycorrhizal fungi elicit a novel intracellular apparatus in *Medicago truncatula* root epidermal cells before infection. Plant Cell 17:3489–3499

Gianinazzi-Pearson V, Dumas-Gaudot E, Gollotte A, Tahiri-Alaoui A, Gianinazzi S (1996) Cellular and molecular defence-related root response to invasion by arbuscular mycorrhizal fungi. New Phytol 133:45–57

Grunwald U, Nyamsuren O, Tarnasloukht M, Lapopin L, Becker A, Mann P, Gianinazzi-Pearson V, Krajinski F, Franken P (2004) Identification of mycorrhiza-regulated genes with arbuscule development-related expression profile. Plant Mol Biol 55:553–566

Güimil S, Chang HS, Zhu T, Sesma A, Osbourn A, Roux C, Ionnidis V, Oakeley EJ, Docquier M, Descombes P, Briggs SP, Paszkowski U (2005) Comparative transcriptomics of rice reveals an ancient pattern of response to microbial colonization. Proc Natl Acad Sci USA 102:8066–8070

Heller G, Adomas A, Li GS, Osborne J, van Zyl L, Sederoff R, Finlay RD, Stenlid J, Asiegbu FO (2008) Transcriptional analysis of *Pinus sylvestris* roots challenged with the ectomycorrhizal fungus *Laccaria bicolor*. BMC Plant Biol 8:19

Hildebrandt U, Ouziad F, Marner FJ, Bothe H (2006) The bacterium *Paenibacillus validus* stimulates growth of the arbuscular mycorrhizal fungus *Glomus intraradices* up to the formation of fertile spores. FEMS Microbiol Lett 254:258–267

Hirayama T, Kieber JJ, Hirayama N, Kogan M, Guzman P, Nourizadeh S, Alonso JM, Dailey WP, Dancis A, Ecker JR (1999) Responsive-to-antagonist1, a Menkes/Wilson disease-related copper transporter, is required for ethylene signalling in Arabidopsis. Cell 97:383–393

Hohnjec N, Vieweg ME, Puhler A, Becker A, Kuster H (2005) Overlaps in the transcriptional profiles of *Medicago truncatula* roots inoculated with two different *Glomus* fungi provide insights into the genetic program activated during arbuscular mycorrhiza. Plant Physiol 137:1283–1301

Hohnjec N, Henckel K, Bekel T, Gouzy J, Dondrup M, Goesmann A, Kuster H (2006) Transcriptional snapshots provide insights into the molecular basis of arbuscular mycorrhiza in the model legume *Medicago truncatula*. Funct Plant Biol 33:737–748

Jacob C, Courbot ML, Martin F, Brun A, Chalot M (2004) Transcriptomic responses to cadmium in the ectomycorrhizal fungus *Paxillus involutus*. FEBS Lett 576:423–427

Johansson T, Le Quere A, Ahren D, Soderstrom B, Erlandsson R, Lundeberg J, Uhlen M, Tunlid A (2004) Transcriptional responses of *Paxillus involutus* and *Betula pendula* during formation of ectomycorrhizal root tissue. Mol Plant-Microbe Interact 17:202–215

Journet EP, Van Tuinen D, Gouzy J, Crespeau H, Carreau V, Farmer MJ, Niebel A, Schiex T, Jaillon O, Chatagnier O, Godiard L, Micheli F, Kahn D, Gianinazzi-Pearson V, Gamas P (2002) Exploring root symbiotic programs in the model legume *Medicago truncatula* using EST analysis. Nucl Acids Res 30:5579–5592

Kim SJ, Hiremath ST, Podila GK (1999) Cloning and identification of symbiosis-regulated genes from the ectomycorrhizal *Laccaria bicolor*. Mycol Res 103:168–172

Kistner C, Winzer T, Pitzschke A, Mulder L, Sato S, Kaneko T, Tabata S, Sandal N, Stougaard J, Webb KJ, Szczyglowski K, Parniske M (2005) Seven *Lotus japonicus* genes required for transcriptional reprogramming of the root during fungal and bacterial symbiosis. Plant Cell 17:2217–2229

Krause K, Kothe E (2006) Use of RNA fingerprinting to identify fungal genes specifically expressed during ectomycorrhizal interaction. J Basic Microbiol 46:387–399

Krüger A, Peskan-Berghofer T, Frettinger P, Herrmann S, Buscot F, Oelmuller R (2004) Identification of premycorrhiza-related plant genes in the association between *Quercus robur* and *Piloderma croceum*. New Phytol 163:149–157

Lacourt I, Duplessis S, Abba S, Bonfante P, Martin F (2002) Isolation and characterization of differentially expressed genes in the mycelium and fruit body of *Tuber borchii*. Appl Environ Microbiol 68:4574–4582

Lapopin L, Gianinazzi-Pearson V, Franken P (1999) Comparative differential display analysis of arbuscular mycorrhiza in *Pisum sativum* and a mutant defective in late stage development. Plant Mol Biol 41:669–677

Le Quere A, Schutzendubel A, Rajashekar B, Canback B, Hedh J, Erland S, Johansson T, Tunlid A (2004) Divergence in gene expression related to variation in host specificity of an ectomycorrhizal fungus. Mol Ecol 13:3809–3819

Le Quere A, Wright DP, Soderstrom B, Tunlid A, Johansson T (2005) Global patterns of gene regulation associated with the development of ectomycorrhiza between birch (*Betula pendula* Roth.) and fungus *Paxillus involutus* (Batsch) Fr. Mol Plant-Microbe Interact 18:659–673

Liu JY, Blaylock LA, Endre G, Cho J, Town CD, VandenBosch KA, Harrison MJ (2003) Transcript profiling coupled with spatial expression analyses reveals genes involved in distinct developmental stages of an arbuscular mycorrhizal symbiosis. Plant Cell 15:2106–2123

Liu JY, Maldonado-Mendoza I, Lopez-Meyer M, Cheung F, Town CD, Harrison MJ (2007) Arbuscular mycorrhizal symbiosis is accompanied by local and systemic alterations in gene expression and an increase in disease resistance in the shoots. Plant J 50:529–544

Manthey K, Krajinski F, Hohnjec N, Firnhaber C, Puhler A, Perlick AM, Kuster H (2004) Transcriptome profiling in root nodules and arbuscular mycorrhiza identifies a collection of novel genes induced during *Medicago truncatula* root endosymbioses. Mol Plant-Microbe Interact 17:1063–1077

Martin F, Tuskan GA, DiFazio SP, Lammers P, Newcombe G, Podila GK (2004) Symbiotic sequencing for the *Populus* mesocosm. New Phytol 161:330–335

Martin F et al. (2008) The genome of *Laccaria bicolor* provides insights into mycorrhizal symbiosis. Nature 452:88–92

Martin-Laurent F, Van Tuinen D, Dumas-Gaudot E, Gianinazzi-Pearson V, Gianinazzi S, Franken P (1997) Differential display analysis of RNA accumulation in arbuscular mycorrhiza of pea and isolation of a novel symbiosis-regulated plant gene. Mol Gen Genet 256:37–44

Massoumou M, Van Tuinen D, Chatagnier O, Arnould C, Brechenmacher L, Sanchez L, Selim S, Gianinazzi S, Gianinazzi-Pearson V (2007) *Medicago truncatula* gene responses specific to arbuscular mycorrhiza interactions with different species and genera of Glomeromycota. Mycorrhiza 17:223–234

Menotta M, Amicucci A, Sisti D, Gioacchini AM, Stocchi V (2004) Differential gene expression during pre-symbiotic interaction between *Tuber borchii* Vittad. and *Tilia americana* L. Curr Genet 46:158–165

Montanini B, Gabella S, Abba S, Peter M, Kohler A, Bonfante P, Chalot M, Martin F, Ottonello S (2006) Gene expression profiling of the nitrogen starvation stress response in the mycorrhizal ascomycete *Tuber borchii*. Fung Genet Biol 43:630–641

Murphy PJ, Langridge P, Smith SE (1997) Cloning plant genes differentially expressed during colonization of roots of *Hordeum vulgare* by the vesicular-arbuscular mycorrhizal fungus *Glomus intraradices*. New Phytol 135:291–301

Nagy F, Karandashov V, Chague W, Kalinkevich K, Tamasloukht M, Xu GH, Jakobsen I, Levy AA, Amrhein N, Bucher M (2005) The characterization of novel mycorrhiza-specific phosphate transporters from *Lycopersicon esculentum* and *Solanum tuberosum* uncovers functional redundancy in symbiotic phosphate transport in solanaceous species. Plant J 42:236–250

Podila GK, Zheng J, Balasubramanian S, Sundaram S, Hiremath S, Brand JH, Hymes MJ (2002) Fungal gene expression in early symbiotic interactions between *Laccaria bicolor* and red pine. Plant Soil 244:117–128

Polidori E, Agostini D, Zeppa S, Potenza L, Palma F, Sisti D, Stocchi V (2002) Identification of differentially expressed cDNA clones in *Tilia platyphyllos-Tuber borchii* ectomycorrhizae using a differential screening approach. Mol Genet Genom 266:858–864

Requena N, Füller P, Franken P (1999) Molecular characterisation of GmFOX2, an evolutionary highly conserved gene from the mycorrhizal fungus *Glomus mosseae*, down-regulated during interaction with rhizobacteria. Mol Plant-Microbe Interact 12:934–942

Requena N, Mann P, Hampp R, Franken P (2002) Early developmentally regulated genes in the arbuscular mycorrhizal fungus *Glomus mosseae*: Identification of GmGIN1, a novel gene with homology to the C-terminus of metazoan hedgehog proteins. Plant Soil 244:129–139

Roussel H, Van Tuinen D, Franken P, Gianinazzi S, Gianinazzi-Pearson V (2001) Signalling between arbuscular mycorrhizal fungi and plants: identification of a gene expressed during early interactions by differential RNA display analysis. Plant Soil 232:13–19

Schaarschmidt S, Gonzalez MC, Roitsch T, Strack D, Sonnewald U, Hause B (2007) Regulation of arbuscular mycorrhization by carbon. The symbiotic interaction cannot be improved by increased carbon availability accomplished by root-specifically enhanced invertase activity. Plant Physiol 143:1827–1840

Schrey SD, Schellhammer M, Ecke M, Hampp R, Tarkka MT (2005) Mycorrhiza helper bacterium *Streptomyces* AcH 505 induces differential gene expression in the ectomycorrhizal fungus *Amanita muscaria*. New Phytol 168:205–216

Schwarz D, Grosch R (2003) Influence of nutrient solution concentration and a root pathogen (*Pythium aphanidermatum*) on tomato root growth and morphology. Sci Hort 97:109–120

Siciliano V, Genre A, Balestrini R, Cappellazzo G, Dewit PJGM, Bonfante P (2007) Transcriptome analysis of arbuscular mycorrhizal roots during development of the prepenetration apparatus. Plant Physiol 144:1455–1466

Smith SE, Read DJ (1997) Mycorrhizal symbiosis. Academic, London.

Tagu D, Python M, Cretin C, Martin F (1993) Cloning symbiosis-related cDNAs from eucalypt ectomycorrhiza by PCR-assisted differential screening. New Phytol 125:339–343

Tahiri-Alaoui A, Antoniw JF (1996) Cloning of genes associated with the colonization of tomato roots by the arbuscular mycorrhizal fungus *Glomus mosseae*. Agronomie 16:699–707

Tamasloukht M, Séjalon-Delmas N, Kluever A, Jauneau A, Roux C, Bécard G, Franken P (2003) Root factor induce mitochondrial-related-gene expression and fungal respiration during the developmental switch from asymbiosis to presymbiosis in the arbuscular mycorrhizal fungus *Gigaspora rosea*. Plant Physiol 131:1468–1478

Tamasloukht M, Waschke A, Franken P (2007) Root exudate-stimulated RNA accumulation in the arbuscular mycorrhizal fungus *Gigaspora rosea*. Soil Biol Biochem 39:1824–1827

van Buuren ML, Maldonado-Mendoza IE, Trieu AT, Blaylock LA, Harrison MJ (1999) Novel genes induced during an arbuscular mycorrhizal (AM) symbiosis formed between *Medicago truncatula* and *Glomus versiforme*. Mol Plant-Microbe Interact 12:171–181

Voiblet C, Duplessis S, Encelot N, Martin F (2001) Identification of symbiosis-regulated genes in *Eucalyptus globulus-Pisolithus tinctorius* ectomycorrhiza by differential hybridization of arrayed cDNAs. Plant J 25:1–12

Waschke A, Sieh D, Tamasloukht M, Fischer K, Mann P, Franken P (2006) Identification of heavy metal-induced genes encoding glutathione S-transferases in the arbuscular mycorrhizal fungus *Glomus intraradices*. Mycorrhiza 17:1–10

Weidmann S, Sanchez L, Descombin J, Chatagnier O, Gianinazzi S, Gianinazzi-Pearson V (2004) Fungal elicitation of signal transduction-related plant genes precedes mycorrhiza establishment and requires the dmi3 gene in *Medicago truncatula*. Mol Plant-Microbe Interact 17:1385–1393

Wulf A, Manthey K, Doll J, Perlick A, Franken P, Linke B, Meyer F, Küster H, Krajinski F (2003) Detection of highly specific transcriptional changes of the model plant *Medicago truncatula* in response to arbuscular mycorrhiza development. Mol Plant-Microbe Interact 16:306–314

Zeppa S, Guescini M, Potenza L, Agostini D, Polidori E, Stocchi V (2000) Analysis of gene expression in the vegetative and fructification phases of the white truffle, *Tuber borchii* Vittad., by mRNA differential display. Biotechnol Lett 22:307–312

Chapter 5
Protein Profiling Analyses in Arbuscular Mycorrhizal Symbiosis

Ghislaine Recorbet and Eliane Dumas-Gaudot

Abstract Because proteins are well known as key effectors of plant responses to environmental cues including recognition, signaling, transport, and defense reactions, interest has been paid to characterize those involved in the establishment and functioning of arbuscular mycorrhizal (AM) symbiosis. The recent development of high throughput techniques has enabled large-scale analyses of symbiosis-related proteins. Different proteomic strategies have been established depending on the symbiotic stage targeted and on the abundance of mycorrhizal material. In mature mycorrhiza, sub-cellular proteomic approaches have been developed in the model legume *Medicago truncatula* to target symbiosis-related membrane proteins eligible for nutrient transport and signaling between symbionts upon arbuscule formation. Modifications in the *M. truncatula* root proteome during early stages of AM symbiosis have also been investigated by comparing protein patterns of non-inoculated roots and roots synchronized for appressorium formation. Concomitantly, proteomic approaches have been developed on in vitro-grown mycorrhiza to identify extraradical fungal proteins along with endomycorrhizins. The genome sequencing programs launched for *M. truncatula* and *Glomus intraradices* are likely to provide additional knowledge about AM symbiosis-related proteins.

5.1 Introduction

As biotrophic micro-organisms, arbuscular mycorrhizal (AM) fungi have to either avoid or suppress plant defense reactions together with redirecting the host metabolic flow to their benefit without harming the host plant. The mechanisms by which this is achieved are almost unknown, but proteins happen to take the lion's share in the paradigms that currently govern plant–biotrophic microbe interactions by playing key roles in mediating recognition, signaling, nutrient transport, plant

G. Recorbet and E. Dumas-Gaudot
UMR 1088 INRA/CNRS 5184/UB Plante-Microbe-Environnement
INRA/CMSEBP 86510, 21065 Dijon Cedex, France
e-mail: dumas@epoisses.inra.fr

C. Azcón-Aguilar et al. (eds.), *Mycorrhizas - Functional Processes and Ecological Impact*,
DOI: 10/1007-978-3-540-87978-7_5,

cell differentiation, and compatibility (Panstruga 2003; Schulze-Lefert and Panstruga 2003). In this respect, proteomics is likely to be one of the best methodologies to decipher some key features of the AM symbiosis. However, roots interacting with AM fungi only display a limited amenability to proteomic analyses by reason of restricted available amounts of pre-symbiotic and symbiotic structures along with very few genomic resources.

Due to the obligate biotrophic status of the mycorrhizal microsymbiont, the fungal asymbiotic phase is restricted to spore germination and the production of a limited amount of mycelium (Balestrini and Lanfranco 2006). Likewise, studies during initial contact between AM fungi and plant roots suffer from difficulties in synchronizing development in the two symbionts (Weidmann et al. 2004). Finally, mycorrhizal symbiosomes only represent a limited amount of biological material that develops within host roots. Contrasting with haustoria of parasitic rusts and powdery mildews that can be purified from infected plant tissues (O'Connell and Panstruga 2006 and references therein), fungal arbuscules have not yet been isolated in sufficient amounts from root tissues. Protein identification using mass spectrometry (MS) techniques also largely depends on genome sequence information. According to He et al. (2004), a simple peptide mass fingerprint (PMF) is sufficient to identify 50–90% of proteins correctly when a complete genome sequence exists. For organisms lacking a complete sequence, tandem mass spectrometry (MS/MS) must be used, which reduces throughput. In silico MS-based protein identification relies on exact matching of masses of peptides and/or peptides fragments to corresponding masses calculated from sequences from databases entries, and it will fail to identify proteins when no sequence is present. Fortunately, many proteins in different species are homologous, permitting identification by their sequence similarity to known homologues from phylogenetically related species (Liska et al. 2004). However, so far very few genomes of plant-colonizing fungi have been annotated (http://www.broad.mit.edu/annotation/fgi/;http://genome.jgi-psf.org/euk/), and they mainly encompass those from pathogenic fungi including *Botrytis cinerea*, *Fusarium graminearum*, *Magnaporthe grisea*, *Ustilago maydis*, *Sclerotinia sclerotiorum*, *Stagonospora nodorum*, *Mycosphaerella graminicola,* and *Nectria haematococca.* When regarding mutualistic fungi interacting with plants, a complete genome sequence is only available for the ectomycorrhizal fungus *Laccaria bicolor* that forms symbiosis with many northern temperate forest trees (http://genome.jgi-psf.org/euk/). Although the United States Department of Energy Joint Institute (http://genome.jgi-psf.org/euk/) also launched a sequencing program in 2004 for the genome of the AM fungus *G. intraradices*, no complete data are yet available. *G. intraradices* was chosen as a model species for AM fungi due of its apparently small genome (15 Mb), its ubiquity in different ecosystems, and its ability to grow in vitro in dual culture with transformed roots (reviewed in Martin et al. 2004). Several expressed sequence tag (EST) libraries have been constructed using different species and genera of AM fungi (see Chapter 4 by Arlt et al.), but only about 5,000 EST sequences have been deposited in GenBank dbEST (http://www.ncbi.nlm.nih.gov/dbEST). In the face of the lack of database sequence data corresponding to Glomeromycota, MS/MS-based identification of AM fungal proteins remains a challenge.

Since the pioneering work of Bestel-Corre et al. (2002), in which the first MS-based identification of mycorrhiza-related proteins was reported, it has become evident that many plant and fungal symbiosis-related proteins may escape such non-targeted analyses performed on whole plant roots, not only because the corresponding amino-acid sequences are lacking in databases but also because of their low abundance or specific structural features. Much effort has thus been put into methodological devices allowing the enrichment of biological extracts for rare proteins relevant to the establishment and functioning of AM symbiosis. Depending on the symbiotic stage targeted and on the abundance of mycorrhizal material, different proteomic strategies have been set up. In mature mycorrhiza, subcellular proteomic approaches have been developed in the model legume *M. truncatula* to target symbiosis-related membrane proteins eligible for nutrient transport and signaling between symbionts upon arbuscule formation. Modifications in the *M. truncatula* root proteome during early stages of AM symbiosis have also been investigated by comparing protein patterns from non-inoculated roots and roots synchronized for appressorium formation. Concomitantly, proteomic approaches have been developed on in vitro-grown mycorrhiza to identify extraradical fungal proteins along with genuine endomycorrhizins.

5.2 Proteomics in Mature Mycorrhiza

In view of the large surface area in which both plant and fungal membranes are in close contact in arbuscule-containing cells, it is widely admitted that plant–fungus interfaces play a major role in nutritional and signal exchanges between both partners leading to a considerable interest into characterizing the molecules transferred and the mechanisms involved (see Chapter 6 by Ferrol and Pérez-Tienda). Sub-cellular proteomics, defined as the large-scale analysis of proteins purified from a cell compartment, has emerged as a promising tool to enrich extracts containing specific proteins. Regarding mycorrhizal associations, two main approaches have been conducted in an attempt to identify plasma membrane (PM) proteins of plant and fungal origin regulated upon AM symbiosis.

5.2.1 Microsome Enrichment

As a preliminary step towards the identification of AM-responsive root PM proteins, a procedure aimed at gaining access to proteins associated with total membrane structures, i.e., microsomes, was developed by combining a subcellular fractionation process with two-dimensional electrophoresis (2-DE) (Benabdellah et al. 1998; Valot et al. 2005). Using differential solubilization of membrane proteins in a chloroform/methanol (C/M) mixture, both peripheral and partially hydrophobic proteins were precipitated in the insoluble C/M fraction (Valot et al. 2004, 2005; Fig. 5.1).

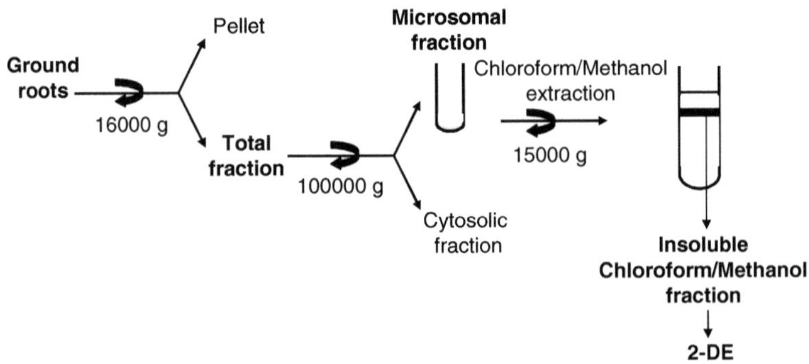

Fig. 5.1 Schematic representation of the procedure used to resolve *Medicago truncatula* root microsomal proteins

Because strictly hydrophobic proteins escape two-dimensional gel-based proteomics, only this latter fraction could be analyzed by 2-DE. Taking advantage of the development of mass spectrometry and of bioinformatic tools, 96 out of 440 well-resolved microsomal root proteins could be identified by matrix-assisted laser desorption ionization-time of flight (MALDI-TOF) peptide mass fingerprinting when mining the *M. truncatula* clustered EST database (Valot et al. 2004). Among them, 21 proteins displayed one possible transmembrane (TM) domain. This procedure was also used to investigate membrane-associated proteins that are regulated in response to the fully established AM symbiosis between *M. truncatula* and *G. intraradices* (Valot et al. 2005). Comparison of 2-DE profiles of the insoluble C/M fraction from non-inoculated and fungus-inoculated roots revealed 36 membrane-associated protein modifications related to the functional AM symbiosis, corresponding to 15 induced, 3 up-regulated and 18 down-regulated polypeptide spots. Among the 25 mycorrhiza-responsive proteins that could be identified following MALDI-TOF and/or tandem mass spectrometry (MS/MS) analyses, 19 were found not to be regulated by a phosphate supply, suggesting they might be candidate markers of the AM symbiosis. Among them were an up-regulated 53-kDa nodulin and an up-regulated acid phosphatase, along with a down-regulated lipoxygenase fragment. This was the first report concerning the over-expression of a nodulin 53-like protein in mycorrhizal roots. Because this nodulin was found to be located in the peribacteroid membrane of root nodules (Saalbach et al. 2002), it was suggested that this protein could also be associated with the periarbuscular membrane. Regarding the acid phosphatase, the increase in the amount and activity of a similar enzyme also found to occur during nodule development, was proposed to be important for efficient nodule metabolism (Penheiter et al. 1997). The decreased abundance in a lipoxygenase fragment suggested that this enzyme might be down-regulated during AM symbiosis. The lipoxygenase pathway is induced during some plant–pathogen interactions leading to the production of signaling molecules, activation of defense genes and phytoalexin accumulation (Blee 2002). Earlier studies have shown that defense-response genes involved in phytoalexin biosynthesis are down-regulated as the symbiosis develops (Harrison and Dixon 1993).

5.2.2 Plasma Membrane Enrichment

Because very little is known about the mycorrhizal periarbuscular membrane, attempts to purify this compartment have relied on its plant-derived PM origin. Due to the under-representation of intrinsic plant membrane proteins on 2-DE gels, strategies have thus been developed in order to ensure a more thorough coverage of PM proteins (Benabdellah et al. 2000; Valot et al. 2006). A qualitative comparative proteomic approach of the PM fractions of *M. truncatula* roots colonized or not with *G. intraradices* was set up (Valot et al. 2006). After root microsome extraction, a discontinuous sucrose gradient-based protocol was used to enrich to a similar extent the PM fractions from control and mycorrhizal roots. To gain access to an enlarged inventory of the proteins present in the PM fractions, two complementary proteomic methodologies were used: (1) liquid trypsin digestion of the PM fraction and separation of the generated peptides by two-dimensional liquid chromatography (2D LC)-MS/MS, and (2) systematic LC-MS/MS analyses of SDS-PAGE resolved proteins (GeLC-MS/MS). The PM fractions of two independent control root experiments were analyzed by the two above-cited methods resulting in the identification of 78 proteins. Fifty-six percent of these were predicted to contain one or more TM domains. Additionally, among the 34 proteins with no TM domain, 5 contained a predicted C-terminal peptide signal for glycosylphosphatidylinositol (GPI)-anchorage to the PM. To identify proteins only present in the PM fraction from *G. intraradices* inoculated roots, the complete list of non-redundant proteins from the two control analyses was used as the reference protein database and only proteins identified with at least four peptides in the PM fraction of inoculated roots were taken into account. Using these criteria, two proteins, corresponding to a PM proton-efflux P-type ATPase (Mtha1) and a blue copper binding protein (Bcp), were found induced in mycorrhizal *M. truncatula* roots as compared to control ones (see Chapter 4 by Arlt et al.). Mtha1 is a highly intrinsic protein containing ten possible TM domains whose activity generates an electrochemical H^+ gradient across the cell plasma membrane using ATP. PM proton-efflux P-type ATPases are believed to produce the driving force necessary for the uptake and efflux of solutes across the plant–microbe interface. Because the expression of *Mtha1* was previously found to be located in arbuscule-containing cells (Krajinski et al. 2002), it was suggested that Mtha1 was likely to be located in the periarbuscular membrane. H^+-ATPases have also been detected in the symbiosome membrane of root nodules (Catalano et al. 2004) and of mycorrhizal roots (Gianinazzi-Pearson et al. 2000). By contrast, ATPase activity is generally lacking from the symbiosome membranes of pathogenic interactions (Parniske 2000). The Bcp is a small protein characterized by a copper binding site together with a putative signal for GPI anchoring at the plasma membrane. GPI-anchored proteins are also known to localize in cholesterol- and glycosphingolipid-rich domains, called lipid rafts, on the cell surface and have been implicated in microbe–host cell interactions and infections (Bhat and Panstruga 2005). A GPI-anchored nodulin (ENOD16) has been detected in the plant-derived symbiosome membrane of *M. truncatula* interacting with *Sinorhizobium meliloti* (Catalano et al. 2004). The Bcp encoding gene has been previously reported as induced in AM-inoculated roots by in silico EST analyses, cDNA arrays and in a subtractive cDNA library (Journet et al. 2002; Manthey et al. 2004; see Chapter 4 by Arlt et al.). Using reporter gene expression, it was also demonstrated that *Bcp* expression is specifically up-regulated in arbuscule-containing regions of mycorrhizal *M. truncatula* roots, pointing to a possible location in the periarbuscular membrane (Hohnjec et al. 2005).

5.3 Proteomics at the Appressorium Stage

In contrast to many plant–pathogenic fungus interactions (Tucker and Talbot 2001), very little is known about the molecular events occurring prior to and during initial contact between the two symbionts in mycorrhiza. Modifications in the *M. truncatula* root proteome during early stages of AM interactions was recently investigated by 2-DE comparisons of protein patterns from non-inoculated roots and roots synchronized for *G. intraradices* appressorium formation on wild-type (J5), penetration-defective (TRV25, *dmi3*), and autoregulation-defective (TR122, *sunn*) genotypes (Amiour et al. 2006). Statistically significant changes in protein abundance were recorded between inoculated and non-inoculated roots, showing that pre-penetration triggered root proteome modifications (Fig. 5.2). The use of MALDI-TOF mass spectrometry enabled the first identification of root proteins whose accumulation was altered in response to early stages of AM symbiosis. Among six root proteins that displayed an increase in abundance in *G. intraradices*-inoculated J5 wild-type plants, a chalcone reductase (CHR), a glutathione-dependent dehydroascorbate reductase (DHAR), a cyclophilin (CYP), and an actin depolymerization factor (ADF) were identified. Root proteins that decreased in density upon appressorium formation corresponded to two glutathione transferases (GST). The identification in wild-type *M. truncatula* of a CHR that increases in abundance prior to root penetration is consistent with the hypothesis that activation of the phenylpropanoid pathway is mediated by early recognition events between the two symbionts (García-Garrido and Ocampo 2002). In view of the decreased abundance of two GSTs in response to *G. intraradices*, it was also suggested that some plant defense reactions might be lowered upon symbiont contact. Because a role in signal transduction has been proposed for the DHAR, CYP, and ADF which displayed an increase in accumulation in *G. intraradices*-inoculated wild-type roots, this study also pointed to the elicitation of new signaling proteins in response to appressorium formation. Although *G. intraradices* differentiated a similar number of appressoria on the roots of the three *M. truncatula* genotypes 5 days after inoculation, major differences with few overlaps were found when comparing the groups of proteins that responded to appressorium formation between wild-type and mutant genotypes.

Among the modifications in the appressorium-responsive root proteome triggered by mutation in *DMI3* were the *DMI3*-dependent responses of DHAR, CYP, and ADF, which was consistent with a role for *DMI3* in decoding and transducing a calcium spiking signal to downstream responses. Among the differences observed

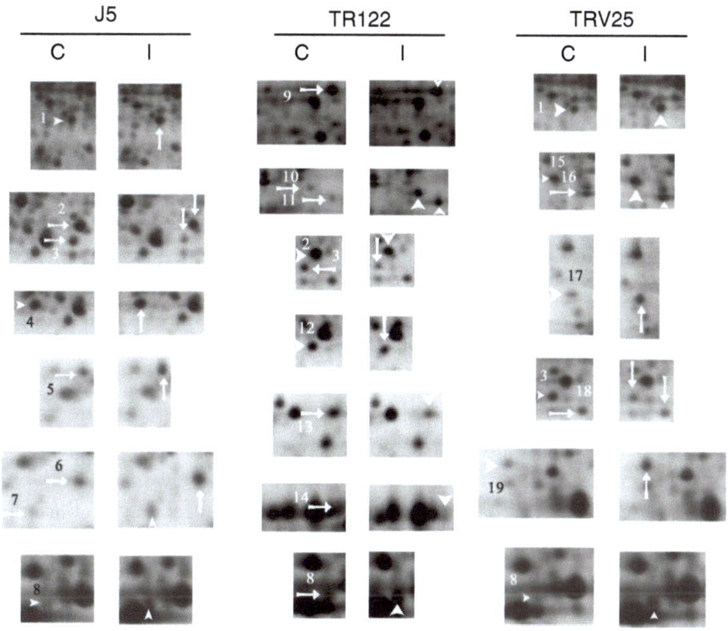

Fig. 5.2 Comparison of root protein abundance in 5-day old wild-type (*J5*), hyper-mycorrhizal (*TR122*) and mycorrhiza-defective (*TRV25*) *M. truncatula* plants either *Glomus intraradices*-inoculated (*I*) or not (*C*). *Horizontal* and *vertical arrows* label root proteins from control and *G. intraradices*-inoculated plants, respectively. *Up*- and *down-headed arrows* indicate increase and decrease in root protein abundance upon appressorium formation in inoculated plants compared to controls, respectively

when comparing the proteins that responded to appressorium formation between the wild-type genotype and the *sunn* mutant defective for the autoregulation of arbuscule formation (hyper-mycorrhizal phenotype) were changes in the abundance of a CHR, a narbonin, two annexins (MtAnn1), a quinone reductase, a protease inhibitor, and DHAR, CYP, and ADF proteins. This indicates that differences in protein accumulation occurring prior to root penetration preceded the over-development of fungal arbuscules in the hyper-mycorrhizal mutant. Because earlier studies have reported increased *MtAnn1* expression (Carvalho-Niebel et al. 1998), and decreased levels of narbonin, Kunitz-type trypsin inhibitor, and quinone reductase transcripts, in roots of the *sunn* mutant upon *S. meliloti* inoculation (El Yahyaoui et al. 2004), these results also point to common early regulatory pathways that may be shared between nodulation and mycorrhizal symbioses. Based on the identification of proteins which showed a modified response to appressorium formation in the *M. truncatula sunn* mutant, it was also suggested that defense reactions, cytoskeleton rearrangement, and/or membrane trafficking events, and auxin signaling might be involved in the early control of mycorrhization (Amiour et al. 2006).

5.4 Fungal Proteomics

To understand the functionality of the AM mycosymbionts, fungal proteins involved in the interaction have to be analyzed. Such studies have been hindered due to the lack of appropriate methodology to obtain sufficient amounts of fungal material with which experiments could be carried out. This has been overcome in recent years after the development of AM monoxenic cultures allowing large amounts of contaminant-free extraradical fungal material to be obtained. Using in vitro Ri T-DNA transformed roots of carrot (*Daucus carota* L.) inoculated with *G. intraradices*, the first reference map for the extraradical proteome of an AM fungus was established, in which over 450 spots were displayed within the window of p*I* 3–10 and molecular mass 10–100 kDa (Fig. 5.3) (Dumas-Gaudot et al. 2004). To obtain information about the *G. intraradices* extraradical proteome, 14 spots were selected on the basis of their signal intensity under UV light exposure and subjected to tandem mass spectrometry. Protein sequences could be obtained for eight proteins (Gi24, Gi90, Gi176, Gi193, Gi275, Gi378, Gi391, and Gi431). Among the spots that gave a hit, Gi 431, representing the second most abundant protein spot, matched with a sequence retrieved from a root-induced cDNA library from *G. intraradices*. Only one sequence was obtained for spot Gi391 that matched with a cDNA from spores of *Glomus versiforme*, another species of Glomeromycota often used as a model

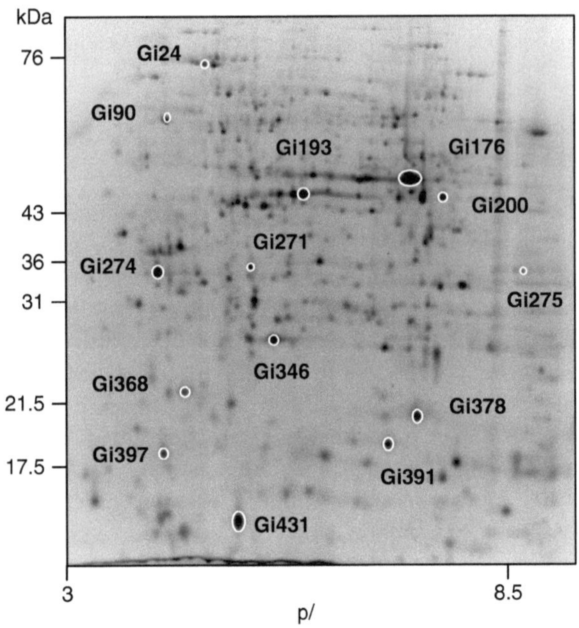

Fig. 5.3 Two-dimensional electrophoretic map of *G. intraradices* extra-radical proteins (300 μg) following sypomhy staining. *Circled spots* indicate proteins analysed by tandem mass spectrometry

fungus. However, for these two spots, no homology with protein having an assigned function was found. Among the extra-radical proteins from *G. intraradices* were identified enzymes of central cell metabolism for spots Gi176 and Gi90. The most abundant spot, Gi176, matched with a sequence retrieved from a library of cDNAs from *G. intraradices* germinating spores, for which an oxydo-reductase function was predicted. Several sequences were obtained for spot Gi90, among which all the homologies found converged to a *Neurospora crassa* ATP synthase beta mitochondrial precursor protein. Spot Gi24 matched with a heat shock protein (HSP70). Finally, concerning Gi193, matches were obtained with sequences from a cDNA library corresponding to *M. truncatula* roots inoculated for 3 weeks with *G. intraradices*. For this spot, homology was found with NmrA-like proteins that are transcription repressors involved in the regulation of nitrogen metabolism in *Aspergillus nidulans*. In an attempt to identify fungal proteins in mycorrhizal roots of carrots, the protein patterns from in vitro-grown mycorrhizal and non-mycorrhizal carrot roots were also compared, resulting in the detection of four induced spots designated m1 to m4. Protein sequences could only be obtained for spots m3 and m4. Because similar mass spectra were obtained for these two spots, it was concluded that they corresponded to the same protein. The sequences of the m3/m4 peptide showed 83% of identity and 88% of similarity with MyK15, a protein detected in mycorrhizal wheat roots but still of unknown function (Fester and Strack 2002). The Myk15 protein was also shown to exhibit sequence homology to the MtBC37G05 EST clone from *G. intraradices*-inoculated *M. truncatula* roots.

Although 2-DE still remains the most convenient workflow for quantitative approaches, shotgun proteomic methods, including GeLC-MS/MS and 2D LC-MS/MS, are more efficient in proteome coverage than qualitative 2-DE in that they allow to gain access to low abundance and/or membrane proteins. The development of such methods to decipher the symbiotic proteome of AM fungi is likely to increase our knowledge about the functional genome of these symbionts.

5.5 Conclusions and Future Perspectives

The use of proteomic approaches more specifically directed towards well-defined stages/tissues/subcellular fractions of roots interacting with AM fungi has enabled the identification of additional symbiosis-related proteins. Interestingly, many of the above-listed proteins tend to support the existence of overlaps between the developmental programs of AM and nodulation symbioses. Despite the importance of protein phosphorylation in signal transduction during cell defense responses in plant–microbe interactions, phosphoproteomic approaches have so far only been developed for pathogenic associations (Thurston et al. 2005). The progress made in the analysis of phosphopeptides should make it possible to identify new components of the regulatory pathways involved in the establishment of AM symbiosis. Likewise, changes in membrane and cell wall proteins are also likely to occur when micro-symbionts contact with their hosts (see Genre et al. 2005), so that the development of subcellular proteomics targeting the early stages of symbiosis is expected to be very fruitful.

When regarding proteomic studies devoted to symbiosomes, recent technical developments in mass spectrometry analyses are expected to provide new insights about the intrinsic proteins that are regulated in mycorrhiza. Methods such as Isotope Coded Affinity Tag (ICAT) labeling that permit the comparison of peptide abundance between two samples appear well suited to enable a quantitative comparison of membrane proteins between symbiotic and non-symbiotic roots. The possibility to detect GPI-anchored proteins by their enzymatic release from PM fractions using phospholipases will also allow their characterization during symbiosis functioning (Rossignol et al. 2006). The use of laser-capture micro-dissection-based techniques that can be coupled to proteome profiling will also help to enrich samples in fungal structures. Regarding proteins of plant origin, transformation methods coupled to protein labeling and microscopy are among the strategies that will help to validate their putative location. Finally, the genome sequencing programs launched for *M. truncatula* and *G. intraradices* are likely to provide additional knowledge about AM symbiosis-related proteins.

References

Amiour N, Recorbet G, Robert F, Gianinazzi S, Dumas-Gaudot E (2006) Mutations in *DMI3* and *SUNN* modify the appressorium-responsive root proteome in arbuscular mycorrhiza. Mol Plant Microbe Interact 19:988–997

Balestrini R, Lanfranco L (2006) Fungal and plant gene expression in arbuscular mycorrhizal symbiosis. Mycorrhiza 16:509–524

Benabdellah K, Azcon-Aguilar C, Ferrol N (1998) Soluble and membrane symbiosis-related polypeptides associated with the development of arbuscular mycorrhizas in tomato (*Lycopersicon esculentum*). New Phytol 140:135–143

Benabdellah K, Azcon-Aguilar C, Ferrol N (2000) Alterations in the plasma membrane polypeptide pattern of tomato roots (*Lycopersicon esculentum*) during the development of arbuscular mycorrhiza. J Exp Bot 51:747–754

Bestel-Corre G, Dumas-Gaudot E, Poinsot V, Dieu M, Dierick J, van Tuinen D, Remacle J, Gianinazzi-Pearson V, Gianinazzi S (2002) Proteome analysis and identification of symbiosis-related proteins from *Medicago truncatula* Gaertn. by two-dimensional electrophoresis and mass spectrometry. Electrophoresis 23:122–137

Bhat RA, Panstruga R (2005) Lipid rafts in plants. Planta 223:5–19

Blee E (2002) Impact of phyto-oxylipins in plant defense. Trends Plant Sci 7:315–321

Carvalho-Niebel F, Lescure N, Cullimore JV, Gamas P (1998) The *Medicago truncatula MtAnn1* gene encoding an annexin is induced by Nod factors and during the symbiotic interaction with *Rhizobium meliloti*. Mol Plant Microbe Interact 11:504–513

Catalano CM, Lane WS, Sherrier DJ (2004) Biochemical characterization of symbiosome membrane proteins from *Medicago truncatula* root nodules. Electrophoresis 25:519–531

Dumas-Gaudot E, Valot B, Bestel-Corre G, Recorbet G, St-Arnaud M, Fontaine B, Dieu M, Raes M, Saravanan R Gianinazzi S, (2004) Proteomics as a way to identify extra-radicular fungal proteins from *Glomus intraradices*-RiT-DNA carrot root mycorrhizas. FEMS Microbiol Ecol 48:401–411

El Yahyaoui F, Kuster H, Ben Amor B, Hohnjec N, Puhler A, Becker A, Gouzy J, Vernie T, Gough C, Niebel A, Godiard L, Gamas P (2004) Expression profiling in *Medicago truncatula* identi-

fies more than 750 genes differentially expressed during nodulation, including many potential regulators of the symbiotic program. Plant Physiol 136:3159–3176
Fester T, Strack D (2002) A mycorrhiza-responsive protein in wheat roots. Mycorrhiza 12:219–222
García-Garrido JM, Ocampo JA (2002). Regulation of the plant defence response in arbuscular mycorrhizal symbiosis. J Exp Bot 53:1377–1386
Genre A, Chabaud M, Timmers T, Bonfante P, Barker DG (2005) Arbuscular mycorrhizal fungi elicit a novel intracellular apparatus in *Medicago truncatula* root epidermal cells before infection. Plant Cell 17:3489–3499
Gianinazzi-Pearson V, Arnould C, Oufattole M, Arango M, Gianinazzi S (2000) Differential activation of H+-ATPase genes by an arbuscular mycorrhizal fungus in root cells of transgenic tobacco. Planta 211:609–613
Harrison M, Dixon RA (1993) Isoflavonoid accumulation and expression of defense gene transcripts during the establishment of vesicular-arbuscular mycorrhizal associations in roots of *Medicago truncatula*. Mol Plant Microbe Interact 6:643–654
He F, Emmet MR, Håkansson K, Hendrikson CL, Marshall A (2004) Theoretical and experimental prospects for protein identification based solely on accurate mass measurement. J Proteome Res 3:61–67
Hohnjec N, Vieweg A, Puhler A, Becker A, Kuster H (2005) Overlaps in the transcriptional profiles of *Medicago truncatula* roots inoculated with two different *Glomus* fungi provide insights into the genetic program activated during arbuscular mycorrhiza. Plant Physiol 137:1283–1301
Journet EP, van Tuinen D, Gouzy J, Crespeau H, Carreau V, Farmer MJ, Niebel A, Chiex T, Saillon O, Chatagnier O, Godiard L, Micheli F, Kahn D, Gianinazzi-Pearson V, Gamas P (2002) Exploring root symbiotic programs in the model legume *Medicago trunctula* using EST analysis. Nuc Acids Res 15:5579–5592
Krajinski F, Hause B, Gianinazzi-Pearson V, Franken P (2002) *Mtha1*, a plasma membrane H^+-ATPase gene from *Medicago truncatula*, shows arbuscule-specific induced expression in mycorrhizal tissue. Plant Biol 4:754–761
Liska AJ, Popov A, Sunyaev S, Coughlin P, Habermann B, Shevchenko A, Bork P, Eric Karsenti E, Shevchenko A (2004) Homology-based functional proteomics by mass spectrometry: application to the *Xenopus* microtubule-associated proteome. Proteomics 4:S2707–2721
Manthey K, Krajinski F, Hohnjec N, Firnhaber C, Puhler A, Perlick AM, Kuster H (2004) Transcriptome profiling in root nodules and arbuscular mycorrhiza identifies a collection of novel genes induced during *Medicago truncatula* root endosymbioses. Mol Plant Microbe Interact 17:1063–1077
Martin F, Tuskan GA, DiFazio SP, Lammers P, Newcombe G, Podila GK (2004) Symbiotic sequencing for the *Populus* mesocosm. New Phytol 161:330–335
O'Connell RJ, Panstruga R (2006) Tête a tête inside a plant cell: establishing compatibility between plants and biotrophic fungi and oomycetes. New Phytol 171:699–718
Panstruga R (2003) Establishing compatibility between plants and obligate biotrophic pathogens. Curr Opin Plant Biol 6:320–326
Parniske M (2000) Intracellular accommodation of microbes by plants: a common developmental program for symbiosis and disease? Curr Opin Plant Biol 3:320–328
Penheiter AR, Duff SM, Sarath G (1997) Soybean root nodule acid phosphatase. Plant Physiol 114:597–604
Rossignol M, Peltier JB, Mock HP, Matros A, Maldonado AM, Jorrín J (2006) Plant proteome analysis: a 2004–2006 update. Proteomics 6:5529–5548
Saalbach G, Erik P, Wienkoop S (2002) Characterisation by proteomics of peribacteroid space and peribacteroid membrane preparations from pea (*Pisum sativum*) symbiosomes. Proteomics 2:325–337
Schulze-Lefert P, Panstruga R (2003) Establishment of biotrophy by parasitic fungi and reprogramming of host cells for disease resistance. Ann Rev Phytopathol 41:641–667
Thurston G, Regan S, Rampitsch C, Xing T (2005) Proteomic and phosphoproteomic approaches to understand plant–pathogen interactions. Physiol Mol Plant Pathol 66:3–11

Tucker SL, Talbot NJ (2001) Surface attachment and pre-penetration stage development by plant pathogenic fungi. Ann Rev Phytopathol 39:385–417

Valot B, Dumas-Gaudot E, Gianinazzi S (2004) Sub-cellular proteomic analysis of a *Medicago truncatula* root fraction. Phytochemistry 65:1721–1732

Valot B, Dieu M, Recorbet G, Raes M, Gianinazzi S, Dumas-Gaudot E (2005) Identification of membrane-associated proteins regulated by the arbuscular mycorrhizal symbiosis. Plant Mol Biol 59:565–580

Valot B, Negroni L, Zivy M, Gianinazzi S, Dumas-Gaudot E (2006) A mass spectrometric approach to identify arbuscular mycorrhiza-related proteins in root plasma membrane fractions. Proteomics 6:S145–155

Weidmann S, Sanchez L, Descombin J, Chatagnier O, Gianinazzi S, Gianinazzi-Pearson V (2004) Fungal elicitation of signal transduction-related plant genes precedes mycorrhiza establishment and requires the *dmi3* gene in *Medicago truncatula*. Mol Plant Microbe Interact 17:1385–1393

Chapter 6
Coordinated Nutrient Exchange in Arbuscular Mycorrhiza

Nuria Ferrol and Jacob Pérez-Tienda

Abstract Arbuscular mycorrhizal (AM) fungi, as obligate biotrophs, rely for their growth and activity on carbon provided by their host plant and, in exchange, they improve the mineral nutrition of the plant, in particular the acquisition of phosphorus and to some extent of nitrogen and other minor nutrients. This nutrient exchange takes place across the symbiotic interfaces that are developed as the fungus colonizes the root system. In this chapter, we provide an overview of the biochemical and molecular mechanisms involved in nutrient transport processes in AM, with special emphasis on those underlying the bidirectional nutrient exchange between symbionts at the symbiotic interfaces.

6.1 Introduction

For approximately 80% of all known plant species, arbuscular mycorrhizal (AM) roots dominate nutrient uptake at the soil–plant interface. This symbiosis is extremely important for the uptake of inorganic phosphate (Pi), but also contributes to the uptake of nitrogen (N) and various trace elements such as copper and zinc (Smith and Read 1997). In return for the improved nutrient supply, up to 20% of the assimilated carbon from the plant is translocated to the root for the formation, maintenance, and function of the AM fungal structures (Graham 2000). The mutualistic interaction between AM fungi and plants is, therefore, based on a bidirectional transfer of mineral nutrients and carbohydrates. Since there is no symplastic continuity between both partners, nutrients must pass an interfacial apoplast before they can be absorbed. In order to have a full picture of the functioning of this mutualism, we need to understand the mechanisms underlying nutrient transport processes across AM interfaces. It is important to know the molecular forms in which nutrients are transported, where they are transferred, and lastly what regulates these transport systems.

N. Ferrol(✉) and J. Pérez-Tienda
Departamento de Microbiología del Suelo y Sistemas Simbióticos
Estación Experimental del Zaidín, CSIC, Profesor Albareda 1, 18008 Granada, Spain
e-mail: nuria.ferrol@eez.csic.es

C. Azcón-Aguilar et al. (eds.), *Mycorrhizas - Functional Processes and Ecological Impact*,
DOI: 10/1007-978-3-540-87978-7_6,

Considerable progress towards understanding the molecular basis of Pi transport in AM has been made during recent years. Presently, less information is available on the transport mechanisms and metabolism of carbon compounds and other mineral nutrients in the symbiosis. However, a survey of recent literature suggests that the gaps in our knowledge are being progressively filled. In this chapter, we focus on the biochemical and molecular mechanisms involved in nutrient transport processes in AM and in the bidirectional nutrient exchange between symbionts at the symbiotic interfaces.

6.2 Creation of a New Compartment: The Symbiotic Interface

During their intraradical growth, AM fungi develop an extended contact area with the root cells, which changes structurally depending on the intercellular or intracellular location of the fungus. The arbuscular interface developed in colonized cortical root cells is believed to function as a key trading place for the exchange of signals and nutrients between the plant and the fungus. Although the fungal hypha penetrates the cell wall of the cortical cell to form the arbuscule, all fungal branches develop inside the cell lumen but they always remain outside the plant plasma membrane. The interface formed between the two symbionts comprises, therefore, the plasma membrane and cell wall of the fungus, a plant-derived membrane which surrounds the arbuscule (the periarbuscular membrane) and an interfacial matrix which separates the periarbuscular membrane from the fungal cell wall. For details about the structural organization and composition of the AM interfaces, readers are referred to Bonfante (2001), Ferrol et al. (2002a) and Balestrini and Bonfante (2005).

6.3 Phosphate Transport in Arbuscular Mycorrhiza

As defined by Smith et al. (2003), mycorrhizal plants have two pathways for Pi uptake: the "direct" pathway at the plant-soil interface, through the root epidermal cells, and the "mycorrhizal" pathway via the fungal mycelium (see also Chapter 7 by Grace et al.). For some plant species–AM fungi combinations, it has been found that AM colonization results in complete inactivation of the direct Pi uptake pathway, and 100% of the phosphorus (P) in plant tissues is provided by the AM fungus (Smith et al. 2004).

6.3.1 Phosphate Uptake from the Soil Solution

Increased Pi acquisition by mycorrhizal plants is mainly due to the extension of the extraradical mycelium into the soil over distances up to several centimeters from the root to take up and translocate Pi from the soil solution to the plant

(Jakobsen et al. 1992). Recent physiological and molecular data suggest that when plants form the AM symbiosis, they alter their Pi acquisition pathways significantly. Gene expression analyses in mycorrhizal roots of different plant species have shown that Pi transporters whose expression decreases with an increase in the P status of the plant, particularly those located at the epidermis, were also down-regulated in the AM symbiosis. To date, down-regulation of Pi transporters has been observed in potato (Rausch et al. 2001), *Medicago truncatula* (Liu et al. 1998), *Lotus japonicus* (Maeda et al. 2006), rice (Paszkowski et al. 2002), barley (Glassop et al. 2005), and maize (Wright et al. 2005). Down-regulation of Pi transporters operating at the plant–soil interface in mycorrhizal roots indicates that the plant relies largely on Pi delivered by the fungal symbiont.

Transport proteins putatively involved in Pi uptake by AM fungi have been identified. In particular, genes encoding high-affinity Pi transporters that are preferentially expressed in the extraradical mycelium, and that are, therefore, likely to be involved in Pi acquisition from the soil solution, have been identified in *Glomus versiforme* (*GvPT*; Harrison and van Buuren 1995), *G. intraradices* (*GiPT*; Maldonado-Mendoza et al. 2001) and *G. mosseae* (*GmosPT*; Benedetto et al. 2005). Functional expression of *GvPT* in yeast showed that its gene product is a high-affinity Pi transporter that operates by proton-coupled symport (Harrison and van Buuren 1995). The H^+-ATPases responsible for the generation of the proton motive force driving the uptake of Pi across the plasma membrane of the extraradical hyphae have been identified in *G. mosseae* (Ferrol et al. 2000; Requena et al. 2003). Detailed expression analyses of *GiPT* and *GmosPT* have revealed that both genes are regulated in a manner typical of genes encoding high-affinity Pi transporters, with maximal expression at the micromolar Pi concentrations usually found in the soil solution. The findings that *GiPT* expression is modulated by the overall P status of the plant (Maldonado-Mendoza et al. 2001), and that *GmosPT* was highly expressed in the intraradical fungal structures (Benedetto et al. 2005), suggest that the fungus may exert control over the amount of Pi that is delivered to the plant.

6.3.2 Phosphorus Metabolism and Translocation in the Fungus

Pi taken up by the fungus is first incorporated into the cytosolic P pool. Excess cytosolic Pi is transported into the fungal vacuoles and condensed into polyphosphates (polyP), linear polymers of three to thousands of Pi residues connected by high-energy bonds, which are believed to play a central role in Pi supply to the plant (Ezawa et al. 2002). The mechanism of polyP translocation from the extraradical to the intraradical hyphae has not yet been elucidated, but it may be driven by cytoplasmic streaming and/or a motile tubular vacuole system (Uetake et al. 2002). The observation that the length of polyP polymers is shorter in mycorrhizal roots than in extraradical hyphae supports the hypothesis that polyP is hydrolyzed in the intraradical fungal structures before Pi is supplied to the host plant (Ohtomo and Saito 2005). Recently, by using a dual-labeling method for the simultaneous *in situ* detection of alkaline phosphatase activity and polyP, it has been shown that this enzyme activity was mainly observed in mature arbuscules where polyP was rarely observed, which supports the idea that polyP is hydrolyzed in the arbuscules and that the alkaline phosphatase is involved in polyP metabolism (Funamoto et al. 2007). As will be presented in the next section, gene expression analyses and RNAi technology have shown that the *L. japonicus* gene *LjPT3* encodes the Pi transporter responsible for the uptake of Pi that is released by the fungus at the symbiotic interface (Maeda et al. 2006). The dual-labeling method described above revealed that in the *LjPT3* knockdown plants the proportion of arbuscules showing high alkaline phosphatase activity was lower than in the vector-control transformants, and that such arbuscules with low activity tended to have high levels of polyP (Funamoto et al. 2007). Furthermore, the expression of *GiALP*, a gene putatively encoding an alkaline phosphatase in the AM fungus *G. intraradices* (Aono et al. 2004), whose transcript levels were found to be higher in the intraradical than in the extraradical hyphae, was suppressed in the *LjPT3* knockdown plants. These data suggest that both polyP metabolism and alkaline phosphatase activity in the arbuscules are controled, in part, by plant Pi uptake via the AM symbiosis. However, it is clear that further studies, such as determining the substrate specificity of the *ALP* gene products, are needed to uncover the mechanisms of polyP breakdown in the intraradical hyphae.

6.3.3 Phosphate Transfer: The Symbiotic Interface at Work

Once Pi is transferred to the arbuscules, it is released from the fungus and transported across the periarbuscular membrane into the cortical cell. The mechanism involved in the release of Pi from the arbuscules is still unknown. Because Pi efflux across the fungal plasma membrane probably follows a concentration gradient, it could be facilitated by an as yet unidentified anion channel, carrier, or pump. In the last few years, the plant transporters implicated in the uptake of the Pi exported across the membrane of the arbuscule have been identified in several plant species. For a detailed review of the mycorrhiza-regulated Pi transporters readers are referred to Bucher (2007) and Javot et al. (2007a).

These transporters are H^+ symporters, which emphasizes the importance of H^+-ATPases and the pH conditions in the interface for nutrient uptake through the mycorrhizal pathway. Up-regulation of H^+-ATPase genes has been reported in different plant species (Murphy et al. 1997; Ferrol et al. 2002b), and induction of plasma membrane H^+-ATPase genes in arbuscule-containing tissues has been demonstrated in tobacco and *M. truncatula* (Gianinazzi-Pearson et al. 2000; Krajinski et al. 2002; see also Chapter 5 by Recorbet and Dumas-Gaudot). Gene expression studies of the plant Pi transporters have revealed that development of the symbiosis induces the novo expression of the so-called mycorrhiza-specific Pi transporters, transporters which are exclusively expressed in mycorrhizal roots, and up-regulation of Pi transporters which have a basal expression in non-mycorrhizal roots, the mycorrhiza up-regulated transporters. Mycorrhiza-specific Pi transporters have been identified in *M. truncatula* (Harrison et al. 2002), rice (Paszkowski et al. 2002), tomato (Nagy et al. 2005), potato, and wheat (Glassop et al. 2005), and mycorrhiza up-regulated Pi transporters in potato (Rausch et al. 2001; Nagy et al. 2005), *L. japonicus* (Maeda et al. 2006), rice (Güimil et al. 2005), barley, maize, (Glassop et al. 2005) and tomato (Nagi et al. 2005; Xu et al. 2007). It is interesting that these genes have divergent sequences despite similarities in their putative physiological functions (Chen et al. 2007).

In situ hybridization and/or *in vivo* promoter analyses of *LePT4*, *StPT3,* and *MtPT4*, the mycorrhiza-induced Pi transporters of tomato, potato, and *M. truncatula*, respectively, demonstrated their predominant, or exclusive, expression in cortical cells colonized by arbuscules, which suggests that the encoded Pi transporters are likely involved in taking up the Pi that is released by the fungus to the interfacial matrix of the arbuscular interface (Rausch et al. 2001; Harrison et al. 2002; Nagy et al. 2005). However, the only protein identified so far in the plant periabuscular membrane surrounding the fine branches of developing and mature arbuscules is the gene product of *MtPT4* (Harrison et al. 2002).

Functional characterization *in planta* of the mycorrhiza-induced Pi transporters is essential to understanding their role in Pi transport in the symbiosis. Knockout mutants of these genes will greatly facilitate their functional analysis. In *L. japonicus*, the knockdown mutant of *LjPT3* exhibited a reduction in mycorrhizal colonization, in Pi uptake via the mycorrhizal pathway, and in growth in Pi-limiting conditions (Maeda et al. 2006). Similarly, it has also been shown that complete silencing of *MtPT4* expression results in premature death of arbuscules and no improvement in Pi acquisition by inoculation with mycorrhizal fungi (Javot et al. 2007b). Based on these data, it was proposed that Pi uptake through MtPT4 serves, either directly or indirectly, as a signal to the plant cell of the presence of a beneficial symbiont. This hypothesis would explain why the symbiosis does not develop under extremely low Pi conditions and why colonization is dramatically stimulated by the application of small amounts of Pi.

6.4 Nitrogen Transport

Recently, advances have also been made in understanding the mechanisms of N transport, another essential nutrient that is believed to play a role in the symbiosis. AM fungi have been shown to directly take up both organic and inorganic N from the soil solution and to transfer it to their host plants. Nitrogen can be taken up as NH_4^+ (Johansen et al. 1992; Frey and Schüepp 1993), NO_3^- (Tobar et al. 1994; Bago et al. 1996), and amino acids (Hawkins et al. 2000). However, a clear preference for NH_4^+ has been demonstrated (Hawkins et al. 2000; Toussaint et al. 2004) which is explained, at least in part, by the extra energy the fungus must expend in reducing NO_3^- to NH_4^+ before it can be incorporated into organic compounds (Marzluf 1996). AM fungi have also been shown to accelerate the decomposition of organic matter, thereby increasing N availability and uptake (Hodge et al. 2001).

Currently, we have little information about the mechanisms of N uptake by AM fungi. Identification and characterization of the fungal genes encoding the proteins operating at the extraradical hyphae together with *in vivo* measurements of the uptake kinetics of different nitrogenous compounds will lead to a better understanding of these processes. Only genes encoding ammonium transporters have been identified so far in *G. intraradices: GintAMT1* (López-Pedrosa et al. 2006), *GintAMT2* (J. Pérez-Tienda et al., unpublished results), plus a third one that is preferentially expressed in intraradical hyphae (Govindarajulu et al. 2005). Functional expression of *GintAMT1* in yeast showed that it encodes a high-affinity NH_4^+ transporter, and gene expression analyses revealed that is transcriptionally up-regulated by micromolar concentrations of NH_4^+ (López-Pedrosa et al. 2006). These data suggest that *GintAMT1* is involved in NH_4^+ uptake by the extraradical mycelia when it is present at micromolar concentrations. Further studies, such as functional characterization, detailed gene expression analyses, and location of the encoded proteins are needed to understand the precise role of the other two *G. intraradices* NH_4^+ transporters.

Recently, elegant isotope labeling experiments in combination with gene expression and enzyme activity data confirmed a model already proposed by Bago et al. (2001) to explain nitrogen flow from the soil to the host plant via the fungus. The model proposes that inorganic N taken up by extraradical hyphae, either as NH_4^+ or NO_3^-, is assimilated in the fungal cytoplasm into arginine, transferred via the tubular vacuoles to the intraradical hyphae, and released to the plant as NH_4^+. The importance of arginine in N transfer is reflected by the fact that this amino acid represents at least 70–90% of the total amino acids in the extraradical hyphae (Johansen et al. 1996; Jin et al. 2005). This model requires that enzymes for N assimilation are expressed differentially in the extraradical and intraradical hyphae. Govindarajulu et al. (2005) showed that in *G. intraradices* a gene of primary N assimilation (glutamine synthetase) was preferentially expressed in the extraradical hyphae while genes involved in arginine breakdown (urease accessory protein and ornithine amino transferase) were more highly expressed in the intraradical hyphae. More recently, in support of this model, it has been shown that the activity of argininosuccinate synthase, an enzyme of the catabolic arm of the urea cycle, in mycorrhizal roots is synchronized with the activity of enzymes of the anabolic arm (arginase and urease) in the extraradical hyphae, and that this synchronization was the result of arginine translocation from the extraradical to the intraradical hyphae (Cruz et al. 2007).

The mechanisms of NH_4^+ transport across the plant and fungal membranes of the symbiotic interface remain unknown, although it was proposed to involve an Ato-like fungal NH_4^+ efflux system and plant NH_4^+ transporters. Non-specific channels such as aquaporins or voltage-dependent cation systems might also contribute to NH_4^+ transfer across the symbiotic interface (Chalot et al. 2006). Plant mycorrhiza-induced NO_3^- and NH_4^+ transporters have been identified in tomato (Hildebrandt et al. 2002), *M. truncatula* (Frenzel et al. 2005; Hohnjec et al. 2005), rice (Güimil et al. 2005), and *L. japonicus* (Deguchi et al. 2007). Up-regulation of NO_3^- transporters suggests mechanisms not only supporting the uptake of NH_4^+ but also the acquisition of nitrate via the mycorrhizal pathway through the fungus. Down-regulation of a *M. truncatula* NO_3^- transporter has been also reported in mycorrhizal roots (Burleigh 2001). It will be interesting to identify the precise expression patterns of these mycorrhiza-regulated NO_3^- and NH_4^+ transporters to explore possible parallels with mycorrhiza-induced Pi transporters.

6.5 Carbon Partitioning in Arbuscular Mycorrhiza

As obligate biotrophs, AM fungi are totally dependent on their hosts to complete their life cycle. Therefore, mechanisms must exist to ensure that mycorrhizal roots receive an adequate supply of sugars for the formation, maintenance, and function of the fungal structures. Since sucrose is the major transport form of photoassimilates in higher plants, this sugar should play an important role in carbon transfer in the symbiosis. Both the mechanism and site of carbon transfer remain unclear. It is likely that sucrose delivered into the apoplast at the arbuscular interface is hydrolyzed via a cell wall invertase, and that the resulting hexoses, principally glucose, are then taken up by the plasma membrane of the fungus. An alternative hypothesis is that the intercellular hyphae growing among the cortical cells are the main sites of fungal hexose uptake. In this section, we examine *in planta* nutrition of AM fungi as established by biochemical and molecular tools. The central questions refer to the compounds acquired by the intraradical fungal structures and to the regulation of host carbohydrate metabolism in response to the symbiosis.

6.5.1 The Fungal Partner

During spore germination, AM fungi are completely reliant on nutrients derived from internal stores, specifically on their lipid (triacylglycerides) stocks (Bago et al. 1999). Although not experimentally proved, it is likely that, during the early stages of colonization as it occurs during the penetration phase of pathogenic biotrophs, the fungus is also reliant on nutrients derived from internal stores. Once inside the host plant, and with the internal reserves likely exhausted, the fungus has to establish itself and to mobilize mechanisms that ensure both an adequate nutrient uptake from the host and an entrainment of plant metabolism to increase the root sink strength to suit the needs of the fungus.

Studies using isotopic labeling with nuclear magnetic resonance spectrometry in AM roots (Shachar-Hill et al. 1995) and radiorespirometry measurements on isolated intraradical hyphae (Solaiman and Saito 1997) have shown that the internal mycelium of the fungus can take up and use hexoses, mainly glucose, but not sucrose. These labeling experiments together with gene expression analyses have revealed that cytoplasmic hexoses are metabolized via the pentose phosphate pathway and that a fraction of these hexoses is used for lipid (triacylglycerides) synthesis, which also serves as a carbon transport form in the fungus (Bago et al. 2002). Unfortunately, none of the numerous efforts carried out up to now to identify and characterize the hexose transporters involved in the crucial process of carbon uptake by the intraradical hyphae has been successful. Recently, a sugar transporter, that functions as a H^+ cotransporter and shows the highest affinity for glucose, has been isolated from the bladders of *Geosiphon pyriformis*, a glomeromycotan fungus that establishes an AM-like symbiosis with cyanobacteria (Schüβler et al. 2006). Isolation of orthologous genes from AM fungi would contribute to a better understanding of carbon transfer processes in AM.

6.5.2 The Plant Partner

Measurements of carbon flux indicate that mycorrhizal plants direct from 4 to 20% more photoassimilates to the root system than non-mycorrhizal plants (Graham 2000). While the drain of carbohydrates by the fungus can be seen as the "cost" of the symbiosis, there is evidence that the net photosynthesis rate of the host plant is enhanced to meet the increased carbohydrate demand of the additional (fungal) carbon sink (Tinker et al. 1994; Wright et al. 1998a). Utilization of sucrose, the form in which photosynthetically fixed carbon is translocated to the plant sink organs, as a source of carbon and energy by mycorrhizal roots depends on its cleavage by either a cytosolic sucrose synthase or an invertase, enzymes that have also been shown to be involved in the generation of the sink strength of plant tissues (Sturm and Tang 1999). To get some insights into the mechanisms governing carbon transfer processes in the symbiosis, biochemical and molecular studies of these sucrolytic enzymes have been performed in mycorrhizal roots. Regarding enzymatic activities, increases have been described for the cytoplasmic and cell wall invertases and sucrose synthase in mycorrhizal clover roots (Wright et al. 1998b) and for cytosolic invertase in mycorrhizal soybean (Schubert et al. 2004). Enhanced transcript levels of sucrose synthase encoding genes were described in mycorrhizal maize (Ravnskov et al. 2003) and of sucrose synthase and vacuolar and cell wall invertases in tomato (García-Rodríguez et al. 2007). Accumulation of sucrose synthase and of vacuolar and cell wall invertase transcripts has also been localized in phloem tissues and in the arbuscule-colonized cortical cells of *M. truncatula* (Hohnjec et al. 2003), *Phaseolus vulgaris* (Blee and Anderson 2002), and tomato (Schaarschmidt et al. 2006). These data suggest that sucrose-cleaving enzymes are involved in the generation of the sink strength of mycorrhizal roots and in providing metabolites

Fig. 6.1 Model of nutrient transport in AM. Fungal and plant transport proteins identified experimentally (●) or as yet unidentified (O) are indicated. H^+-ATPases generate the H^+ motive force driving nutrient transport across the plant or fungal plasma membranes. Pi is taken up by the extraradical mycelium from the soil solution by a high-affinity Pi transporter that operates by H^+-Pi symport. Pi is then transported from extraradical to intraradical mycelium as polyP in the fungal vacuoles and Pi resulting from polyP hydrolysis is transferred across the fungal plasma membrane to the apoplast of the symbiotic interface. Finally, Pi is taken up by the colonized

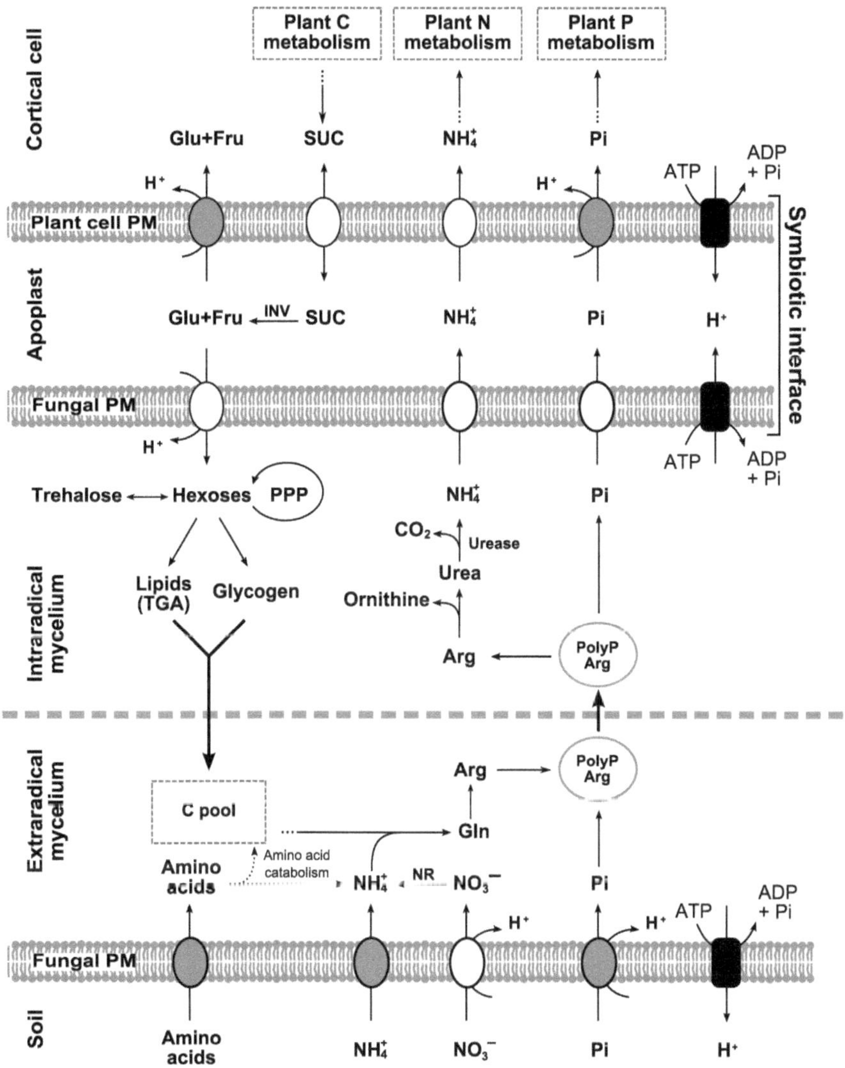

Fig. 6.1 (continued) cortical root cell by a H^+-Pi symporter. Concerning N transport, the extraradical mycelium can take up both inorganic (NH_4^+, NO_3^-) and organic N through plasma membrane NH_4^+ transporters, NO_3^- transporters, and amino acid permeases, respectively. N is assimilated via nitrate reductase (*NR*) and the glutamine synthetase/GOGAT cycle and then converted into arginine (*Arg*), which is loaded into the fungal vacuoles and translocated along the coenocytic fungal hyphae from the extraradical to the intraradical mycelium. Arginine is broken down in the intraradical mycelium, releasing urea and ornithine. NH_4^+ released from urea breakdown by the action of urease is transferred from the fungus to the apoplast of the symbiotic interface and then taken up by the host plant. Regarding carbon fluxes, sucrose from the phloem is either imported into cortical root cells or cleaved at the symbiotic interface by apoplastic invertases (*INV*). The resulting hexoses are then taken up by plant or fungal hexose transporters. Fungal cytoplasmic hexoses are transformed into trehalose, metabolized via the pentose phosphate pathway (*PPP*) or used for the biosynthesis of glycogen and lipids, compounds that are then exported to the extraradical mycelium

(hexoses) to either the colonized cell to cope with its higher metabolic activity or to the fungus.

Recently, the role of apoplastic invertases, enzymes that are believed to make available to the fungus the carbon compounds present in the symbiotic interface, has been addressed by using a functional approach (Schaarschmidt et al. 2007a, b). Root-specific elevated apoplastic invertase activity, achieved by expressing in tobacco a chimeric gene encoding an apoplast-located yeast-derived invertase, resulted in increased root hexose levels but had no effect on mycorrhizal colonization by *G. intraradices*. Furthermore, over-expression of the yeast gene at three different compartments (cytosol, vacuole, and cell wall) in root cells of *M. truncatula* did not affect mycorrhizal colonization by either *G. mosseae* or *G. intraradices*, in spite of the enhanced hexose to sucrose ratio of the transformed roots. In contrast, it was observed that decreased root hexose content, achieved by either reducing assimilate supply of roots as a consequence of reduced phloem loading of photoassimilates, by root-specific expression of the *Arabidopsis* apoplastic invertase inhibitor or by over-expression of the yeast invertase in the leaf apoplast, strongly inhibited AM colonization. These data indicate that an undersupply of carbohydrates leads to a decreased development of the fungus and that carbon supply in the arbuscular mycorrhizal interaction depends on the activity of apoplastic invertases.

Sugar transporters also play a pivotal role in sugar distribution throughout the plant (Sauer 2007), but very little is known about their regulation in AM. Up-regulation in *M. truncatula* mycorrhizal roots, especially in highly colonized regions of the root, of a hexose transporter likely involved in sugar uptake suggests a role for its encoded protein in supplying sugars to root cells critically involved in the symbiotic interaction (Harrison 1996). These data suggest that both the plant and the fungus compete for hexoses present in the apoplast of the symbiotic interface and that the plant exerts some control on carbohydrate drain to the fungus. Transcriptional activation of putative hexose transporters has been also reported in maize mycorrhizal roots (Wright et al. 2005) and in leaves of tomato mycorrhizal plants (García-Rodríguez et al. 2005). Up-regulation of a sugar transporter in the leaves of the colonized plants suggests a role for its encoded protein in carbon allocation from source to sink tissues. Further studies are needed to better understand how the fungus induces metabolic sinks at the colonization sites within the root to ensure an adequate carbohydrate flux from the host plant.

Figure 6.1 summarizes the mechanisms described in this chapter and illustrates the currently accepted model of nutrient exchange between the plant and the fungus in the AM symbiosis.

6.6 Conclusions

In recent years, progress in the understanding of nutrient transport processes in AM has been made at the molecular level. Pi transport and carbon metabolism have been extensively studied and several key plant and fungal genes have been isolated.

However, gene isolation is just at the beginning, and defining the function of these proteins is a major challenge in the future. Since mycorrhizal associations are also important for supplying the plant with other mineral nutrients, the consideration of an exchange of Pi alone against carbohydrates at the plant–fungus interface gives an incomplete view of the transfer processes occurring in AM. Efforts must be made, therefore, to isolate genes encoding the proteins that mediate the transport of other nutrients, such as N, K, Zn, or Cu. Recently, progress has also been made on the understanding of N transfer in the symbiosis, although further studies are needed to elucidate the mechanisms of these transport processes and their contribution to the mineral nutrition of the plant in specific physiological situations.

We currently have little information about the regulation of the transfer processes across the specialized interfaces in AM or about the mechanisms involved in polarizing the transfer of nutrients between the symbionts. Since nutrient exchange between the plant and the fungus is a requirement for a functional symbiosis, it is not surprising that the exchanged molecules, mainly but not exclusively Pi and carbon, also act as regulatory components of the symbiosis. Further research will determine whether carbon allocation to the fungus is linked to the symbiotic delivery of Pi or other nutrients, and will enable full understanding of the basis of a successful AM symbiosis.

Acknowledgements Financial support was provided by the CICyT project AGL2006-08218/AGR, Spain.

References

Aono T, Maldonado-Mendoza IE, Dewbre GR, Harrison MJ, Saito M (2004) Expression of alkaline phosphatase genes in arbuscular mycorrhizas. New Phytol 162:525–534

Bago B, Vierheilig H, Piché Y, Azcón-Aguilar C (1996) Nitrate depletion and pH changes induced by the extraradical mycelium of the arbuscular mycorrhizal fungus *Glomus intraradices* grown in monoxenic culture. New Phytol 133:273–280

Bago B, Pfeffer PE, Douds DD, Brouillette J, Bécard G, Shachar-Hill Y (1999) Carbon metabolism in spores of the arbuscular mycorrhizal fungus *Glomus intraradices* as revealed by nuclear magnetic resonance spectroscopy. Plant Physiol 121:263–271

Bago B, Pfeffer P, Shachar-Hill Y (2001) Could the urea cycle be translocating nitrogen in the arbuscular mycorrhizal symbiosis? New Phytol 149:4–8

Bago B, Zipfel W, Williams RM, Jun J, Arreola R, Lammers PJ, Pfeffer PE, Shachar-Hill Y (2002) Translocation and utilization of fungal storage lipid in the arbuscular mycorrhizal symbiosis. Plant Physiol 128:108–124

Balestrini R, Bonfante P (2005) The interface compartment in arbuscular mycorrhizae: a special type of plant cell wall. Plant Biosystems 139:8–15

Benedetto A, Magurno F, Bonfante P, Lanfranco L (2005) Expression profiles of a phosphate transporter gene (*GmosPT*) from the endomycorrhizal fungus *Glomus mosseae*. Mycorrhiza 15:620–627

Blee KA, Anderson AJ (2002) Transcripts for genes encoding soluble acid invertase and sucrose synthase accumulate in root tip and cortical cells containing mycorrhizal arbuscules. Plant Mol Biol 50:197–211

Bonfante P (2001) At the interface between mycorrhizal fungi and plants: the structural organization of cell wall, plasma membrane and cytoskeleton. In: Hock B (ed) The Mycota: Fungal Associations, Springer, Berlin, pp 45–61

Bucher M (2007) Functional biology of plant phosphate uptake at root and mycorrhiza interfaces. New Phytol 173:11–26
Burleigh SH (2001) Relative quantitative RT-PCR to study the expression of plant nutrient transporters in arbuscular mycorrhizas. Plant Sci 160:899–904
Chalot M, Blaudez D, Brun A (2006) Ammonia: a candidate for nitrogen transfer at the mycorrhizal interface. Trends Plant Sci 11:263–266
Chen AQ, Hu J, Sun SB, Xu GH (2007) Conservation and divergence of both phosphate- and mycorrhiza-regulated physiological responses and expression patterns of phosphate transporters in solanaceous species. New Phytol 173:817–831
Cruz C, Egsgaard H, Trujillo C, Ambus P, Requena N, Martins-Loucao MA, Jakobsen I (2007) Enzymatic evidence for the key role of arginine in nitrogen translocation by arbuscular mycorrhizal fungi. Plant Physiol 144:782–792
Deguchi Y, Banba M, Shimoda Y, Chechetka SA, Suzuri R, Okusako Y, Ooki Y, Toyokura K, Suzuki A, Uchiumi T, Higashi S, Abe M, Kouchi H, Izui K, Hata S (2007) Transcriptome profiling of *Lotus japonicus* roots during arbuscular mycorrhiza development and comparison with that of nodulation. DNA Res 14:117–133
Ezawa T, Smith SE, Smith FA (2002) P metabolism and transport in AM fungi. Plant Soil 244:221–230
Ferrol N, Barea JM, Azcón-Aguilar C (2000) The plasma membrane H^+-ATPase gene family in the arbuscular mycorrhizal fungus *Glomus mosseae*. Current Genet 37:112–118
Ferrol N, Barea JM, Azcón-Aguilar C (2002a) Mechanisms of nutrient transport across interfaces in arbuscular mycorrhizas. Plant Soil 244:231–237
Ferrol N, Pozo MJ, Antelo M, Azcón-Aguilar C (2002b) Arbuscular mycorrhizal symbiosis regulates plasma membrane H^+-ATPase gene expression in tomato plants. J Exp Bot 53:1683–1687
Frenzel A, Manthey K, Perlick AM, Meyer F, Puhler A, Kuster H, Krajinski F (2005) Combined transcriptome profiling reveals a novel family of arbuscular mycorrhizal-specific *Medicago truncatula* lectin genes. Mol Plant Microbe Interact 18:771–782
Frey B, Schuepp H (1993) Acquisition of nitrogen by external hyphae of arbuscular mycorrhizal fungi associated with *Zea mays* L. New Phytol 124:221–230
Funamoto R, Saito K, Oyaizu H, Saito M, Aono T (2007) Simultaneous *in situ* detection of alkaline phosphatase activity and polyphosphate in arbuscules within arbuscular mycorrhizal roots. Funct Plant Biol 34:803–810
García-Rodríguez S, Pozo MJ, Azcón-Aguilar C, Ferrol N (2005) Expression of a tomato sugar transporter is increased in leaves of mycorrhizal or *Phytophthora parasitica*-infected plants. Mycorrhiza 15:489–496
García-Rodríguez S, Azcón-Aguilar C, Ferrol N (2007) Transcriptional regulation of host enzymes involved in the cleavage of sucrose during arbuscular mycorrhizal symbiosis. Physiol Plant 129:737–746
Gianinazzi-Pearson V, Arnould C, Oufattole M, Arango M, Gianinazzi S (2000) Differential activation of H^+-ATPase genes by an arbuscular mycorrhizal fungus in root cells of transgenic tobacco. Planta 211:609–613
Glassop D, Smith SE, Smith FW (2005) Cereal phosphate transporters associated with the mycorrhizal pathway of phosphate uptake into roots. Planta 222:688–698
Govindarajulu M, Pfeffer PE, Jin HR, Abubaker J, Douds DD, Allen JW, Bucking H, Lammers PJ, Shachar-Hill Y (2005) Nitrogen transfer in the arbuscular mycorrhizal symbiosis. Nature 435:819–823
Graham JH (2000) Assessing costs of arbuscular mycorrhizal symbiosis agroecosystems fungi. In: Podila GK, Douds DD Jr. (eds) Current Advances in Mycorrhizae Research. APS Press, St. Paul, MN, pp 127–140
Güimil S, Chang HS, Zhu T, Sesma A, Osbourn A, Roux C, Ionnidis V, Oakeley EJ, Docquier M, Descombes P, Briggs SP, Paszkowski U (2005) Comparative transcriptomics of rice reveals an ancient pattern of response to microbial colonization. Proc Natl Acad Sci USA 102:8066–8070

Harrison MJ (1996) A sugar transporter from *Medicago truncatula*: Altered expression pattern in roots during vesicular-arbuscular (VA) mycorrhizal associations. Plant J 9:491–503

Harrison MJ, van Buuren ML (1995) A phosphate transporter from the mycorrhizal fungus *Glomus versiforme*. Nature 378:626–629

Harrison MJ, Dewbre GR, Liu JY (2002) A phosphate transporter from *Medicago truncatula* involved in the acquisition of phosphate released by arbuscular mycorrhizal fungi. Plant Cell 14:2413–2429

Hawkins HJ, Johansen A, George E (2000) Uptake and transport of organic and inorganic nitrogen by arbuscular mycorrhizal fungi. Plant Soil 226:275–285

Hildebrandt U, Schmelzer E, Bothe H (2002) Expression of nitrate transporter genes in tomato colonized by an arbuscular mycorrhizal fungus. Physiol Plant 115:125–136

Hodge A, Campbell CD, Fitter AH (2001) An arbuscular mycorrhizal fungus accelerates decomposition and acquires nitrogen directly from organic material. Nature 413:297–299

Hohnjec N, Perlick AM, Pühler A, Kuster H (2003) The *Medicago truncatula* sucrose synthase gene MtSucS1 is activated both in the infected region of root nodules and in the cortex of roots colonized by arbuscular mycorrhizal fungi. Mol Plant Microbe Interact 16:903–915

Hohnjec N, Vieweg ME, Pühler A, Becker A, Kuster H (2005) Overlaps in the transcriptional profiles of *Medicago truncatula* roots inoculated with two different *Glomus* fungi provide insights into the genetic program activated during arbuscular mycorrhiza. Plant Physiol 137:1283–1301

Jakobsen I, Abbott LK, Robson AD (1992) External hyphae of vesicular arbuscular-mycorrhizal fungi associated with *Trifolium subterraneum* L. 2. Hyphal transport of ^{32}P over defined distances. New Phytol 120:509–516

Javot H, Pumplin N, Harrison MJ (2007a) Phosphate in the arbuscular mycorrhizal symbiosis: transport properties and regulatory roles. Plant Cell Environ 30:310–322

Javot H, Penmetsa RV, Terzaghi N, Cook DR, Harrison MJ (2007b) A *Medicago truncatula* phosphate transporter indispensable for the arbuscular mycorrhizal symbiosis. Proc Natl Acad Sci USA 104:1720–1725

Jin H, Pfeffer PE, Douds DD, Piotrowski E, Lammers PJ, Shachar-Hill Y (2005) The uptake, metabolism, transport and transfer of nitrogen in an arbuscular mycorrhizal symbiosis. New Phytol 168:687–696

Johansen A, Jakobsen I, Jensen ES (1992) Hyphal transport of ^{15}N-labeled nitrogen by a vesicular-arbuscular mycorrhizal fungus and its effect on depletion of inorganic soil N. New Phytol 122:281–288

Johansen A, Finlay RD, Olsson PA (1996) Nitrogen metabolism of external hyphae of the arbuscular mycorrhizal fungus *Glomus intraradices*. New Phytol 133:705–712

Krajinski F, Hause B, Gianinazzi-Pearson V, Franken P (2002) *Mtha1*, a plasma membrane H^+-ATPase gene from *Medicago truncatula*, shows arbuscule-specific induced expression in mycorrhizal tissue. Plant Biol 4:754–761

Liu H, Trieu AT, Blaylock LA, Harrison MJ (1998) Cloning and characterization of two phosphate transporters from *Medicago truncatula* roots: Regulation in response to phosphate and to colonization by arbuscular mycorrhizal (AM) fungi. Mol Plant Microbe Interact 11:14–22

López-Pedrosa A, González-Guerrero M, Valderas A, Azcón-Aguilar C, Ferrol N (2006) *GintAMT1* encodes a functional high-affinity ammonium transporter that is expressed in the extraradical mycelium of *Glomus intraradices*. Fungal Genet Biol 43:102–110

Maeda D, Ashida K, Iguchi K, Chechetka SA, Hijikata A, Okusako Y, Deguchi Y, Izui K, Hata S (2006) Knockdown of an arbuscular mycorrhiza-inducible phosphate transporter gene of *Lotus japonicus* suppresses mutualistic symbiosis. Plant Cell Physiol 47:807–817

Maldonado-Mendoza IE, Dewbre GR, Harrison MJ (2001) A phosphate transporter gene from the extra-radical mycelium of an arbuscular mycorrhizal fungus *Glomus intraradices* is regulated in response to phosphate in the environment. Mol Plant Microbe Interact 14:1140–1148

Marzluf GA (1996) Regulation of nitrogen metabolism in mycelial fungi. In: Brambl R, Marzluf GA (eds) The Mycota III: Biochemistry and Molecular Biology. Springer, Berlin, Heidelberg, pp 357–368

Murphy PJ, Langridge P, Smith SE (1997) Cloning plant genes differentially expressed during colonization of roots of *Hordeum vulgare* by the vesicular-arbuscular mycorrhizal fungus *Glomus intraradices*. New Phytol 135:291–301

Nagy R, Karandashov V, Chague W, Kalinkevich K, Tamasloukht M, Xu GH, Jakobsen I, Levy AA, Amrhein N, Bucher M (2005) The characterization of novel mycorrhiza-specific phosphate transporters from *Lycopersicon esculentum* and *Solanum tuberosum* uncovers functional redundancy in symbiotic phosphate transport in solanaceous species. Plant J 42:236–250

Ohtomo R, Saito M (2005) Polyphosphate dynamics in mycorrhizal roots during colonization of an arbuscular mycorrhizal fungus. New Phytol 167:571–578

Paszkowski U, Kroken S, Roux C, Briggs SP (2002) Rice phosphate transporters include an evolutionarily divergent gene specifically activated in arbuscular mycorrhizal symbiosis. Proc Natl Acad Sci USA 99:13324–13329

Rausch C, Daram P, Brunner S, Jansa J, Laloi M, Leggewie G, Amrhein N, Bucher M (2001) A phosphate transporter expressed in arbuscule-containing cells in potato. Nature 414:462–466

Ravnskov S, Wu Y, Graham JH (2003) Arbuscular mycorrhizal fungi differentially affect expression of genes coding for sucrose synthases in maize roots. New Phytol 157:539–545

Requena N, Breuninger M, Franken P, Ocon A (2003) Symbiotic status, phosphate, and sucrose regulate the expression of two plasma membrane H^+-ATPase genes from the mycorrhizal fungus *Glomus mosseae*. Plant Physiol 132:1540–1549

Sauer N (2007) Molecular physiology of higher plant sucrose transporters. FEBS Lett 581:2309–2317

Schaarschmidt S, Roitsch T, Hause B (2006) Arbuscular mycorrhiza induces gene expression of the apoplastic invertase LIN6 in tomato (*Lycopersicon esculentum*) roots. J Exp Bot 57:4015–4023

Schaarschmidt S, González MC, Roitsch T, Strack D, Sonnewald U, Hause B (2007a) Regulation of arbuscular mycorrhization by carbon. The symbiotic interaction cannot be improved by increased carbon availability accomplished by root-specifically enhanced invertase activity. Plant Physiol 143:1827–1840

Schaarschmidt S, Kopka J, Ludwig-Muller J, Hause B (2007b) Regulation of arbuscular mycorrhization by apoplastic invertases: enhanced invertase activity in the leaf apoplast affects the symbiotic interaction. Plant J 51:390–405

Schubert A, Allara P, Morte A (2004) Cleavage of sucrose in roots of soybean (*Glycine max*) colonized by an arbuscular mycorrhizal fungus. New Phytol 161:495–501

Schüβler A, Martin H, Cohen D, Fitz M, Wipf D (2006) Characterization of a carbohydrate transporter from symbiotic glomeromycotan fungi. Nature 444:933–936

Shachar-Hill Y, Pfeffer PE, Douds D, Osman SF, Doner LW, Ratcliffe RG (1995) Partitioning of intermediary carbon metabolism in vesicular-arbuscular mycorrhizal leek. Plant Physiol 108:7–15

Smith SE, Read DK (1997) Mycorrhizal Symbiosis. Academic, London

Smith SE, Smith FA, Jakobsen I (2003) Mycorrhizal fungi can dominate phosphate supply to plants irrespective of growth responses. Plant Physiol 133:16–20

Smith SE, Smith FA, Jakobsen I (2004) Functional diversity in arbuscular mycorrhizal (AM) symbioses: the contribution of the mycorrhizal P uptake pathway is not correlated with mycorrhizal responses in growth or total P uptake. New Phytol 162:511–524

Solaiman MDZ, Saito M (1997) Use of sugars by intraradical hyphae of arbuscular mycorrhizal fungi revealed by radiorespirometry. New Phytol 136:533–538

Sturm A, Tang GQ (1999) The sucrose-cleaving enzymes of plants are crucial for development, growth and carbon partitioning. Trends Plant Sci 4:401–407

Tinker PB, Durall DM, Jones MD (1994) Carbon efficiency in mycorrhizas - Theory and sample calculations. New Phytol 128:115–122

Tobar R, Azcón R, Barea JM (1994) Improved nitrogen uptake and transport from ^{15}N-labeled nitrate by external hyphae of arbuscular mycorrhiza under water-stressed conditions. New Phytol 126:119–122

Toussaint JP, St-Arnaud M, Charest C (2004) Nitrogen transfer and assimilation between the arbuscular mycorrhizal fungus *Glomus intraradices* Schenck and Smith and Ri T-DNA roots of *Daucus carota* L. in an *in vitro* compartmented system. Can J Microbiol 50:251–260

Uetake Y, Kojima T, Ezawa T, Saito M (2002) Extensive tubular vacuole system in an arbuscular mycorrhizal fungus, *Gigaspora margarita*. New Phytol 154:761–768

Wright DP, Scholes JD, Read DJ (1998a) Effects of VA mycorrhizal colonization on photosynthesis and biomass production of *Trifolium repens* L. Plant Cell Environ 21:209–216

Wright DP, Read DJ, Scholes JD (1998b) Mycorrhizal sink strength influences whole plant carbon balance of *Trifolium repens* L. Plant Cell Environ 21:881–891

Wright DP, Scholes JD, Read DJ, Rolfe SA (2005) European and African maize cultivars differ in their physiological and molecular responses to mycorrhizal infection. New Phytol 167:881–896

Xu GH, Chague V, Melamed-Bessudo C, Kapulnik Y, Jain A, Raghothama KG, Levy AA, Silber A (2007) Functional characterization of LePT4: a phosphate transporter in tomato with mycorrhiza-enhanced expression. J Exp Bot 58:2491–2501

Chapter 7
Deciphering the Arbuscular Mycorrhizal Pathway of P Uptake in Non-responsive Plant Species

Emily J. Grace, F. Andrew Smith, and Sally E. Smith

Abstract Although arbuscular mycorrhizal (AM) symbioses are considered to be mutualistic, plant benefit is not always immediately obvious. Non-responsiveness in terms of growth and phosphorus (P) nutrition is observed in a wide variety of plant species, including natives and some widely cultivated crops (e.g. cereals). Non-responsiveness is primarily attributed to variations in the exchange of carbon (C) and P between the symbionts. Here, we explore recent insights into P uptake in non-responsive plants. The AM pathway of P uptake can be functional in non-responsive plants, as shown by fungal $^{32/33}$P uptake, which has raised questions regarding functionality of the direct uptake pathway. As the mechanisms for P uptake via AM and direct uptake pathways are revealed, we can begin to explore functional differences at the molecular level. Identifying factors which influence AM responsiveness will provide critical insights for future crop breeding efforts.

7.1 Introduction

Symbiotic interactions of AM fungi with roots of land plants are widespread. They occur in both natural and agricultural ecosystems and probably involve ~80% of land plants (Smith and Read 1997). Conventionally, the symbiosis is considered to be a mutualism, based upon the reciprocal exchange of nutrients. The fungi are obligate symbionts, relying on plants as sole sources of carbon (C). In return, plants receive nutrients such as phosphorus (P) and nitrogen, which are taken up from the soil by external hyphae of the fungi (see Chapter 6 by Ferrol and Pérez-Tienda). The evolutionary conservation and widespread occurrence of AM symbioses are testaments to the importance of AM in plant function. Indeed, AM symbioses have been demonstrated to improve disease tolerance (see Chapter 9 by Pozo et al.),

E.J. Grace (✉), F.A. Smith, and S.E. Smith
Soil and Land Systems, School of Earth and Environmental Sciences, Waite Campus
The University of Adelaide, Adelaide, South Australia, 5005
e-mail: emily.grace@adelaide.edu.au

C. Azcón-Aguilar et al. (eds.), *Mycorrhizas - Functional Processes and Ecological Impact*, 89
DOI: 10/1007-978-3-540-87978-7_7, © Springer-Verlag Berlin Heidelberg 2009

increase drought resistance and decrease the accumulation of heavy metals (see Chapter 8 by González-Guerrero et al.), although it has been suggested that these benefits have evolved relatively recently (Fitter 2006). Improved P nutrition is still considered the primary benefit of AM symbioses, and total plant P uptake and growth response are the most commonly reported measures of AM function. Plant responses to AM colonisation are diverse, ranging from large positive increases in growth and P uptake to growth depressions (Tawaraya 2003). This diversity in responsiveness is considered to reflect the diversity in function of different plant–AM fungal combinations. This review brings together recent advances in measurement of the contributions of the AM fungus to plant P nutrition and current knowledge of molecular processes of P uptake in AM plants, with particular emphasis on situations where plants show low or no positive responses to the symbiosis.

7.2 The Challenge of P Uptake from Soil and Symbiotic Interfaces

AM plants have two possible pathways for the acquisition of P from the soil solution, the direct uptake pathway via the root epidermis and root hairs, and the AM pathway via external fungal hyphae and colonised cortical cells (Fig. 7.1a). The total P content of an AM plant results from the combined operation of these path-

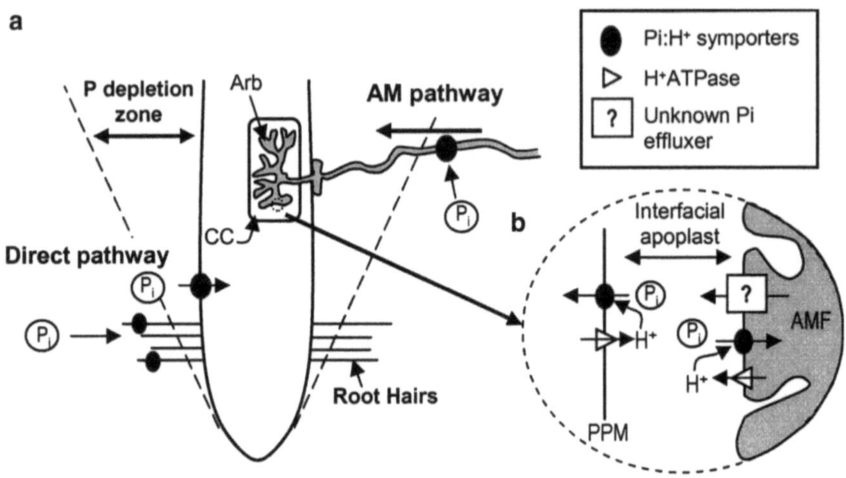

Fig. 7.1 Schematic representation of P uptake in an AM root. **(a)** P uptake from soil solution is mediated by plant and fungal P_i:H^+ symporters expressed in external hyphae or root epidermis and root hairs. Rapid uptake of P by roots leads to formation of a depletion zone, fungal hyphae extend beyond depletion zones taking up P and transporting it to intracellular arbuscules (*Arb*) in colonised cortical cells (*CC*). **(b)** P taken up by the fungus is released into the interfacial apoplast via an unknown mechanism. Plant P uptake across the plant plasma membrane (*PPM*) at symbiotic interfaces is also mediated by P_i:H^+ symporters in competition with AM fungal retrieval

ways. Although the direct pathway has been studied extensively in model systems such as *Arabidopsis thaliana* [which is constitutively non-mycorrhizal (NM)], the prevalence of AM in most natural and agricultural environments suggests that an understanding of the integration of the two pathways is essential to provide a realistic picture of plant nutrient uptake. The following section briefly outlines the processes involved in P uptake in an AM plant and the interfaces at which P transfer occurs.

Both plants and AM fungi acquire P from the soil solution as inorganic orthophosphate (P_i). Although the total quantity of P in the soil can be high, the concentration of P_i in the soil solution rarely exceeds 10 μM (Bieleski 1973). In contrast, cytosolic P concentrations are within the millimolar range (Mimura 1999). Uptake of P_i from soil therefore requires an active, high-affinity transportation system. Over the last decade, plant and fungal P_i transporters involved in P uptake from soil have been identified (see Smith et al. 2003a and references therein). These transporters share high sequence similarity and together are members of the P_i:H^+ symporter family. Functional characterisation predicts high-affinity kinetics, consistent with uptake of P_i from the low concentrations in the soil solution. P_i transporters have been localised to the root epidermis and root hairs and to the extraradical hyphae (Fig. 7.1a). Both plant and fungal P_i transporters respond to P_i concentrations in the environment and to the P status of the organism (Mimura 1999; Maldonado-Mendoza et al. 2001).

The slow rate of diffusion of P_i in soil (10^{-8} to 10^{-11} cm^2 s^{-1}; Bieleski 1973) compared with rapid plant uptake results in the formation of a depletion zone adjacent to the root surface. P uptake by NM plants is therefore limited by the rate of diffusion of P_i into this zone and/or the ability of the plant to extend roots into undepleted soil. In a mycorrhizal plant, AM fungal hyphae extend beyond this depletion zone (Fig. 7.1a), their small diameter (2–15 μm) enables access to smaller soil pores than plant roots, whilst P translocation to the plant through fungal hyphae occurs at rates very much faster than the diffusion of P_i through soil. Following P uptake by extraradical hyphae, amounts in excess of metabolic requirements are transferred to the vacuoles and stored as polyphosphate (polyP). Numerous studies have investigated the dynamics of polyP sequestration, storage and transport in AM hyphae, but the mechanism of P release to the symbiotic interfaces remains obscure (see Ezawa et al. 2002).

The precise location of P transfer to the plant has received considerable attention. Intracellular arbuscules have long been considered the primary site for P transfer. These interfaces are formed when fungal hyphae penetrate the walls of root cortical cells and invaginate the plasma membranes to form the highly branched structures once considered characteristic of the symbiosis. Extension of the plasma membrane around the arbuscule provides a large increase in surface area and results in the formation of an extensive and intimate interface between the symbionts. However, not all AM fungi form arbuscules. Intracellular hyphal coils can provide an equivalent surface area to arbuscules (Dickson and Kolesik 1999) and there is now recognition that these too may play an important role in P transfer. Localisation of plant P_i transporter expression to cortical cells containing arbuscules and hyphal coils provides strong evidence for P transfer via intracellular interfaces (Harrison et al. 2002; Karandashov et al. 2004; Glassop et al. 2005). However, it is noteworthy that the fungus remains within an apoplastic compartment, with no cytoplasmic continuity between the symbionts. Transfer of P between the symbionts therefore requires both efflux from the fungus and uptake by the plant (Fig. 7.1b).

The mechanisms for efflux of P from AM fungal hyphae to the interfacial apoplast remain obscure, although specific efflux channels have been invoked to account for the considerable P release (Smith et al. 2001). P in the interfacial apoplast must be actively transported across the plasma membrane into the plant cell, presumably presenting a challenge similar to uptake from the soil, although Pi concentrations in the apoplast have not been determined. Localisation of H^+-ATPases to plant membranes surrounding arbuscules (Gianinazzi-Pearson et al. 2000) and the expression of plant P_i transporters supports this model (Harrison et al. 2002; see also Chapter 6 by Ferrol and Pérez-Tienda). The plant P_i transporters expressed at symbiotic interfaces are members of the same family as the root epidermal P_i transporters; insights from their expression and functional characterisation are discussed below.

7.3 Diversity in Plant Responses to AM Colonisation

AM-induced increases in plant growth are often accompanied by significant increases in P concentration. These data provided the first evidence for the key role of P in symbiotic nutrient exchange. Typically, plants which respond positively to AM colonisation have limited capacity to grow at low P, and AM colonisation improves P uptake and results in increased growth over a broad range of external P concentrations. However, responses of plants to AM colonisation are diverse, ranging from dramatic increases in growth and P nutrition to negligible changes in growth and even growth depressions. The term mycorrhizal responsiveness (MR), calculated according to Eq. 7.1, has been developed to quantify AM-induced changes in plant growth (MGR) (Baon et al. 1993) or P uptake (MPR) (Li et al. 2008), where AM and NM refer to either dry matter or total plant P, respectively.

$$\mathrm{MR} = \frac{(\mathrm{AM} - \mathrm{NM})}{\mathrm{NM}} \times 100 \tag{7.1}$$

In the context of this review, non-responsive plants are those which demonstrate negligible or negative growth responses to AM colonisation (zero or negative MR) even at low P supply. In an extensive literature survey of growth responses of 250 AM plant species including field and forage crops, wild grasses and forbs, and trees, Tawaraya (2003) showed that non-responsive species are less common than positively responsive species, but that some were identified in all groups investigated.

Non-responsiveness of AM plants has been related to physiological traits which increase P uptake under P-limiting conditions, thereby reducing reliance on AM fungi for acquisition of P. Such traits include extensive, finely branched root systems and long root hairs (Jakobsen et al. 2005) or the ability of the plant to modify the rhizosphere thus increasing access to soil P. In *Triticum aestivum* and *Hordeum vulgare*, cultivar differences in AM responsiveness are correlated with efficiency of P acquisition or P utilisation, leading to the suggestion that breeding programs targeting improved P efficiency may have led to a decline in AM responsiveness of modern cereal cultivars (Baon et al. 1993; Zhu et al. 2001).

Whereas it is possible to make broad generalisations regarding the tendency for certain plant species to be positively responsive or non-responsive to colonization, it is important to recognise that AM responsiveness is not dependent solely upon the plant genotype. Responsiveness is also influenced by the fungal genotype (Munkvold et al. 2004) and the environmental conditions which influence the C and P status in the plant; primarily light, temperature, soil P and soil pH (Johnson et al. 1997). Diversity in plant responses is therefore considered to result from diversity in function of different AM fungus–plant interactions and is conventionally attributed to perturbations in the balance between benefit derived from increased access to growth-limiting nutrients and cost of supplying C to the fungal symbiont. C allocation to the AM fungus can account for 20% of plant photosynthates (Jakobsen and Rosendahl 1990) and, in many instances, growth depressions are attributed to this C drain (see Jakobsen 1995 and references therein).

Both transient and persistent growth depressions are occasionally observed amongst positively responsive plant species. In both instances, these growth depressions can be explained with respect to C demand and P supply by the AM fungus. For example, persistent growth depressions are frequently reported at high P supply and are attributed to a shift in the balance between symbiotic cost and benefit. At high P, growth of NM plants is not limited by P and there is little benefit of increased supply via AM fungi, whilst the C cost of the fungal symbiont is maintained. This theory is supported by observations of AM citrus in which growth depression at high P was overcome by elevated CO_2 supply (Jifon et al. 2002) (for further discussion see Jakobsen 1995).

As with growth depressions at high P, growth depressions at low P in non-responsive plants are often attributed to fungal C drain. Therefore, it might be expected that the magnitude of growth depression will be correlated with the extent of colonisation and, hence, C demand. However, there is an increasing body of evidence suggesting that this is not the case. As is apparent from Table 7.1, growth depressions are frequently unrelated to the degree of colonisation. The mycorrhizal growth response (MGR) of *H. vulgare* colonised by two AM fungi was −36% and −34% whilst colonisation was 72% and 19%, respectively (Grace et al. 2008). In *T. aestivum* cv. Newton, growth depressions due to *Glomus mosseae* or *Glomus versiforme* were equal (−50%, Table 7.1), but these fungi colonised 5% and 61% of the root length, respectively (Hetrick et al. 1992). Such findings have been largely ignored, but they suggest that the traditional model used to explain non-responsiveness should be re-examined.

Whereas calculation of MGR or MPR has proved useful for comparison of whole-plant responses to AM, it is important to clearly distinguish MPR from

Table 7.1 Mycorrhizal growth response (*MGR*)[a] and corresponding % colonisation in non-responsive AM symbioses of cereals

Plant species	Fungal species	MGR (%)	% colonisation	References
Hordeum vulgare	*Glomus intraradices*	–36.0	72	Grace et al. (2008)
	G. geosporum	–34.1	19	
Triticum aestivum cv. Newton	*G. versiforme*	–49.0	61	Hetrick et al. (1992)
	G. mosseae	–50.6	5	
T. aestivum cv. Kanzler	*G. mosseae*	–60.0	42	Hetrick et al. (1992)
	G. versiforme	–44.5	91	
	G. monosporum	–66.1	17	
	G. etunicatum	–35.5	18	
T. aestivum cv. Brookton	*G. intraradices*	–64.0	47	Li et al. (2008)
	Gigaspora margarita	–82.0	11	

[a]Calculated according to Eq. 7.1 (Sect. 7.3)

estimates of the actual contribution of the AM pathway to P uptake. Most estimates have been based on the difference in P uptake between AM and NM plants (e.g. Smith et al. 1994) and rely on the assumption that AM colonisation has no effect on the direct uptake of nutrients into plant roots. As positive values are only obtained in plants that are positively responsive to AM, such calculations have led to the notion that the AM pathway is non-functional in terms of P uptake in non-responsive plants. However, an increasing body of evidence from both radiotracer experiments and molecular studies (see below) demonstrates that the contribution of the AM pathway is not necessarily related to plant responses. The assumption that non-responsive plants receive no nutritional benefit from AM symbionts must therefore be re-examined, particularly with respect to roles and management of AM in agricultural systems and natural ecosystems.

7.4 Radiotracer Studies Enable Measurement of AM Contribution Regardless of Whole Plant Responses

Experimental designs using compartmented pots in which radiotracer is supplied in a hyphal compartment (HC) accessible only to AM fungal hyphae have been highly effective at demonstrating nutrient (P, Zn, NO_3^- and NH_4^+) uptake and transfer to plants via the AM pathway (Johansen et al. 1993; Pearson and Jakobsen 1993; Bürkert and Robson 1994; Tobar et al. 1994; Smith et al. 2000). As is evident from Table 7.2, $^{32/33}P$ transfer via the AM pathway has been demonstrated for a range of plant and fungal combinations including both positively responsive and non-responsive interactions.

Early radiotracer experiments enabled comparison of the relative AM contributions in different plant–fungus combinations. Comparing ^{32}P uptake in two native grasses, Hetrick et al. (1994) demonstrated that positively responsive *Andropogon*

Table 7.2 Mycorrhizal growth response (*MGR*) or P response (*MPR*) of plant and AM fungal combinations in which P transfer via the AM pathway has been detected using radiotracers. In those instances where fungal contribution has been quantified the percent contribution is reported

Plant species	AM fungal species[a]	MGR[b]	MPR[b]	^{32}P or ^{33}P Transfer[c]	Contribution[d] (%)	References
Native grasses						
Andropogon gerardii	*Glomus etunicatum*	+	+	Y	?	Hetrick et al. (1994)
Bromus inermis	*G. etunicatum*	=	=	Y	?	Hetrick et al. (1994)
Cultivated spp.						
Cucumis sativus	*G. caledonium* BEG15	−18	−13	Y	?	Ravnskov and Jakobsen (1995)
	G. invermaium WUM10	−7	−1	Negligible	?	Ravnskov and Jakobsen (1995)
	G. caledonium BEG15	9	−6	Y	109[e]	Pearson and Jakobsen (1993)
	G. invermaium WUM10	15	17	Y	21[e]	Pearson and Jakobsen (1993)
	Scutellospora calospora	10	−1	Y	7[e]	Pearson and Jakobsen (1993)
Hordeum vulgare	*G. intraradices*	−36	−40	Y	60	Grace et al. (2008)
	G. geosporum	−34	−40	Very low	?	Grace et al. (2008)
	G. intraradices	–	+	Y	?	Zhu et al. (2003)
Linum usitatissimum	*G. caledonium*	1,100	1,075	Y	^{6}30	Ravnskov and Jakobsen (1995), Smith et al. (2003b),
	G. intraradices	1,425	1,850	Y	100	Smith et al. (2003b, 2004)
	G. invermaium	662	1,272	Y	?	Ravnskov and Jakobsen (1995)
	Gigaspora rosea	100	200	Y	Very low	Smith et al. (2003b, 2004)
Medicago truncatula	*G. caledonium*	220	200	Y	^{6}70	Smith et al. (2003b, 2004), Smith et al. (2000)
	G. intraradices	220	290	Y	80	Smith et al. (2003b, 2004)
	Gigaspora rosea	−33	5	Y	Very low	Smith et al. (2003b, 2004)
	S. calospora	+	+	Y	?	Smith et al. (2000)
Solanum lycopersicum	*G. caledonium*	−12	−10	Y	20	Smith et al. (2003b, 2004)
	G. intraradices BEG 87	−18	−4	Y	100	Smith et al. (2003b, 2004)
	Gigaspora rosea	−32	−27	Y	Very low	Smith et al. (2003b, 2004)

(continued)

Table 7.2 (continued)

Plant species	AM fungal species[a]	MGR[b]	MPR[b]	^{32}P or ^{33}P Transfer[c]	Contribution[d] (%)	References
	G. intraradices BEG 87	100	116	Y	20	Poulsen et al. (2005)
	G. intraradices WFVAM 23	100	100	Y	70	Poulsen et al. (2005)
	G. versiforme	62	60	Y	70	Poulsen et al. (2005)
Trifolium repens	*G. mosseae*	+		Y	?	Li et al. (1991)
Trifolium subterraneum	*Acaulospora laevis*	−36	−12	Y	?	Jakobsen et al. (1992)
	G. invermaium	−27	6	Y	?	Jakobsen et al. (1992)
	S. calospora	−44	−62	Y	?	Jakobsen et al. (1992)
Triticum aestivum	*G. caledonium*	−8	−5	Y	?	Ravnskov and Jakobsen (1995)
	G. intraradices	=	=	Y	80	Li et al. (2006)
	G. invermaium	−1	27	Negligible	?	Ravnskov and Jakobsen (1995)
T. aestivum cv. Newton	*G. etunicatum* and *G. mosseae*	−14	179	Y	?	Hetrick et al. (1996)
T. aestivum cv. Turkey	*G. etunicatum* and *G. mosseae*	81	250	Y	?	Hetrick et al. (1996)

[a]Where relevant the AM fungal isolate number follows the species name.
[b]Mycorrhizal growth response (MGR) or mycorrhizal P response (MPR); positive (+) or no change (=) relative to NM control plants; where possible the MGR/MPR has been calculated according to Eq. 7.1 (Sect. 7.3).
[c]Was ^{32}P or ^{33}P transfer observed? Y = yes.
[d]Contribution of the mycorrhizal pathway to plant P uptake as calculated by: % contribution = (specific activity P plant/specific activity P pot) × (soil P pot/soil P HC) × (hyphal length density pot/hyphal length density HC) × 100. ? = not known
[e]See reference for calculation method.

gerardii received less ^{32}P than non-responsive *Bromus inermis*. Furthermore, in *T. aestivum*, the specific activity (kBq ^{32}P mg^{-1} P) was lower in the positively responsive landrace Turkey, than in the non-responsive modern cultivar, Newton, indicating that the AM pathway made a greater contribution to the non-responsive cultivar (Hetrick et al. 1996). The overwhelming conclusion must be that the response, in terms of MGR or MPR, in non-responsive plants is entirely unrelated to the contribution of the AM pathway to P uptake.

Tracer studies also provide insights into the role of the fungal symbiont in determining diversity in plant responsiveness. In *Cucumis sativus*, Pearson and Jakobsen (1993) observed differences in plant P content (and hence MPR) depending on the identity of the AM fungal symbiont, in the order *Glomus caledonium* < *Scutellospora calospora* << *Glomus invermaium*. P content was unrelated to the amount of ^{32}P transported from the

HC, which increased in the order *S. calospora* < *G. invermaium* << *G. caledonium*. ^{32}P transfer was not correlated with hyphal length in the HC or % colonisation, leading the authors to suggest that hyphal P uptake, P translocation and P release to the plant are likely to be important control points governing the diversity in P contribution of different AM fungal species (see Munkvold et al. 2004 for further discussion).

Although these early experiments with radiotracers demonstrated transfer of ^{32}P via the AM pathway and enabled comparison of relative transfer under particular experimental conditions, attempts to quantify the contribution of the AM pathway were confounded by large HCs distant from the plant roots, which favoured the AM plants and fungal symbionts with extensive hyphal networks (Li et al. 1991; Pearson and Jakobsen 1993). A recent advance, depending on determination of the specific activity of $^{32/33}$P in the plant and in a small radiolabeled HC, and hyphal length densities in the main pot and HC, has largely overcome these issues (Smith et al. 2003b, 2004). The experiment demonstrated significant differences in contribution of the AM pathway to *Medicago truncatula*, *Solanum lycopersicum* and *Linum usitatissimum* (see Table 7.2). The contribution also varied with fungal species and, as had been previously observed, was not related to percent colonisation, plant growth response or total P uptake. Of particular significance was the finding that *Glomus intraradices* contributed up to 100% of total plant P to both *L. usitatissimum*, which showed a positive growth response (MGR 1,425%), and *S. lycopersicum* which showed a growth depression (MGR −18%). These results clearly demonstrate that the AM pathway can be functional in non-responsive plants.

Whilst re-emphasising that responsiveness is not an appropriate measure of mycorrhizal contribution to P uptake, these results also show that the direct, epidermal P uptake pathway can be suppressed during AM symbiosis in favour of the AM pathway. One explanation for the apparent down-regulation of direct P uptake in AM plants is that the formation of the depletion zone (Fig. 7.1) occurs more rapidly in AM plants. Competition between fungal hyphae and roots for P uptake in zones close to the root could lead to rapid depletion of P_i pools accessible to root transporters and decreased overall contribution of the direct pathway. However, an alternative explanation involving changes in the expression of plant P_i transporters has also been invoked. The discovery of the transporters responsible for P uptake by each pathway has enabled further investigation of this phenomenon.

7.5 The Role of P_i Transporters in P Uptake by an AM Plant

Identification of P_i transporters involved in AM fungal and plant P uptake provides tools for further investigation of processes controlling P influx via the direct and AM pathways. Numerous reviews have focussed on the identification and functional characterisation of these transporters (Smith et al. 2003a; Karandashov and Bucher 2005; Javot et al. 2007b; see also Chapter 6 by Ferrol and Pérez-Tienda). Accordingly, we highlight significant points related to variability in P fluxes via the plant and fungal P uptake pathways. Although plant and fungal P_i transporters have been identified in both responsive and non-responsive species, there has, so far, been limited focus on differences between these species and the resultant effects on symbiotic function.

7.5.1 Expression of P_i Transporters in the AM Pathway

7.5.1.1 AM Fungal P_i Transporters

The AM fungal P_i:H^+ symporter of *G. versiforme*, *GvPT*, is predominantly expressed in the external mycelium, with only low levels of expression in roots of the plant symbiont, *M. truncatula* (Harrison and van Buuren 1995). This pattern was taken as evidence for unidirectional flux of P from the fungal symbiont to the plant. In contrast, high levels of expression of the *GvPT* orthologue *GmosPT*, from *G. mosseae,* were observed not only in external mycelium but also in colonised roots of *C. sativus* (Benedetto et al. 2005). Using laser micro-dissection associated with gene expression analyses, Balestrini et al. (2007) localised intraradical expression of *GmosPT* in arbuscules in cortical cells of *S. lycopersicum*. Expression of the AM fungal H^+-ATPase, *GmHA5*, was also identified exclusively in arbuscules in this study. The expression of both a fungal P_i transporter and a fungal H^+-ATPase in arbuscules provides the first evidence for potential active P transport by fungal structures at intracellular interfaces and suggests that the fungal symbiont may compete with the plant for P in the interfacial apoplast. The differences in expression of P_i transporters from different fungi and at different locations raise important questions relating to symbiotic function. If AM fungi have different abilities to compete with plants for uptake from the interfacial apoplast this could have significant implications for overall efficiency of symbiotic P transfer in different plant–fungus combinations. Further investigation is clearly necessary. It will be interesting to determine whether the expression of fungal P_i transporters in roots is influenced by identity of host species and/or environmental conditions.

7.5.1.2 AM-Inducible Plant P_i Transporters

Plant transporters involved in P_i uptake from the interfacial apoplast have been identified in a range of plant species, including the non-responsive (or poorly responsive) *H. vulgare*, *T. aestivum* and *S. lycopersicum* (Glassop et al. 2005; Nagy et al. 2005). These AM-inducible P_i transporters fall into two categories, based upon their expression patterns: AM-specific P_i transporters expressed exclusively in AM roots, and AM-upregulated P_i transporters which also show low level expression in NM roots and/or shoots. RNA hybridisation and reporter-gene activity consistently demonstrate expression of AM-inducible P_i transporters in root cortical cells containing intracellular structures including arbuscules, hyphal coils and arbusculate coils (Karandashov et al. 2004; Glassop et al. 2005). In *M. truncatula*, immunolocalisation of the AM-specific P_i transporter MtPT4 revealed expression of protein exclusively on the plant plasma membrane surrounding arbuscules, it being strongest around mature arbuscules and coordinated with arbuscule development and decay (Harrison et al. 2002).

In *H. vulgare*, the AM-inducible P_i transporter *HvPT8* is expressed at high levels in AM roots regardless of plant responses to AM colonisation (Grace et al. 2008), corroborating ^{32}P data which suggested a 60% contribution of the AM pathway (Table 7.2). In *S. lycopersicum*, investigations with a *reduced mycorrhizal colonisation* mutant (*rmc*) and its wild type progenitor demonstrated that the AM-inducible P_i transporters, *LePT3* and *LePT4* are only expressed in symbiotic interactions capable of ^{32}P transfer to the plant (Poulsen et al. 2005). These data indicate that AM-inducible P_i transporters may provide useful molecular markers signalling formation of a functional symbiosis. However, there is no evidence at present to support suggestions that expression of these transporters can be correlated with P flux via the AM pathway.

The heterologous expression of AM-inducible P_i transporters in yeast has been used to functionally characterise these transporters and investigate transport kinetics. Although this system is not ideal (Smith et al. 2003a), results suggest that there may be considerable variation among different plants. The apparent K_m (64 μM) for the AM-upregulated transporter StPT3 suggests fairly high-affinity transport characteristics, whereas MtPT4 demonstrated low-affinity kinetics (K_m = 493–685 μM) (Harrison et al. 2002). The possible existence of both high and low-affinity AM inducible transporters suggests that P concentrations in the interfacial apoplast may be variable. This is conceivable if different AM fungi differ in their efficiency of P release to the interfacial apoplast and/or reabsorption (see above), and is supported by observations that P retention time in fungal hyphae differs between AM species (Jakobsen et al. 1992; Smith et al. 2000) and that some AM fungi accumulate polyP in hyphae rather than releasing it directly to the plant (R. Shibata, personal communication). Thus, the plant transport system will need to be adaptable and capable of P uptake over a range of P concentrations. Such requirements may explain the existence of multiple AM-inducible transporters in some plant species (see below). Indeed, these transporters may have evolved from others involved in scavenging P from the plant apoplast during plant maturation and senescence. It will be interesting to see whether these transporters differ in their uptake capacities and whether their expression is differentially regulated by different AM fungi or at different times during development of the symbiosis.

Multiple AM-inducible P_i transporters have been identified in the Solanaceae and in *Oryza sativa* (Nagy et al. 2005; Glassop et al. 2007). The targeted expression analyses of Balestrini et al. (2007) in *S. lycopersicum* demonstrated expression of the AM-inducible *LePT3* and AM-specific *LePT4* transporters exclusively in arbuscule-containing cortical cells. The AM-inducible *LePT5* was also expressed in non-colonised cells of AM roots and was assumed to be associated with the presence of intercellular hyphae. Nagy et al. (2005) used a transposon insertion mutant to investigate the mycorrhizal phenotype of *S. lycopersicum* mutants with loss-of-function of LePT4. Mycorrhizal colonisation and ^{33}P transfer via the mycorrhizal pathway were unaffected in *lept4–1* mutants, suggesting that LePT3 and LePT5 are able to compensate for loss of function of LePT4 and indicating functional overlap amongst these transporters. This is in stark contrast to recent reports of mutants from the Fabaceae, in which only a single AM-inducible transporter has been identified to date. Partial RNAi knockdown of the *LjPT3* transporter of *Lotus* reduced both growth response and colonisation levels of mutant plants compared with vector control plants (Maeda et al. 2006), whereas in *Medicago* complete silencing of *MtPT4* using RNAi resulted in total inhibition of the positive growth and P response usually observed for this highly responsive species (Javot et al. 2007a). Detailed observation of colonisation patterns in the *MtPT4* RNAi mutant and an MtPT4 loss-of-function mutant revealed the premature collapse and senescence of arbuscules (Javot et al. 2007a). Both LjPT3 and MtPT4 appear to be crucial for transfer of P to the plant via mycorrhizal interfaces. In addition, the data of Javot et al. (2007a) suggest that fungal P supply to the plant is essential for maintenance of a compatible AM interaction. These authors hypothesised that insufficient C transfer to the fungus may be responsible for arbuscule senescence; this was supported by the observation that fungal hyphae did not proliferate outside the root of the *mtpt4–1* mutant. The suggestion that P and C transfer may be intrinsically linked has been invoked by a number of authors (Woolhouse 1975; Fitter 2006). However, such linkage does not account for variations in P transfer via the AM pathway (Table 7.1), and hence with the notion that some AM fungi ‘cheat’ their hosts by acquiring C without donating P (Johnson et al. 1997; Kiers and van der Heijden 2006).

These data suggest that symbiotic function may be more tightly controlled in some plant families than in others. It will be interesting to determine whether double or triple knockout mutants of the AM-inducible P_i transporters in members of the Solanaceae display the same reduced mycorrhization phenotype as members of the Fabaceae. It is clear that, in the Fabaceae at least, AM-inducible P_i transporters play a crucial role in the integration of plant and fungal processes which lead to a compatible AM interaction. The mechanisms of this control remain to be elucidated.

7.5.2 P Uptake via the Direct Pathway in an AM Plant

The high affinity P_i transporters involved in P_i uptake at the root epidermis are down-regulated at high external P and responsive to the P status of the plant (Liu et al. 1998a; Rausch and Bucher 2002). Down-regulation of these transporters has also been observed during colonisation by AM fungi. In non-responsive *O. sativa*, six of the ten P_i transporters expressed in roots were down-regulated by AM colonisation at low P supply (Paszkowski et al. 2002). In responsive *M. truncatula*, a steady decline in expression of the epidermal transporters *MtPT1* and *MtPT2* was observed with increasing AM colonisation (Liu et al. 1998b), and MtPT1 protein levels mirrored transcript levels in this response (Chiou et al. 2001). Such concomitant changes in transcript and protein abundance provide evidence for transcriptional control of P_i transporter regulation.

It has been suggested that the down-regulation of epidermal P_i transporters in roots upon AM colonisation is primarily a function of improved P status of the plant (Burleigh and Bechmann 2002). However, few gene expression studies have included the physiological measurements that are necessary to provide further insight into this phenomenon. In a more extensive study of *M. truncatula* colonised by seven AM fungi, down-regulation of *MtPT2* varied depending on AM fungal species (Burleigh et al. 2002). In this experiment, a low level correlation was observed between shoot P concentration in AM plants and *MtPT2* expression. In *H. vulgare*, down-regulation of the root epidermal P_i transporters *HvPT1*, *HvPT2* and *HvPT3* was observed in conjunction with increases in tissue P content resulting from P fertilisation (Glassop et al. 2005). Expression of *HvPT1* and *HvPT2* was also lower in roots of AM than NM plants grown at low P, despite similar shoot and root P concentrations, whereas *HvPT3* transcript levels remained quite high in AM roots. In contrast to data from responsive *M. truncatula*, these results suggest an AM-specific signalling pathway involved in the down-regulation of epidermal P_i transporters that is independent of the P response pathways in the plant.

Down-regulation of epidermal P_i transporters in AM roots provides a significant link with ^{32}P uptake data indicating that the direct uptake pathway is switched off (at least partially) in some AM symbioses. A number of studies have begun to investigate this further. Colonisation of *H. vulgare* by *G. intraradices* resulted in reduced growth and P uptake relative to NM controls (Grace et al. 2008). However, plants were extensively colonised and the AM pathway contributed 60% of shoot P (Table 7.2). In this experiment, the root epidermal P_i transporters were not down-regulated (as had been previously reported; Glassop et al. 2005) and the decrease in contribution of the direct pathway could not be correlated with changes in their expression. Poulsen et al. (2005) reached a similar conclusion using *S. lycopersicum*, although in that case AM colonisation resulted in increases in plant growth and P uptake (Table 7.2). Interestingly, in the interaction between *S. lycopersicum* and *G. intraradices* BEG 87, the AM pathway accounted for only 20% of plant P uptake, indicating that the MPR of 116% was due to an increase in P uptake via the direct pathway. However, there was no clear correlation between the contribution of the direct uptake pathway and changes in expression of the epidermal P_i transporters. Although transcriptional regulation has been identified as an important primary control point for plant P_i transport, recent advances suggest that both post-transcriptional and post-translational modification may also be involved (Bucher 2007; see Bucher 2007 and references there in). The role of these processes in determining P fluxes by both the direct and AM uptake pathways remains to be determined.

A confounding factor in gene expression studies is the use of whole root samples for analysis of changes in gene expression. AM colonisation is non-synchronous: the plant root system is patchily colonised and colonisation units vary in age and stage of development. Sampling whole roots may mask cell-specific changes in transcript accumulation or changes in the localisation of gene expression. A number of studies have reported that expression of StPT1 and StPT2 of *Solanum tuberosum* is not altered in AM roots (Karandashov et al. 2004; Nagy et al. 2005). However, using a split-root system, Rausch et al. (2001) demonstrated localised down-regulation of these genes in the colonised half of the root system only. These observations, together with those of Gordon-Weeks et al. (2003) demonstrating differential expression of StPT1 and StPT2 during root development, highlight the need for targeted sampling in gene expression studies. New technologies, such as laser micro-dissection, will be critical in furthering our understanding of the role that P transporter expression plays in governing P fluxes via the plant and AM pathways. If AM fungi have differential ability to directly regulate the expression of plant P_i transporters, this may be pivotal to understanding the observed diversity in plant responses to colonisation.

7.6 Conclusions

P uptake in AM plants results from the combined operation of the direct root uptake pathway and the AM pathway. In non-responsive plants, it has been assumed that the AM pathway is non-functional and that growth depressions result from C supply to an AM fungus that confers little benefit to plant P uptake. However, it is clear from experiments utilising radiotracers that the AM pathway is not only functional in many non-responsive associations, but in some instances can take over from the direct uptake pathway in P supply to the plant. In addition, the notion that growth depressions result from AM fungal C drain is not entirely upheld. Growth depressions in non-responsive species are observed even at very low levels of AM colonisation when the C demand of the AM fungal symbiont is likely to be quite low. It has been suggested (Li et al. 2008) that this is a result of the epidermal P uptake switching off in response to fungus–plant recognition, even though P flux through the AM pathway is small. However, the down-regulation of epidermal P_i transporters is not consistently observed in AM roots. Whether this relates to methodological difficulties or real differences in plant and fungal control of gene expression remains to be determined. Improving our understanding of the signal pathways will be critical.

Non-responsive AM plants include widely cultivated crop species. We propose that non-responsive species present the greatest potential for increasing productivity and/or yield with AM symbioses, whether by selective manipulation of growth conditions or engineered traits which increase their reliance on and responsiveness to AM. The challenge now is to discover the processes, conditions, signals or mechanisms involved in these complex interactions and to apply them to future breeding efforts. The question of why the direct and AM pathways of P uptake are not additive in non-responsive plants should be a key research focus. If the epidermal pathway is not switched off when the AM pathway is operating in a non-responsive species, we hypothesise that a positive MGR would result and agricultural benefits follow. Exploring this issue should be a focus for future research. Detailed quantitative analyses of gene expression and transporter regulation, coupled with thorough physiological measurements, will be crucial in improving our understanding of these processes and will play an important role in guiding future efforts to improve the efficiency and utilisation of P by plant systems.

Acknowledgments We thank Mark Tester and Olivier Cotsaftis for supervision of the project, helpful discussions and ongoing support. Emily Grace is grateful for a Commonwealth Hill post-graduate research scholarship and to the Australian Centre for Plant Functional Genomics for research support and infrastructure. Part of our research program is also supported by the Australian Research Council.

References

Balestrini R, Gomez-Ariza J, Lanfranco L, Bonfante P (2007) Laser microdissection reveals that transcripts for five plant and one fungal phosphate transporter genes are contemporaneously present in arbusculated cells. Mol Plant Microbe Interact 20:1055–1062

Baon JB, Smith SE, Alston AM (1993) Mycorrhizal responses of barley cultivars differing in P-efficiency. Plant Soil 157:97–105

Benedetto A, Magurno F, Bonfante P, Lanfranco L (2005) Expression profiles of a phosphate transporter gene (*GmosPT*) from the endomycorrhizal fungus *Glomus mosseae*. Mycorrhiza 15:620–627

Bieleski RL (1973) Phosphate pools, phosphate transport, and phosphate availability. Annu Rev Plant Phys 24:225–252

Bucher M (2007) Functional biology of plant phosphate uptake at root and mycorrhiza interfaces. New Phytol 173:11–26

Bürkert B, Robson A (1994) 65Zn uptake in subterranean clover (*Trifolium subterraneum* L.) by three vesicular-arbuscular mycorrhizal fungi in a root-free sandy soil. Soil Biol Biochem 26:1117–1124

Burleigh SH, Bechmann IE (2002) Plant nutrient transporter regulation in arbuscular mycorrhizas. Plant Soil 244:247–251

Burleigh SH, Cavagnaro TR, Jakobsen I (2002) Functional diversity of arbuscular mycorrhizas extends to the expression of plant genes involved in P nutrition. J Exp Bot 53:1593–1601

Chiou T-J, Liu H, Harrison MJ (2001) The spatial expression patterns of a phosphate transporter (MtPT1) from *Medicago truncatula* indicate a role in phosphate transport at the root/soil interface. Plant J 25:281–293

Dickson S, Kolesik P (1999) Visualisation of mycorrhizal fungal structures and quantification of their surface area and volume using laser scanning confocal microscopy. Mycorrhiza 9:205–213

Ezawa T, Smith SE, Smith FA (2002) P metabolism and transport in AM fungi. Plant Soil 244:221–230

Fitter AH (2006) What is the link between carbon and phosphorus fluxes in arbuscular mycorrhizas? A null hypothesis for symbiotic function. New Phytol 172:3–6

Gianinazzi-Pearson V, Arnould C, Oufattole M, Arango M, Gianinazzi S (2000) Differential activation of H+-ATPase genes by an arbuscular mycorrhizal fungus in root cells of transgenic tobacco. Planta 211:609–613

Glassop D, Smith SE, Smith F (2005) Cereal phosphate transporters associated with the mycorrhizal pathway of phosphate uptake into roots. Planta 222:688–698

Glassop D, Godwin RM, Smith SE, Smith FW (2007) Rice phosphate transporters associated with phosphate uptake in rice roots colonised with arbuscular mycorrhizal fungi. Can J Bot 85:644–651

Gordon-Weeks R, Tong YP, Davies TGE, Leggewie G (2003) Restricted spatial expression of a high-affinity phosphate transporter in potato roots. J Cell Sci 116:3135–3144

Grace EJ, Cotsaftis O, Tester M, Smith FA, Smith SE (2008) Arbuscular Mycorrhizal inhibition of growth in barley cannot be attributed to extent of colonisation, fungal puptake or effects on expression of plant phosphate transporter genes. New Phytologist, in press.

Harrison MJ, van Buuren ML (1995) A phosphate transporter from the mycorrhizal fungus *Glomus versiforme*. Nature 378:626–632

Harrison MJ, Dewbre GR, Liu JY (2002) A phosphate transporter from *Medicago truncatula* involved in the acquisition of phosphate released by arbuscular mycorrhizal fungi. Plant Cell 14:2413–2429
Hetrick BAD, Wilson GWT, Cox TS (1992) Mycorrhizal dependence of modern wheat varieties, landraces, and ancestors. Can J Bot 70:2032–2040
Hetrick BAD, Wilson GWT, Schwab AP (1994) Mycorrhizal activity in warm-season and cool-season grasses: variation in nutrient-uptake strategies. Can J Bot 72:1002–1008
Hetrick BAD, Wilson GWT, Todd TC (1996) Mycorrhizal response in wheat cultivars: relationship to phosphorus. Can J Bot 74:19–25
Jakobsen I (1995) Transport of phosphorus and carbon in VA mycorrhizas. In: Varma A, Hock B (eds) Mycorrhiza; structure, function, molecular biology and biotechnology. Springer, Berlin, pp 297–324
Jakobsen I, Rosendahl L (1990) Carbon flow into soil and external hyphae from roots of mycorrhizal cucumber plants. New Phytol 115:77–83
Jakobsen I, Abbott LK, Robson AD (1992) External hyphae of vesicular-arbuscular mycorrhizal fungi associated with *Trifolium subterraneum* L. 2. Hyphal transport of 32P over defined distances. New Phytol 120:509–516
Jakobsen I, Chen B, Munkvold L, Lundsgaard T, Zhu YG (2005) Contrasting phosphate acquisition strategies of mycorrhizal fungi with that of root hairs using the root hairless barley mutant. Plant Cell Environ 28:928–938
Javot H, Penmetsa RV, Terzaghi N, Cook DR, Harrison MJ (2007a) A *Medicago truncatula* phosphate transporter indispensable for the arbuscular mycorrhizal symbiosis. Proc Natl Acad Sci USA 104:1720–1725
Javot H, Pumplin N, Harrison MJ (2007b) Phosphate in the arbuscular mycorrhizal symbiosis: transport properties and regulatory roles. Plant Cell Environ 30:310–322
Jifon JL, Graham JH, Drouillard DL, Syvertsen JP (2002) Growth depression of mycorrhizal citrus seedlings grown at high phosphorus supply is mitigated by elevated CO2. New Phytol 153:133–142
Johansen A, Jakobsen I, Jensen ES (1993) External hyphae of vesicular-arbuscular mycorrhizal fungi associated with *Trifolium subterraneum* L. 3. Hyphal transport of 32P and 15N. New Phytol 124:61–68
Johnson NC, Graham JH, Smith FA (1997) Functioning of mycorrhizal associations along the mutualism-parasitism continuum. New Phytol 135:575–586
Karandashov V, Bucher M (2005) Symbiotic phosphate transport in arbuscular mycorrhizas. Trends Plant Sci 10:22–29
Karandashov V, Nagy R, Wegmuller S, Amrhein N, Bucher M (2004) Evolutionary conservation of a phosphate transporter in the arbuscular mycorrhizal symbiosis. Proc Natl Acad Sci USA 101:6285–6290
Kiers ET, van der Heijden MGA (2006) Mutualistic stability in the arbuscular mycorrhizal symbiosis: exploring hypotheses of evolutionary cooperation. Ecology 87:1627–1636
Li HY, Smith SE, Holloway RE, Zhu YG, Smith FA (2006) Arbuscular mycorrhizal fungi contribute to phosphorus uptake by wheat grown in a phosphorus-fixing soil even in the absence of positive growth responses. New Phytol 172:536–543
Li HY, Smith FA, Dickson S, Holloway RE, Smith SE (2008) Plant growth depressions in arbuscular mycorrhizal symbioses: not just caused by carbon drain? New Phytol 178:852–862
Li X-L, Marschner H, George E (1991) Acquisition of phosphorus and copper by VA-mycorrhizal hyphae and root-to-shoot transport in white clover. Plant Soil 136:49–57
Liu C, Muchhal US, Uthappa M, Kononowicz AK, Raghothama KG (1998a) Tomato phosphate transporter genes are differentially regulated in plant tissues by phosphorus. Plant Physiol 116:91–99
Liu H, Trieu AT, Blaylock LA, Harrison MJ (1998b) Cloning and characterization of two phosphate transporters from *Medicago truncatula* roots: regulation in response to phosphate and response to colonization by arbuscular mycorrhizal (AM) fungi. Mol Plant Microbe Interact 11:14–22

Maeda D, Ashida K, Iguchi K, Chechetka SA, Hijikata A, Okusako Y, Deguchi Y, Izui K, Hata S (2006) Knockdown of an arbuscular mycorrhiza-inducible phosphate transporter gene of *Lotus japonicus* suppresses mutualistic symbiosis. Plant Cell Physiol 47:807–817

Maldonado-Mendoza IE, Dewbre GR, Harrison MJ (2001) A phosphate transporter gene from the extra-radical mycelium of an arbuscular mycorrhizal fungus *Glomus intraradices* is regulated in response to phosphate in the environment. Mol Plant Microbe Interact 14:1140–1148

Mimura T (1999) Regulation of phosphate transport and homeostasis in plant cells. Int Rev Cytol 191:149–200

Munkvold L, Kjoller R, Vestberg M, Rosendahl S, Jakobsen I (2004) High functional diversity within species of arbuscular mycorrhizal fungi. New Phytol 164:357–364

Nagy R, Karandashov V, Chague V, Kalinkevich K, Tamasloukht MB, Xu G, Jakobsen I, Levy AA, Amrhein N, Bucher M (2005) The characterization of novel mycorrhiza-specific phosphate transporters from *Lycopersicon esculentum* and *Solanum tuberosum* uncovers functional redundancy in symbiotic phosphate transport in solanaceous species. Plant J 42:236–250

Paszkowski U, Kroken S, Roux C, Briggs SP (2002) Rice phosphate transporters include an evolutionarily divergent gene specifically activated in arbuscular mycorrhizal symbiosis. Proc Natl Acad Sci USA 99:13324–13329

Pearson JN, Jakobsen I (1993) The relative contribution of hyphae and roots to phosphorus uptake by arbuscular mycorrhizal plants measured by dual labelling with 32P and 33P. New Phytol 124:489–494

Poulsen KH, Nagy R, Gao L-L, Smith SE, Bucher M, Smith FA, Jakobsen I (2005) Physiological and molecular evidence for Pi uptake via the symbiotic pathway in a reduced mycorrhizal colonization mutant in tomato associated with a compatible fungus. New Phytol 168:445–454

Rausch C, Bucher M (2002) Molecular mechanisms of phosphate transport in plants. Planta 216:23–37

Rausch C, Daram P, Brunner S, Jansa J, Lalol M, Leggewie G, Amrhein N, Bucher M (2001) A phosphate transporter expressed in arbuscule-containing cells in potato. Nature 414:462–466

Ravnskov S, Jakobsen I (1995) Functional compatibility in arbuscular mycorrhizas measured as hyphal P transport to the plant. New Phytol 129:611–618

Smith FA, Jakobsen I, Smith SE (2000) Spatial differences in acquisition of soil phosphate between two arbuscular mycorrhizal fungi in symbiosis with *Medicago truncatula*. New Phytol 147:357–366

Smith FW, Mudge SR, Rae AL, Glassop D (2003a) Phosphate transport in plants. Plant Soil 248:71–83

Smith SE, Read DJ (1997) Mycorrhizal Symbiosis. Academic, Cambridge,

Smith SE, Dickson S, Morris C, Smith FA (1994) Transfer of phosphate from fungus to plant in VA mycorrhizas: calculation of the area of symbiotic interface and of fluxes of P from two different fungi to *Allium porrum* L. New Phytol 127:93–99

Smith SE, Dickson S, Smith FA (2001) Nutrient transfer in arbuscular mycorrhizas: how are fungal and plant processes integrated? Aust J Plant Physiol 28:683–694

Smith SE, Smith FA, Jakobsen I (2003b) Mycorrhizal fungi can dominate phosphate supply to plants irrespective of growth responses. Plant Physiol 133:16–20

Smith SE, Smith FA, Jakobsen I (2004) Functional diversity in arbuscular mycorrhizal (AM) symbioses: the contribution of the mycorrhizal P uptake pathway is not correlated with mycorrhizal responses in growth or total P uptake. New Phytol 162:511–524

Tawaraya K (2003) Arbuscular mycorrhizal dependency of different plant species and cultivars. Soil Sci Plant Nutr 49:655–668

Tobar R, Azcón R, Barea JM (1994) Improved nitrogen uptake and transport from 15N-labelled nitrate by external hyphae of arbuscular mycorrhiza under water-stressed conditions. New Phytol 126:119–122

Woolhouse H (1975) Membrane structure and transport problems considered in relation to phosphorus and carbohydrate movement and the regulation of the endotrophic mycorrhizal associations. In: Sanders F, Mosse B, Tinker P (eds) Endomycorrhizas. Academic, London, pp 209–223

Zhu YG, Smith SE, Barritt AR, Smith FA (2001) Phosphorus (P) efficiencies and mycorrhizal responsiveness of old and modern wheat cultivars. Plant Soil 237:249–255

Zhu YG, Smith FA, Smith SE (2003) Phosphorus efficiencies and responses of barley (*Hordeum vulgare* L.) to arbuscular mycorrhizal fungi grown in highly calcareous soil. Mycorrhiza 13:93–100

Chapter 8
Mechanisms Underlying Heavy Metal Tolerance in Arbuscular Mycorrhizas

Manuel González-Guerrero, Karim Benabdellah, Nuria Ferrol, and Concepción Azcón-Aguilar

Abstract Arbuscular mycorrhizal fungi are able to tolerate a wide range of metal concentrations in soils. A number of passive and active molecular processes are employed by these fungi to maintain metal homeostasis. The main passive mechanism is the binding of metals to the fungal walls, responsible for a significant percentage of the metal retained. Meanwhile in the cytosol, a number of chelators (metallothioneins, glutathione) bind the metals very efficiently. Heavy metal transporters collaborate with the intracellular chelators to actively reduce the levels of metal by pumping metal out of the cytosol. Additionally, the fungus strives to reduce the free radicals produced by heavy metals. In this chapter, we discuss the most recent progress in the identification and characterization of the elements involved in maintaining metal homeostasis in arbuscular mycorrhizal fungi, as well as how the heavy metal control systems of the plant are affected by the development of the symbiosis.

8.1 Introduction

Although present in low amounts in living beings, some metals are essential for life. As many industrial chemical reactions which are catalyzed by metals, many enzymes also need transition metals in their catalytical core (Fraústro da Silva and Williams 2001). Transition metals, such as zinc, copper, iron, or cobalt, are present in every organism. They play structural and catalytic roles, such as the zinc finger domains and oxygen transport, or are involved in enzymatic reactions in processes

M. González-Guerrero(✉), K. Benabdellah, N. Ferrol, and C. Azcón-Aguilar
Department of Chemistry and Biochemistry, Worcester Polytechnic Institute, 100 Institute Rd, Worcester MA, 01609 USA
e-mail: manuelgg@wpi.edu

C. Azcón-Aguilar et al. (eds.), *Mycorrhizas - Functional Processes and Ecological Impact*,
DOI: 10/1007-978-3-540-87978-7_8,

such as photosynthesis, oxidative respiration, protection against free radicals, or lipid biosynthesis (Fraústro da Silva and Williams 2001). They are essential and are considered as mineral micronutrients. Other metals, such as cadmium, mercury, or lead, are highly toxic due to their chemical nature that can mimic essential metals. Excessive levels of mineral micronutrients are also highly toxic, since they can alter the levels of other nutrients or produce highly toxic free radicals (Fenton reactions) (Halliwell and Gutteridge 1989). In order to precisely regulate the levels of these elements, while at the same time being able to remove the non-essential ones, living beings have developed a number of redundant systems to monitor and control metal homeostasis. At a molecular level, these involve transporters, metallothioneins, and metallochaperones (Palmiter 1998; Guerinot 2000; O'Halloran and Culotta 2000; Eide 2004; Argüello et al. 2007).

Although life most probably originated in hot vents in the depth of the oceans, an acidic environments rich in metals (Segerer et al. 1993), most ecosystems nowadays present low levels of metals. This is more so in soils, especially basic soils, in which metal mobility is very limited (Hodgson 1963). In fact, one of the main factors limiting crop production is the deficiency of soils in essential micronutrients (Ruel and Bouis 1998). Many are the strategies by which plants strive to acquire essential metals from soils, which range from acidifying the soil to secreting chelators (Marschner and Römheld 1994), but probably one of the most extensive is the same that most plants use to ensure phosphate acquisition, the establishment of associations with soil fungi to form arbuscular mycorrhizas (AM). The establishment of mycorrhizal symbioses allows plants to supplement the function of the root system and to incorporate low mobility nutrients, such as certain forms of phosphate, nitrogen compounds, and metals (Bago et al. 1996; Díaz et al. 1996; Smith and Read 1997; Hawkins et al. 2000; Rufyikiri et al. 2002). An extensive body of knowledge exists on the physiology and molecular biology of phosphate metabolism in AM (Maldonado-Mendoza et al. 2001; Harrison et al. 2002; Karandashov and Bucher 2005; Javot et al. 2007; see also Chapters 6 and 7 by Ferrol and Pérez-Tienda and Grace et al., respectively), but much less is known about how metal homeostasis is controlled in mycorrhizal fungi or how the metals are delivered to the plant. The aim of this chapter is to summarize our current knowledge on this aspect in AM.

8.2 Mycorrhizas and Plant Metal Homeostasis

Although now overshadowed by studies on phosphate metabolism, the first published effect of mycorrhizas on plant mineral nutrition revealed that mycorrhizal apple seedlings had a higher Cu content than non-mycorrhizal ones (Mosse 1957). This finding was followed by similar observations in many plant–fungus associations and for different metals (Weissenhorn et al. 1995; Rivera-Becerril et al. 2002; Rufyikiri et al. 2003). The increased metal uptake is not due to a possible up-regulation of plant metal transporters, but to the uptake capability of the extraradical hyphae of AM fungi, as was clearly shown by radiolabeling studies (Cooper and

Tinker 1978; Entry et al. 1999; Rufyikiri et al. 2004). Consequently, AM fungi are able to improve micronutrient assimilation when plants are growing in deficient soils (Díaz et al. 1996; Chen et al. 2003). The mechanisms responsible could involve the higher soil volume that is accessible to the mycorrhizal plant, exudates from the AM fungi that help to mobilize essential metals, or the existence of transporters with higher affinity than those of plant origin (Smith and Read 1997). However, it has also been observed that AM fungi are able to confer increased metal tolerance to host plants (Entry et al. 1999; Rivera-Becerril et al. 2002). This has been shown in several types of soils, for several metals, and with different fungus–plant combinations. Element analysis of mycorrhizal and non-mycorrhizal plants reveals that mycorrhizal plants usually have lower levels of metals in the shoots than non-mycorrhizal ones (Rivera-Becerril et al. 2002; Chen et al. 2003). Microanalytical studies of metal localization in mycorrhizal roots have shown that metals accumulate preferentially in the fungal structures, thus preserving the plant tissues (Weiersbye et al. 1999; Rivera-Becerril et al. 2002). Therefore, it has been postulated that AM function as a 'buffer' that protects the plant against damaging alterations in metal levels in the soil (Díaz et al. 1996). As a result of the increased metal tolerance conferred by mycorrhizas, some authors have proposed that mycorrhizal plants could be used as an additional tool in bioremediation of soils contaminated with high levels of metals (Jacquot et al. 2000; Leyval et al. 2002).

The possibility of using mycorrhizal plants in bioremediation practices was further reinforced by the isolation of mycorrhizal fungi from soils with high levels of metals (del Val et al. 1999a; Malcová et al. 2003). These isolates are able to increase plant metal tolerance to heavy metals to a higher extent than isolates from unpolluted sites (del Val et al. 1999b; Malcová et al. 2003). In order to persist in environments with high metal content, AM fungi adapt through changes in the mechanisms that control intracellular metal homeostasis.

8.3 Mechanisms of Metal Homeostasis in AM Fungi

8.3.1 Changes in Fungal Development

AM fungi are able to alter their developmental pattern in order to avoid stress. It has been shown that in order to avoid unfavorable conditions in a heterogeneous environment, the mycelium develops more extensively in some sites than in others, which suggests that the fungal colony is able to detect distressing conditions and to avoid them (Bago et al. 2004). A similar strategy for metal avoidance has been shown in AM and other fungi (Fomina et al. 2000; Pawlowska and Charvat 2004). In the presence of metals, the fungal colony tends to extend the exploratory phase (either germinating or exploratory hyphae), probably trying to find less stressful environments (Pawlowska and Charvat 2004). In addition, it has been shown that spores in contaminated environments contain higher levels of metals than the rest of the fungal colony (González-Guerrero et al. 2008), a mechanism that would allow the mycelium to continue functioning in the presence of toxic substances. Similarly, vesicles of the intraradical mycelium accumulate more metal than the hyphae (Weiersbye et al. 1999).

8.3.2 Extracellular Chelation and Cell Wall Binding

In highly polluted sites, the possibilities for mycelium to avoid contaminants are very limited. Therefore, strategies have been developed to withstand the stress. The easiest and the most economical in terms of energy use is the immobilization of the cations in soil. This is achieved by the release of metal chelators. The nature of these chelators is diverse, but they are mainly organic acids such as citrate or oxalate (Green and Clausen 2003). Although no compound playing this role has been detected in AM fungi, ericoid and ectomycorrhizal fungi are able to produce iron chelators (Szaniszlo et al. 1981; Haselwandter et al. 1992), suggesting that AM fungi could synthesize molecules that would play the same role. The secretion of these compounds can result in up to 85% reduction of heavy metal incorporation, as described for the ectomycorrrhizal fungus *Paxillus involutus* (Bellion et al. 2006). Regardless of the possibility of the secretion of organic acids as chelators, a glycoprotein exuded by AM fungi, glomalin, has been postulated to play a role in metal immobilization (González-Chávez et al. 2004). In soils, glomalin has been proven to bind up to 4 mg of Cu per gram of protein (González-Chávez et al. 2004). It also binds Fe, Mn, Pb, and Cd though with lower affinity (Chern et al. 2007). The broad range of metal sequestered by glomalin has made it a candidate for strategies of biostabilization (Khan 2005).

However, the production and secretion of fungal exudates still requires a relatively high investment of energy. As an alternative, the fungal cell wall has a high content of chitin with potential metal binding sites, such as hydroxyls, carboxyls, or aminos (Strandberg et al. 1981). In terms of energy spent, it would be more economical for the mycorrhizal fungi to use this negatively charged structure to specifically adsorb the metals. In fact, certain studies indicate that the fungal wall is responsible for 50% of the metal retained by AM fungi (Joner et al. 2000). This adsorption is extremely fast and it takes less than 30 min to reach saturation (Joner et al. 2000). Heavy metal localization studies by EDXS have revealed that the binding capacity of the cell wall of spores and hyphae are different (González-Guerrero et al. 2008); in general, the spore wall retained a higher concentration of metals than the hyphal wall. Additionally, it has to be taken into account that the wall is not necessarily a static structure, but could be remodeled and its chemical composition altered. In this regard, fungi forming ericoid mycorrhiza present a differential chitin synthase expression pattern when exposed to high levels of Zn and a different wall arrangement (Lanfranco et al. 2002a, 2004), which can result in a higher metal binding capacity. Further modifications, such as the incorporation of melanin, can further increase the biosorption capacity of the cell wall (Fogarty and Tobin 1996). In ectomycorrhizal fungi, laccase activity and other enzymes involved in the production of polyphenolic compounds are induced under exposure to Cu or Cd (Palmieri et al. 2000; Faraco et al. 2003). It is tempting to speculate that a similar mechanism may also exist in other mycorrhizal fungi.

8.3.3 Cytosol Complexation

Most of the metabolic reactions occur in the cytosol or in the lumen of cell organelles. Therefore, it has to be a highly buffered environment in order to avoid any possible imbalance that might alter the chemical reactions that take place within. In this regard, free metals in the cytosol could represent a serious threat to the fitness of the cell. For this reason, organisms have developed mechanisms which sequester and deliver metals in a bound form, keeping the cytosolic levels of metals in the fM-pM range (Outten and O'Halloran 2001), with no free metals at all for practical purposes. This is achieved by a number of cytosolic chelators, of which the best known are metallothioneins (MTs).

MTs are low molecular weight polypeptides, typically of 70–80 amino acids. They share no specific structure, but all have a high number of cysteine residues which are involved in metal coordination and sequestration (Hamer 1986; Vasak and Hasler 2000). Different studies have shown MTs to have a complex role, being involved not only in metal homeostasis but also in oxidative stress protection (Palmiter 1998; Maret 2000, 2003).

Three MTs have been identified so far in AM fungi (Stommel et al. 2001; Lanfranco et al. 2002b; González-Guerrero et al. 2007). All of them have been isolated from EST libraries obtained from fungal samples grown under standard conditions. This observation hints at a role of MTs which does not involve sequestration of high levels of metals. Yeast complementation studies showed that proteins coded by *GmarMT1* and *GintMT1* are able to bind Cu and Cd (Lanfranco et al. 2002b; González-Guerrero et al. 2007). Both genes appear to be induced by high levels of Cu in the media, while Cd exposure results in no effect or even a down-regulation of gene expression. More detailed studies of *GintMT1* revealed that the signal for induction is an increase in the oxidative stress caused by the sudden increase in Cu levels rather than Cu per se (González-Guerrero et al. 2007), highlighting the role of MTs in oxidative stress protection. Likewise, MTs isolated from ectomycorrhizal fungi present a similar role, being induced by Cu and Cd and able to restore the Cu and Cd tolerance levels of a MT-deficient yeast strain (Bellion et al. 2007).

Overall, it seems that MTs play a role in metal homeostasis, although they probably do not constitute the primary control point for metal detoxification in the long term. This is not unusual, since the production of a molecule that would just bind metal and accumulate in the cytoplasm (no transporter for MTs is known) is not an energy-wise strategy. In fact, when a full-scale screening was performed in *Arabidopsis thaliana* for mutants resulting in higher levels of metals in the tissues, none of them was affected in MT genes (Lahner et al. 2003). Therefore, other mechanisms must exist which ensure the sequestration of the metal and the turnover of the chelator.

Glutathione (GSH) and GSH oligomers, such as phytochelatins, have also been associated with metal resistance (Zhu et al. 1999; Cobbett and Goldsbrough 2002). They can bind metals, mainly Cu and Cd, and the resulting complex is removed from the cytosol by specific transporters (see Sect. 8.3.4) that translocate the metal into the vacuole (Ortiz et al. 1995; Marin et al. 1998). As will be discussed in the next section, the isolation of multidrug resistance protein (MRP) transporters in AM fungi suggests that this is a viable strategy in these organisms. In ectomycorrhizal fungi, GSH concentrations and the activity of the GSH synthesis pathway are elevated upon Cd exposure (Howe et al. 1997; Courbot et al. 2004). However, no evidence of phytochelatins nor the enzyme that produces phytochelatins, phytochelatin synthase, have been found in AM fungi, although a number of different strategies (direct phytochelatin detection, PCR with degenerated primers, probes from different organisms, screening of available genomes, and gene libraries) have been attempted to detect them (M. González-Guerrrero, C. Azcón-Aguilar and N. Ferrol, unpublished results). The detection of GSH dimers in some fungi, the synthesis of which is not catalyzed by a phytochelatin synthase (Canovas et al. 2004; Wünschmann et al. 2007), leaves the possibility open for a similar process occurring in AM fungi.

In any case, metal trafficking in a cell is possible due to the concerted action of transporters and metallochaperones (O'Halloran and Culotta 2000; Argüello et al. 2007). Metallochaperones are metal shuttles that collect the metal from the influx transporters and provide it to efflux transporters or to accepting apo-proteins (O'Halloran and Culotta 2000). In the genome of the ectomycorrhizal fungus *Laccaria bicolor*, six genes putatively encoding metallochaperones have been identified. However, to date, no metallochaperone has been identified in AM fungi. Future work on the genomes or with SSH libraries of AM fungi will provide information about the presence or not of these molecules.

Overall, the relative importance of metal chelation in the cytosol greatly differs in the different types of mycorrhizal fungi. Ectomycorrhizal fungi retain a significant proportion of the incorporated metal in the cytosol, up to 20% of the total Cd in certain cases (Blaudez et al. 2000). However, cytosolic levels of metals in AM fungi remain barely detectable even after a week of metal exposure, especially Cd (González-Guerrero et al. 2008).

8.3.4 Metal Trafficking Across Membranes

As previously stated, the overall strategy of survival in the presence of high levels of metals is the reduction of their levels in the cytosol. Sequestration and immobilization in the cytosol is not a feasible long-term strategy of metal tolerance. For this reason, the control of heavy metal transport across biological membranes is the key element of an active control of heavy metal homeostasis. There is a great variety of metal transporters with different specificities and mechanisms of energy transduction (Puig and Thiele 2002; Eide 2004; Argüello et al. 2007). Most of them, except the Ctr which has only been detected in eukaryotes, are present in all domains of life. This hints at the evolutionary importance of metal transporters and the necessary redundancy.

Cation diffusion facilitator (CDF) transporters are a family of metal transporters involved in Zn, Fe, and Co homeostasis. Their biological role is ion efflux from the cytosol, either outside the cell or into an organelle (MacDiarmid et al. 2003; Kim et al. 2004). A member of this family, *GintZnT1*, was the first heavy metal transporter identified in AM fungi (González-Guerrero et al. 2005). This transporter was able to decrease cytosolic Zn levels in yeast, although not to an extent able to restore the phenotype of a *zrc1 cot1* yeast mutant, affected in the two vacuolar CDF transporters (MacDiarmid et al. 2002). The expression pattern of the gene reveals a role in Zn detoxification in the early moments of Zn exposure. This hints at the action of other Zn transporters working at later stages, which might complement the action of GintZnT1. For instance, ZIP transporters might be involved in this process since they are responsible for the influx of Zn or Fe into the cytosol (Eide 2004). It is interesting to note that, in yeast, long-term acclimation to high levels of Zn is achieved through down-regulation of uptake systems rather than up-regulation of efflux ones (Zhao and Eide 1996; Zhao et al. 1998), since it is a more conservative strategy in energetic terms. A gene with homology to members of the ZIP transporter family has been identified in *Glomus intraradices*, although a role at the plant–fungus interface rather than in Zn tolerance mechanisms of the extraradical mycelium was proposed for its gene product (Hildebrandt et al. 2007).

The subcellular location of GintZnT1 has not been determined yet. EDXS analysis of *G. intraradices* mycelium treated with Zn revealed that this metal tends to accumulate in vacuoles during the first 8 h of exposure, with no subsequent increase in vacuolar Zn levels from 8 h to the 7 days point (González-Guerrero et al. 2008). This is a pattern similar to that of *GintZnT1* expression, which is only up-regulated during the first hours of Zn exposure (González-Guerrero et al. 2005). Both results suggest that GintZnT1, or another CDF with a similar expression profile, is localized in the vacuolar membrane.

As described for Zn, EDXS observations showed that Cu and Cd also accumulate preferentially in the vacuoles of *G. intraradices* (González-Guerrero et al. 2008). Therefore, specific metal transporters must exist to introduce these two metals in these organelles. Recently, we have cloned and characterized an ABC transporter of the MRP subfamily, *GintABC1*, which might be responsible for Cu and Cd trafficking into the vacuole (M. González-Guerrero et al., unpublished results). These are transporters that couple the hydrolysis of ATP to the transport of a wide range of compounds: GSH conjugates into vacuoles in the case of the MRP subfamily (Kolukisaoglu et al. 2002; Kruh and Belinsky 2003). The expression profile of *GintABC1*, upregulated by Cu and by a short pulse of Cd, overlaps with the pattern of Cu and Cd accumulation in vacuoles. The high homology of GintABC1 to the *Saccharomyces cerevisiae* yeast Cd factor, a vacuolar ABC transporter involved in Cd detoxification, together with its transcriptional regulation by metals, suggest a localization for GintABC1 also in the vacuolar membrane.

Once in vacuoles, metals have to be stabilized. EDXS data hint at an association between metals and polyphosphate (González-Guerrero et al. 2008). Our study also shows that vacuoles having the highest metal content are localized in the spores. Given that a direct incorporation of metals through the spore wall is unlikely, and that the metals are probably incorporated into vacuoles in the hyphal cytosol, our observations predict an as yet unknown mechanism of metal-sensing in the fungus that would divert metal-rich vacuoles to spores. In addition, since vacuoles rich in phosphate and metals are diverted to the spore, our model predicts that exposure to high levels of metals would result in inadequate phosphate transfer to the plant. In this context, a change in the vacuolar morphology from tubular to spherical has been shown when the ectomycorrhizal fungus *P. involutus* is exposed to Zn (Tuszynska et al. 2006). This change is relevant, since tubular vacuoles have been associated with long-distance phosphate transport in fungi (Ashford 2002; Darrah et al. 2006).

8.3.5 Antioxidant Mechanisms to Cope with Heavy Metals

As mentioned above, heavy metals partially exert their damaging effects through the induction of oxidative stress. This is particularly true for redox-active metals such as Cu or Fe, which generate reactive oxygen species (ROS) (Halliwell and Gutteridge 1989). In order to live in an aerobic environment, living organisms have developed an impressive array of anti-oxidant metabolites and enzymes that scavenge or prevent the formation of ROS, hence protecting cells from oxidative damage. The antioxidant defences include enzymes such as catalases (CATs), superoxide dismutases (SODs), glutaredoxins (Grxs), and glutathione S-transferases (GSTs), and low molecular weight antioxidants such as GSH and vitamins C, E, and B6. To date, only a few genes encoding proteins putatively involved in ROS homeostasis have been identified and characterized in AM fungi: two *CuZn-SOD*s, ten genes putatively encoding GSTs, one *Grx*, one gene encoding a protein involved in Vitamin B6 biosynthesis, and three MTs (see Sect. 8.3.3). Although antioxidants play multiple roles in diverse physiological processes, we present here an overview of the role of antioxidants in the response to metal-induced stress in AM fungi.

Superoxide dismutases are ubiquitous components of the cellular antioxidant system. These proteins catalyze the dismutation of the superoxide anion (O_2^-) to oxygen (O_2) and hydrogen peroxide (H_2O_2). Two CuZnSOD genes, named *GmarCuZnSOD* and *GintCuZnSOD*, have been described in *Gigaspora margarita* and *G. intraradices*, respectively (Lanfranco et al. 2005; M. González-Guerrero et al., unpublished results).

Glutathione S-tranferases are a family of multifunctional detoxification enzymes. They catalyze the conjugation of the thiol group of GSH to some electrophilic moieties of diverse hydrophobic toxicants. Some GSTs also have GSH-dependent peroxidase activity against lipid hydroperoxide and hydrogen peroxides (Sheehan et al. 2001). Ten genes putatively encoding GSTs were isolated from *G. intraradices* SSH libraries, and most of them were up-regulated by heavy metals (Ouziad et al. 2005; Waschke et al. 2006). Both of these studies provided evidence about the involvement of these fungal genes in the alleviation of the damage caused by metal-induced oxidative stress.

Glutaredoxins play an important role in redox regulation in eukaryotic and prokaryotic cells (Holmgren et al. 2005). The *G. intraradices* glutaredoxin (GintGRX1) belongs to the ‘thioredoxin fold’ family (Bacik and Hazes 2006). Up-regulation of this gene by heavy metals, among other stimuli, reveals its role in the protection against heavy metal-induced oxidative stress (K. Benabdellah et al., 2008).

Finally, a novel gene involved in vitamin B6 biosynthesis has been isolated from *G. intraradices* (*GintPDX1.1*) (K. Benabdellah et al., unpublished results). Vitamin B6, also known as pyridoxine, has been identified as a potent antioxidant with a high ability to quench superoxide ions and prevent lipid peroxidation (Titiz et al. 2006). *GintPDX1.1* regulation studies showed a differential expression in response to Cu and ROS, suggesting a role for vitamin B6 in response to heavy metal exposure.

All the above-mentioned mechanisms of metal homeostasis in AM fungi are summarized in Fig. 8.1.

8.4 The Mycorrhizal Plant

Mycorrhizal establishment results in an increased tolerance to both excess and deficiency of essential heavy metals (Díaz et al. 1996; Chen et al. 2003). At least partially, this is the result of the molecular mechanisms ongoing in the mycorrhizal fungi. However, plants must also accommodate their own metal homeostasis to the presence of the microsymbiont. Effects of Cd on the plant have been reported at the level of protein profiles or nuclear ploidy level (Repetto et al. 2003; Repetto et al. 2007). Given that the mycorrhizal fungus is going to both provide metal to the plant and protect it against metal toxicity, it is to be expected that plant mechanisms responsible for these processes are down-regulated in mycorrhizal plants. In this respect, it has been shown that there is no up-regulation of a metallothionein gene nor any increase in glutathione reductase activity in mycorrhizal pea roots such as occurs in non-mycorrhizal plants (Rivera-Becerril et al. 2005). Additionally, some plant heavy metal transporters are down-regulated. This is the case of *MtZIP2* from *Medicago truncatula* or *LeNramp1* and *LeNramp3* from tomato plants (Burleigh et al. 2003; Ouziad et al. 2005). Moreover, mycorrhization also affects the tissue distribution of plant heavy metal transporters (Ouziad et al. 2005). To date, no up-regulated heavy metal transporter has been identified in mycorrhizal tissues. Such a protein would be a candidate transporter for the acquisition by the plant of metals provided by the mycorrhizal fungi.

8.5 Perspectives

The last 5 years have witnessed the first glimpses into the molecular basis of metal homeostasis in mycorrhizas. Intracellular chelators and transporters are beginning to be identified. However, we still lack a comprehensive view of these

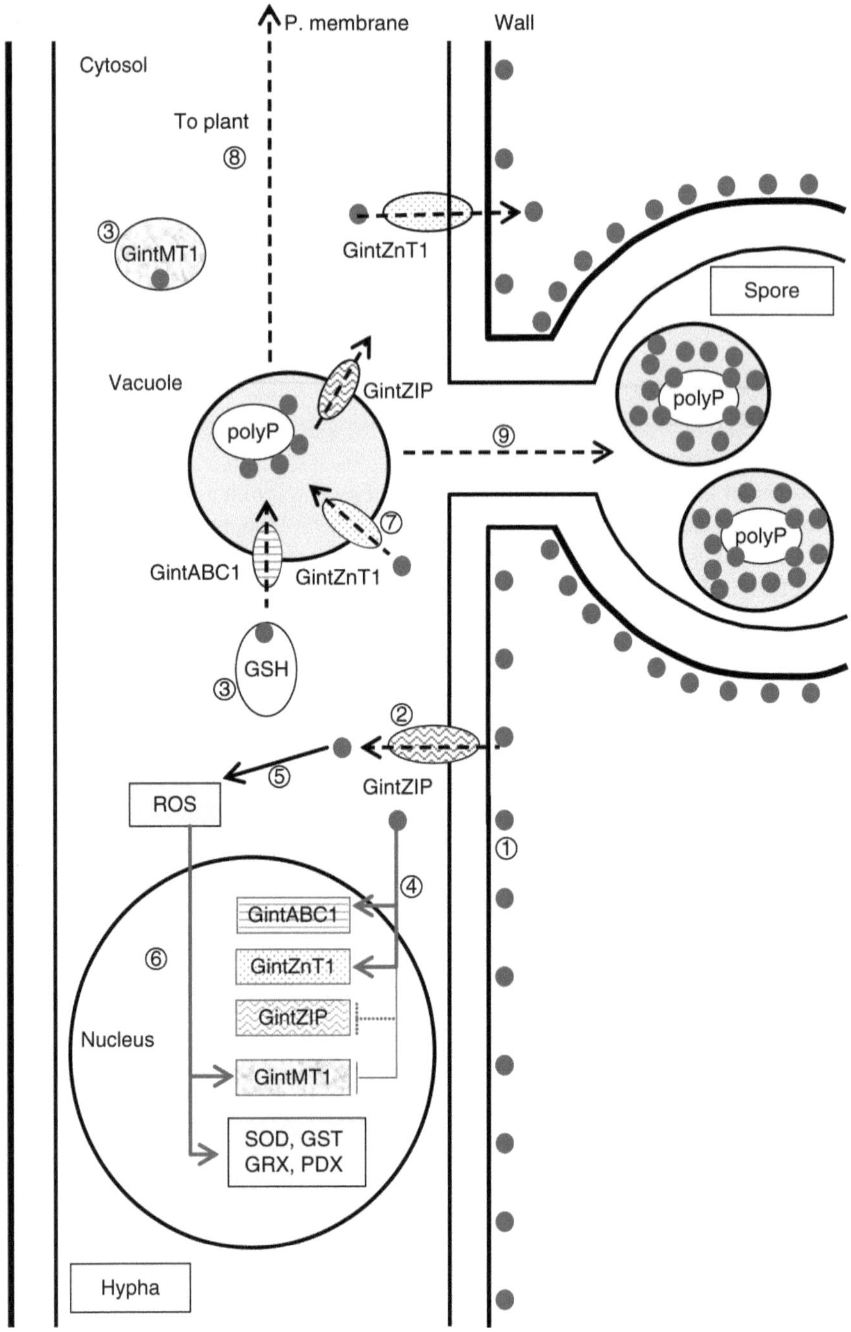
P. membrane
Wall
Cytosol
To plant
⑧
③ GintMT1
GintZnT1
Spore
Vacuole
GintZIP
polyP
⑨
polyP
polyP
⑦
GintABC1
GintZnT1
GSH
③
②
⑤
GintZIP
ROS
①
④
GintABC1
⑥
GintZnT1
GintZIP
Nucleus
GintMT1
SOD, GST
GRX, PDX
Hypha

homeostatic processes. The methodological limitations of working with mycorrhizal fungi, especially AM fungi, are largely responsible for the slowness of progress. Advances in the knowledge of the genome of these fungi will greatly improve our efforts, and new candidate genes will be identified and analyzed. So far, molecular studies have been limited to gene expression and heterologous expression, usually in yeast. Information on the subcellular localization of the transporters is essential to understand their physiological role, whether in intracellular trafficking or in metal transfer to the root. In this regard, identification of the plant transporter responsible for metal uptake at the periarbuscular membrane will be an important keystone to our understanding of plant–metal homeostasis under natural conditions.

Fungal ecotypes adapted to high levels of metals have probably developed specific heavy metal control strategies, such as reinforced fungal walls, more efficient efflux transporters, or transporters with altered specificity. AM fungi isolated from polluted and unpolluted sites should be compared in order to identify the determinants of metal tolerance. Finally, an increased understanding of these processes will provide essential tools for efficient phytoremediation practices and for an increase in the productivity in crop production, both pressing problems of the twenty-first century.

Acknowledgments The authors are grateful to the Consejería de Innovación, Ciencia y Empresa of the Andalusian Autonomic Governement (ref. P06-CVI-02263) for financial support to part of the work reported here. Karim Benabdellah was supported by an I3P contract from the Spanish Council for Scientific Research (CSIC).

Fig. 8.1 Tentative scheme of the mechanisms of metal homeostasis in *Glomus intraradices*. Although the negatively charged fungal wall binds metal ions (*1*), some of them are incorporated into the cytosol (*2*) through specific metal transporters, such as GintZIP. In the cytosol, metals are sequestered by intracellular chelators, such as metallothioneins (GintMT1) or glutathione (GSH) (*3*). At higher concentrations, they induce the expression of efflux systems, such as *GintABC1* or *GintZnT1*, as well as inhibiting *GintMT1* and probably the uptake systems, i.e., *GintZIP* (*4*). Through metal-catalyzed Fenton reactions, or the depletion of the GSH pool, reactive oxygen species (ROS) are produced (*5*), which result in gene activation (*GintMT1* and others) (*6*). The efflux systems will transport metals to the exterior or introduce them into the vacuoles (*7*), where they associate to polyphosphate. Some of these vacuoles will be directed to the plant root, where they will release the phosphate and probably the metals (*8*). However, if the concentration of metals within the vacuoles is too high, they might be diverted to the spores (*9*). Experimentally validated data are shown with *solid lines*, while hypotheses based on other model organisms (*Saccharomyces cerevisiae* or *Arabidopsis thaliana*) are shown as *dashed lines*. Effects on gene regulation are indicated by *grey lines*, both induction (→) and inhibition (—|)

References

Argüello J, Eren E, González-Guerrero M (2007) The structure and function of heavy metal transport P1B-ATPases. Biometals 20:233–248

Ashford AE (2002) Tubular vacuoles in arbuscular mycorrhizas. New Phytol 154:545–547

Bacik JP, Hazes B (2006) Crystal structures of a poxviral glutaredoxin in the oxidized and reduced states show redox-correlated structural changes. J Mol Biol 365:1545–1558

Bago B, Vierheilig H, Piché Y, Azcón-Aguilar C (1996) Nitrate depletion and pH changes induced by the extraradical mycelium of the arbuscular mycorrhizal fungus *Glomus intraradices* grown in monoxenic culture. New Phytol 133:273–280

Bago B, Cano C, Azcón-Aguilar C, Samson J, Coughlan AP, Piché Y (2004) Differential morphogenesis of the extraradical mycelium of an arbuscular mycorrhizal fungus grown monoxenically on spatially heterogeneous culture media. Mycologia 96:452–462

Bellion M, Courbot M, Jacob C, Blaudez D, Chalot M (2006) Extracellular and cellular mechanisms sustaining metal tolerance in ectomycorrhizal fungi. FEMS Microbiol Lett 254:173–181

Bellion M, Courbot M, Jacob C, Guinet F, Blaudez D, Chalot M (2007) Metal induction of a *Paxillus involutus* metallothionein and its heterologous expression in *Hebeloma cylindrosporum*. New Phytol 174:151–158

Benabdellah K, Merlos M, Azcón-Aguilar C, Ferrol N (2008). GintGRX1, the first characterized glomeromycotan glutaredoxin, is a multifunctional enzyme that responds to oxidative stress. Fungal Genet Biol. doi:10.1016/j.fgb.2008.09.019

Blaudez D, Botton B, Chalot M (2000) Cadmium uptake and subcellular compartmentation in the ectomycorrhizal fungus *Paxillus involutus*. Microbiology 146:1109–1117

Burleigh S, Kristensen B, Bechmann IE (2003) A plasma membrane zinc transporter from *Medicago truncatula* is up-regulated in roots by Zn fertilization, yet down-regulated by arbuscular mycorrhizal colonization. Plant Mol Biol 52:1077–1088

Canovas D, Vooijs R, Schat H, De Lorenzo V (2004) The role of thiol species in the hypertolerance of *Aspergillus* sp P37 to arsenic. J Biol Chem 279:51234–51240

Chen B, Li X, Tao H, Christie P, Wong M (2003) The role of arbuscular mycorrhiza in zinc uptake by red clover growing in a calcareous soil spiked with various quantities of zinc. Chemosphere 50:839–846

Chern EC, Tsai DW, Ogunseitan OA (2007) Deposition of glomalin-related soil protein and sequestered toxic metals into watersheds. Environ Sci Technol 41:3566–3572

Cobbett C, Goldsbrough P (2002) Phytochelatins and metallothioneins: Roles in heavy metal detoxification and homeostasis. Annu Rev Plant Biol 53:159–182

Cooper KM, Tinker PB (1978) Translocation and transfer of nutrients in vesicular arbuscular mycorrhizas II. Uptake and translocation of phosphorus, zinc and sulphur. New Phytol 81:43–52

Courbot M, Diez L, Ruotolo R, Chalot M, Leroy P (2004) Cadmium-responsive thiols in the ectomycorrhizal fungus *Paxillus involutus*. Appl Environ Microbiol 70:7413–7417

Darrah PR, Tlalka M, Ashford AE, Watkinson SC, Fricker MD (2006) The vacuole system is a significant intracellular pathway for longitudinal solute transport in basidiomycete fungi. Eukaryot Cell 5:1111–1125

del Val C, Barea JM, Azcón-Aguilar C (1999a) Diversity of arbuscular mycorrhizal fungus populations in heavy-metal-contaminated soils. Appl Environ Microbiol 65:718–723

del Val C, Barea JM, Azcón-Aguilar C (1999b) Assessing the tolerance to heavy metals of arbuscular mycorrhizal fungi isolated from sewage sludge-contaminated soils. Appl Soil Ecol 11:261–269

Díaz G, Azcón-Aguilar C, Honrubia M (1996) Influence of arbuscular mycorrhizae on heavy metals (Zn and Pb) uptake and growth of *Lygeum spartum* and *Anthillis cytisoides*. Plant Soil 180:241–249

Eide DJ (2004) The ABC of solute carriers. The SLC39 family of metal ion transporters. Eur J Physiol 447:796–800
Entry JA, Watrud LS, Reeves M (1999) Accumulation of 137Cs and 90Sr from contaminated soil by three grass species inoculated with mycorrhizal fungi. Environ Pollut 104:449–457
Faraco V, Sannia G, Giardina P (2003) Metal-responsive elements in *Pleurotus ostratus* laccase gene promoters. Microbiology 149:2155–2162
Fogarty RV, Tobin JM (1996) Fungal melanins and their interactions with metals. Enzyme Microb Technol 19:311–317
Fomina M, Ritz K, Gadd GM (2000) Negative fungal chemotropism to toxic metals. FEMS Microbiol Lett 193:207–211
Fraústro da Silva JJR, Williams RJP (2001) The biological chemistry of the elements, 2nd edn. Oxford University Press, New York
González-Chávez MC, Carrillo-González R, Wright SF, Nichols KA (2004) The role of glomalin, a protein produced by arbuscular mycorrhizal fungi, in sequestering potentially toxic elements. Environ Pollut 130:317–323
González-Guerrero M, Azcón-Aguilar C, Mooney M, Valderas A, MacDiarmid CW, Eide DJ, Ferrol N (2005) Characterization of a *Glomus intraradices* gene encoding a putative Zn transporter of the cation diffusion facilitator family. Fungal Genet Biol 42:130–140
González-Guerrero M, Cano C, Azcón-Aguilar C, Ferrol N (2007) *GintMT1* encodes a functional metallothionein in *Glomus intraradices* that responds to oxidative stress. Mycorrhiza 17:327–335
González-Guerrero M, Melville LH, Ferrol N, Lott JNA, Azcón-Aguilar C, Peterson RL (2008) Ultrastructural localization of heavy metals in the extraradical mycelium and spores of the arbuscular mycorrhizal fungus *Glomus intraradices*. Can J Microbiol 54:103–110
Green F, Clausen CA (2003) Copper tolerance of brown-rot fungi: time course of oxalic acid production. Int Biodeterior Biodegrad 51:145–149
Guerinot ML (2000) The ZIP family of metal transporters. Biochim Biophys Acta 1465:190–198
Halliwell B, Gutteridge JMC (1989) Free radicals in biology and medicine. Clanderon Press, Oxford, UK
Hamer DH (1986) Metallothionein. Ann Rev Biochem 55:913–951
Harrison M, Dewbre G, Liu J (2002) A phosphate transporter from *Medicago truncatula* involved in the acquisition of phosphate released by arbuscular mycorrhizal fungi. Plant Cell 14:2413–2429
Haselwandter K, Dobernigg B, Beck W, Jung G, Cansier A, Winkelmann G (1992) Isolation and identification of hydroxamate siderophores of ericoid mycorrhizal fungi. Biometals 5:51–56
Hawkins HJ, Johansen A, George E (2000) Uptake and transport of organic and inorganic nitrogen by arbuscular mycorrhizal fungi. Plant Soil 226:275–285
Hildebrandt U, Regvar M, Bothe H (2007) Arbuscular mycorrhiza and heavy metal tolerance. Phytochemistry 68:139–146
Hodgson JF (1963) Chemistry of the micronutrient elements in soils. Adv Agron 15:119–159
Holmgren A, Johansson C, Berndt C, Lönn ME, Hudemann C, Lillig CH (2005) Thiol redox control via thioredoxin and glutaredoxin systems. Biochem Soc Trans 33:1375–1377
Howe R, Evans RL, Ketteridge SW (1997) Copper-binding proteins in ectomycorrhizal fungi. New Phytol 135:123–131
Jacquot E, van Tuinen D, Gianinazzi S, Gianinazzi-Pearson V (2000) Monitoring species of arbuscular mycorrhizal fungi in planta and in soil by nested PCR: Application to the study of the impact of sewage sludge. Plant Soil 226:179–188
Javot H, Pumplin N, Harrison MJ (2007) Phosphate in the arbuscular mycorrhizal symbiosis: Transport properties and regulatory roles. Plant Cell Environ 30:310–322
Joner EJ, Briones R, Leyval C (2000) Metal-binding capacity of arbuscular mycorrhizal mycelium. Plant Soil 226:227–234
Karandashov V, Bucher M (2005) Symbiotic phosphate transport in arbuscular mycorrhizas. Trends Plant Sci 10:22–29

Khan AG (2005) Role of soil microbes in the rhizospheres of plants growing on trace metal contaminated soils in phytoremediation. J Trace Elem Med Biol 18:355–364

Kim D, Gustin JL, Lahner B, Persans MW, Baek D, Yun DJ, Salt DE (2004) The plant CDF family member TgMTP1 from the Ni/Zn hyperaccumulator *Thlaspi goesingense* acts to enhance efflux of Zn at the plasma membrane when expressed in *Saccharomyces cerevisiae*. Plant J 39:237–251

Kolukisaoglu HU, Bovet L, Klein M, Eggmann T, Geisler M, Wanke D, Martinoia E, Schulz B (2002) Family business: the multidrug-resistance related protein (MRP) ABC transporter genes in *Arabidopsis thaliana*. Planta 216:107–119

Kruh GD, Belinsky MG (2003) The MRP family of drug efflux pumps. Oncogene 22:7537–7552

Lahner B, Gong J, Mahmoudian M, Smith E, Abid K, Rogers E, Guerinot M, Harper J, Ward J, McIntyre L, Schoroeder J, Salt D (2003) Genomic scale profiling of nutrient and trace elements in *Arabidopsis thaliana*. Nat Biotechnol 21:1215–1221

Lanfranco L, Balsamo R, Martino E, Perotto S, Bonfante P (2002a) Zinc ions alter morphology and chitin deposition in an ericoid fungus. Eur J Histochem 46:341–350

Lanfranco L, Bolchi A, Ros E, Ottonello S, Bonfante P (2002b) Differential expression of a metallothionein gene during the presymbiotic versus the symbiotic phase of an arbuscular mycorrhizal fungus. Plant Physiol 130:58–67

Lanfranco L, Balsamo R, Martino P, Bonfante P, Perotto S (2004) Zinc ions differentially affect chitin synthase gene expression in an ericoid mycorrhizal fungus. Plant Biosyst 138:271–277

Lanfranco L, Novero M, Bonfante P (2005) The mycorrhizal fungus *Gigaspora margarita* possesses a CuZn superoxide dismutase that is up-regulated during symbiosis with legume hosts. Plant Physiol 137:1319–1330

Leyval C, Joner EJ, del Val C, Haselwandter K (2002) Potential of arbuscular mycorrhizal fungi for bioremediation. In: Gianinazzi S, Schüepp H, Barea JM, Haselwandter K (eds) Mycorrhizal technology in agriculture. Birkhäuser Verlag, Basel, pp 175–186

MacDiarmid CW, Milanick MA, Eide D (2002) Biochemical properties of vacuolar zinc transport systems of *Saccharomyces cerevisiae*. J Biol Chem 277:39187–39194

MacDiarmid C, Milanick M, Eide D (2003) Induction of the *ZRC1* metal tolerance gene in zinc-limited yeast confers resistance to zinc shock. J Biol Chem 278:15065–15072

Malcová R, Rydlová J, Vosátka M (2003) Metal-free cultivation of *Glomus* sp. BEG 140 isolated from Mn-contaminated soil reduces tolerance to Mn. Mycorrhiza 13:151–157

Maldonado-Mendoza IE, Dewbre GR, Harrison MJ (2001) A phosphate transporter gene from the extra-radical mycelium of an arbuscular mycorrhizal fungus *Glomus intraradices* is regulated in response to phosphate in the environment. Mol Plant Microbe Interact 14:1140–1148

Maret W (2000) The function of zinc metallothionein: a link between cellular zinc and redox state. J Nutr 130:1455S-1458S

Maret W (2003) Cellular zinc and redox states converge in the metallothionein/thionein pair. J Nutr 133:1460S-1462

Marin E, Leonhardt N, Vavasseur A, Forestier C (1998) Cloning of AtMRP1, an *Arabidopsis thaliana* cDNA encoding a homologue of the mammalian multidrug resistance-associated protein. Biochim Biophys Acta 1369:7–13

Marschner H, Römheld V (1994) Strategies of plants for acquisition of iron. Plant Soil 165:261–274

Mosse B (1957) Growth and chemical composition of mycorrhizal and non-mycorrhizal apples. Nature 179:349–362

O'Halloran TV, Culotta VC (2000) Metallochaperones, an intracellular shuttle service for metal ions. J Biol Chem 275:25057–25060

Ortiz DF, Ruscitti T, McCue KF, Ow DW (1995) Transport of metal-binding peptides by HMT1, a fission yeast ABC-type vacuolar membrane protein. J Biol Chem 270:4721–4728

Outten C, O'Halloran T (2001) Femtomolar sensitivity of metalloregulatory proteins controlling zinc homeostasis. Science 292:2488–2492

Ouziad F, Hildebrandt U, Schmelzer E, Bothe H (2005) Differential gene expressions in arbuscular mycorrhizal-colonized tomato grown under heavy metal stress. J Plant Physiol 162:634–649

Palmieri F, Giardina P, Bianco C, Fontanella B, Sannia G (2000) Copper induction of laccase isoenzymes in the ligninolytic fungus *Pleurotus ostreatus*. Appl Environ Microbiol 66:920–924

Palmiter RD (1998) The elusive function of metallothioneins. Proc Natl Acad Sci USA 85:8428–8430

Pawlowska TE, Charvat I (2004) Heavy-metal stress and development patterns of arbuscular mycorrhizal fungi. Appl Environ Microbiol 70:6643–6649

Puig S, Thiele DJ (2002) Molecular mechanisms of copper uptake and distribution. Curr Opin Chem Biol 6:171–180

Repetto O, Bestel-Corre G, Dumas-Gaudot E, Berta G, Gianinazzi-Pearson V, Gianinazzi S (2003) Targeted proteomics to identify cadmium-induced protein modifications in *Glomus mosseae*-inoculated pea roots. New Phytol 157:555–567

Repetto O, Massa N, Gianinazzi-Pearson V, Dumas-Gaudot E, Berta G (2007) Cadmium effects on populations of root nuclei in two pea genotypes inoculated or not with the arbuscular mycorrhizal fungus *Glomus mosseae*. Mycorrhiza 17:111–120

Rivera-Becerril F, Calantzis C, Turnau K, Caussanel JP, Belimov AA, Gianinazzi S, Strasser RJ, Gianinazzi-Pearson V (2002) Cadmium accumulation and buffering of cadmium-induced stress by arbuscular mycorrhiza in three *Pisum sativum* L. genotypes. J Exp Bot 53:1177–1185

Rivera-Becerril F, van Tuinen D, Martin-Laurent F, Metwally A, Dietz K, Gianinazzi S, Gianinazzi-Pearson V (2005) Molecular changes in *Pisum sativum* L. roots during arbuscular mycorrhiza buffering of cadmium stress. Mycorrhiza 16:51–60

Ruel MT, Bouis HE (1998) Plant breeding: A long-term strategy for the control of zinc deficiency in vulnerable populations. Am J Clin Nutr 68:488S-494S

Rufyikiri G, Thiry Y, Wang L, Delvaux B, Declerck S (2002) Uranium uptake and translocation by the arbuscular mycorrhizal fungus *Glomus intraradices* under root-organ culture conditions. New Phytol 156:275–281

Rufyikiri G, Thiry Y, Declerck S (2003) Contribution of hyphae and roots to uranium uptake and translocation by arbuscular mycorrhizal carrot roots under root-organ culture conditions. New Phytol 158:391–399

Rufyikiri G, Declerck S, Thiry Y (2004) Comparison of ^{232}U and ^{33}P uptake and translocation by the arbuscular mycorrhizal fungus *Glomus intraradices* in root organ culture conditions. Mycorrhiza 14:203–207

Segerer AH, Burggraf S, Fiala G, Huber G, Huber R, Pley U, Stetter KO (1993) Life in hot springs and hydrothermal vents. Orig Life Evol Biosph 23:77–90

Sheehan D, Meade G, Foley VM, Dowd CA (2001) Structure, function and evolution of glutathione transferases: implications for classification of non-mammalian members of an ancient enzyme superfamily. Biochem J 360:1–16

Smith SE, Read DJ (1997) Mycorrhizal symbiosis. Academic, London

Stommel M, Mann P, Franken P (2001) EST-library construction using spore RNA of the arbuscular mycorrhizal fungus *Gigaspora rosea*. Mycorrhiza 10:281–285

Strandberg GW, Shumate SE, Parrott JR Jr, (1981) Microbial cells as biosorbents for heavy metals: accumulation of uranium by *Saccharomyces cerevisiae* and *Pseudomonas aeruginosa*. Appl Environ Microbiol 41:237–245

Szaniszlo PJ, Powell PE, Reid CPP, Cline GR (1981) Production of hydroxamate siderophores iron chelators by ectomycorrhizal fungi. Mycologia 73:1158–1174

Titiz O, Tambasco-Studart M, Warzych E, Apel K, Amrhein N, Laloi C, Fitzpatrick TB (2006) PDX1 is essential for vitamin B6 biosynthesis, development and stress tolerance in *Arabidopsis*. Plant J 48:933–946

Tuszynska S, Davies D, Turnau K, Ashford AE (2006) Changes in vacuolar and mitochondrial motility and tubularity in response to zinc in a *Paxillus involutus* isolate from a zinc-rich soil. Fungal Genet Biol 43:155–163

Vasak M, Hasler DW (2000) Metallothioneins: new functional and structural insights. Curr Opin Chem Biol 4:177–183

Waschke A, Sich M, Tamasloukht M, Fischer K, Mann P, Franken P (2006) Identification of heavy metal-induced genes encoding glutathione S-transferase in the arbuscular mycorrhizal fungus *Glomus intraradices*. Mycorrhiza 17:1–10

Weiersbye IM, Straker CJ, Przybylowicz, WJ (1999) Micro-PIXE mapping of elemental distribution in arbuscular mycorrhizal roots of the grass *Cynodon dactylon*, from gold and uranium mine tailings. Nucl Inst Methods Phys Res 158:335–343

Weissenhorn I, Leyval C, Belgy G, Berthelin J (1995) Arbuscular mycorrhizal contribution to heavy metal uptake by maize (*Zea mays*) in pot cultures with contaminated soil. Mycorrhiza 5:245–251

Wünschmann J, Beck A, Meyer L, Letzel T, Grill E, Lendzian KJ (2007) Phytochelatins are synthesized by two vacuolar serine carboxypeptidases in *Saccharomyces cerevisiae*. FEBS Lett 581:1681–1687

Zhao H, Eide D (1996) The yeast *ZRT1* gene encodes the zinc transporter protein of a high-affinity uptake system induced by zinc limitation. Proc Natl Acad Sci USA 93:2454–2458

Zhao H, Butler E, Rodgers J, Spizzo T, Duesterhoeft S, Eide D (1998) Regulation of zinc homeostasis in yeast by binding of the ZAP1 transcriptional activator to zinc-responsive promoter elements. J Biol Chem 273:28713–28720

Zhu YL, Pilon-Smits EA, Tarun AS, Weber SU, Jouanin L, Terry N (1999) Cadmium tolerance and accumulation in Indian mustard is enhanced by overexpressing gamma-glutamylcysteine synthetase. Plant Physiol 121:1169–1178

Chapter 9
Priming Plant Defence Against Pathogens by Arbuscular Mycorrhizal Fungi

María J. Pozo, Adriaan Verhage, Javier García-Andrade, Juan M. García, and Concepción Azcón-Aguilar

Abstract Root colonisation by arbuscular mycorrhizal fungi (AMF) can improve plant resistance/tolerance to biotic stresses. Although this bioprotection has been amply described in different plant systems, the underlying mechanisms remain largely unknown. Besides mechanisms such as improved plant nutrition and competition, experimental evidence supports the involvement of plant defence mechanisms in the observed protection. During mycorrhiza establishment, modulation of plant defence responses occurs upon recognition of the AMF in order to achieve a functional symbiosis. As a consequence of this modulation, a mild, but effective activation of the plant immune responses may occur, not only locally but also systemically. This activation leads to a primed state of the plant that allows a more efficient activation of defence mechanisms in response to attack by potential enemies.

9.1 Introduction

Mutually beneficial interactions between plants and microbes are frequent in nature. Common benefits for the plant are improved plant nutrition and/or increased capability to cope with adverse conditions. In the case of arbuscular mycorrhizal (AM) associations, the symbioses alter plant physiology, leading to a better mineral nutrition and to increased resistance/tolerance to biotic and abiotic stresses. Although it should be noted that the ability to enhance plant resistance/tolerance differs among AM fungal isolates and can be modulated by environmental conditions, general trends emerge from the multiple studies dealing with mycorrhiza-induced resistance in different pathosystems. Enhanced resistance/tolerance to soil-borne pathogens has been widely reported in mycorrhizal plants (Whipps 2004). Although it is clear that the symbiosis may also impact plant interactions

M.J. Pozo (✉), A. Verhage, J. García-Andrade, J.M. García, and C. Azcón-Aguilar
Departamento de Microbiología del Suelo y Sistemas Simbióticos,
Estación Experimental del Zaidín, CSIC, Profesor Albareda 1, 18008 Granada, Spain
e-mail: mariajose.pozo@eez.csic.es

C. Azcón-Aguilar et al. (eds.), *Mycorrhizas - Functional Processes and Ecological Impact*, 123
DOI: 10/1007-978-3-540-87978-7_9, © Springer-Verlag Berlin Heidelberg 2009

with aboveground attackers, the outcome of those interactions is less clear and seems to depend largely on the attacker lifestyle (Pozo and Azcón-Aguilar, 2007). This finding points to a differential regulation of plant defence signalling pathways. In this chapter, we summarise the information available regarding mycorrhiza-induced resistance (MIR) with special emphasis on the involvement of plant defence mechanisms.

9.2 Mycorrhiza-Induced Resistance Against Pathogens

Most studies on protection by AM against soil-borne diseases report reductions in incidence and/or severity of root rot or wilting caused by fungi such as *Rhizoctonia*, *Fusarium* or *Verticillium*, and root rot caused by oomycetes including *Phytophthora*, *Pythium* and *Aphanomyces*. A comprehensive review of these studies was compiled by Whipps (2004). Similarly, a reduction in deleterious effects caused by parasitic nematodes, such as *Pratylenchus* and *Meloidogyne*, is common in mycorrhizal plants (de la Peña et al. 2006; Li et al. 2006). Because of the common localisation in the root of both attackers and AM fungi, it is generally difficult to discern the local or systemic character of the protection. However, the use of split-root experimental systems, allowing physical separation between AM fungi (AMF) and pathogens, has revealed reductions in pathogen infection and in disease symptoms in the non-mycorrhizal parts of root systems of mycorrhizal plants. Systemic protection at the root system level has been demonstrated against *Phytophthora* and *Ralstonia* in tomato (Cordier et al. 1998; Pozo et al. 2002; Zhu and Yao 2004), *Gaeumannomyces* in wheat (Khaosaad et al. 2007) and recently against plant parasitic nematodes in banana plants (Elsen et al. 2008). Such effectiveness against a diverse range of attackers, including nematodes, oomycetes, bacteria and fungi, confirms the broad spectrum character of the induced resistance associated with the AM symbiosis.

Information about mycorrhizal effects on aboveground diseases is scarce and apparently less conclusive. Early reports associated AM symbioses with enhanced susceptibility to viruses (see Whipps 2004 for review). In line with those studies, it was proposed that mycorrhizal plants are more susceptible to shoot pathogens (Shaul et al. 1999). However, recent studies dealing with pathogens of different lifestyles have reported a more complex reality. Biotrophic pathogens, such as powdery mildew and rust fungi (*Blumeria*, *Oidium*, *Uromyces*), seem to perform better in mycorrhizal plants, although increased tolerance was often observed in terms of plant mass and yield (Gernns et al. 2001; Whipps 2004). Concerning hemibiotrophs, the impact of the symbiosis varies from no effect to reduction of the disease, for example against *Colletotrichum orbiculare*, the causal agent of anthracnose in cucumber (Lee et al. 2005; Chandanie et al. 2006). Finally, several studies have observed a positive effect of the symbiosis on plant resistance to other shoot pathogens. In tomato, AM reduced disease symptoms caused by a phytoplasma and conferred protection against the necrotrophic fungus *Alternaria solani* (Lingua et al. 2002; Fritz et al. 2006; de la Noval et al. 2007). Similarly, colonisation by

Glomus mosseae reduces disease symptoms and proliferation of *Pseudomonas syringae* in tomato leaves (J. García-Andrade and M.J. Pozo, unpublished results). Liu and coworkers have recently shown increased resistance of mycorrhizal *Medicago* to the shoot bacterial pathogen *Xanthomonas campestris* (Liu et al. 2007).

9.3 Effects of AM Symbioses on Phytophagous Insects

Interaction with herbivorous insects is also altered in mycorrhizal plants, as the symbiosis has an impact on the growth and/or survival of those insects. Again, while the symbiosis consistently reduces attacks by root-feeding insects, effects on foliar-feeding ones are more variable. A detailed analysis of published studies allowed Gange (2006) to reach important conclusions, despite the diversity in experimental approaches and systems analysed. In general, AM reduce the incidence of generalist chewing insects, while sap-feeding or specialist insects show increases in performance on mycorrhizal plants (Gange 2006). Such a pattern may indicate that the final outcome of the interaction is largely determined by the insect lifestyle and the degree of specialisation. While generalist insects are sensitive to plant defence mechanisms, specialist herbivores are likely to be able to circumvent the defences of their host plant and remain undetected. As a result, generalists may be affected by the enhanced defence capacity of mycorrhizal plants, while specialists will circumvent the defences and may benefit from improved nutritional status of the mycorrhizal host plant. Despite these general trends, the outcome depends on the plants and organisms involved in each specific interaction. For example, although several works reported increased performance of sap-feeding insects in mycorrhizal plants (Gange 2006), colonisation by *G. mosseae* significantly reduced the performance of potato aphids in tomato (Guerrieri et al. 2004). The scheme in Fig. 9.1 summarises the general effects of AM on different plant attackers.

9.4 Mechanisms of Mycorrhiza Induced Resistance

There is experimental evidence that plant protection by AM results from a combination of mechanisms acting at different levels (Azcón-Aguilar and Barea 1996). The most widely accepted mechanisms to explain mycorrhiza-induced protection are the improvement of plant nutrition and the compensation by the symbiosis of the damage caused by the pathogen. However, studies including nutrient-supplemented controls have shown that mycorrhizal effects cannot be regarded as a mere consequence of improved phosphorus nutrition (Shaul et al. 1999; Fritz et al. 2006; Liu et al. 2007). Another important factor is competition between AMF and pathogens for photosynthates and, in the case of root pathogens, for colonisation sites (Cordier et al. 1998). Mycorrhizal establishment also induces changes in root system

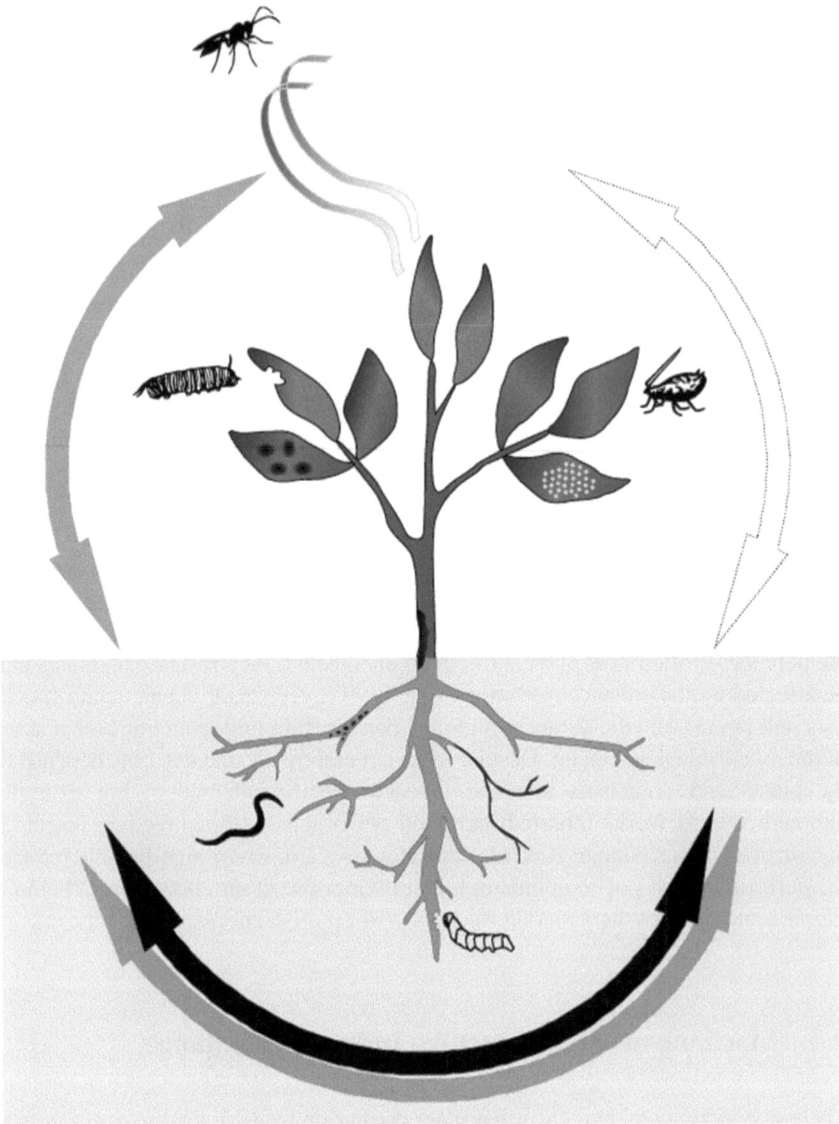

Fig. 9.1 Spectrum of efficiency of mycorrhiza-induced resistance (MIR). AM symbioses generally reduce incidence and/or damage caused by soil-borne pathogens, nematodes and root-chewing insects (*bottom*). The protection results from the combination of local and systemic mechanisms (represented by a *double arrow*). In aboveground tissues, MIR is effective against necrotrophic pathogens and generalist chewing insects (*left*). Indirect defence responses are also enhanced: parasitoids are more attracted by volatiles released by AM plants. Viral and fungal biotrophs, as well as phloem-feeding insects, perform better on mycorrhizal plants (*right*). *Solid* and *open arrows* indicate increased resistance or susceptibility, respectively, of mycorrhizal plants. Drawing by J. Perez-Tienda. Reproduced with permission from Pozo and Azcón-Aguilar (2007)

architecture and morphology. These changes may alter the dynamics of infection by the pathogen, although direct evidence for such a correlation is lacking. An additional level of complexity is the fact that mycorrhization determines important changes in rhizosphere microbial populations, and these changes may lead to the stimulation of components of the microbiota with antagonistic activity toward certain root pathogens (Barea et al. 2005). Finally, the contribution of plant defence mechanisms has been highlighted in different studies, and this will be the main focus of this chapter.

As mentioned above, several of these mechanisms may be operative simultaneously, with individual contributions depending on environmental conditions, timing of the interaction and partners involved (Azcón-Aguilar and Barea 1996; Whipps 2004). A key factor in the induction of resistance seems to be the extension of root colonisation by AMF. Studies comparing different mycorrhizal colonisation levels conclude the requirement of a well-established AM symbiosis for local and systemic induced resistance (Slezack et al. 2000; Khaosaad et al. 2007). However, there are reports on biocontrol of pathogens in non-AM species by co-culture with mycorrhizal plants. Whether this effect is related to induced resistance or to microbial interactions in the soil remains to be elucidated (St-Arnaud et al. 1997).

9.5 Modulation of Plant Defence Responses in Mycorrhizal Plants

During interactions with microorganisms, plants are able to recognise microbe-derived molecules and tailor their defence responses according to the type of micro-organism encountered. The molecular dialogue established between both partners will determine the final outcome of the relationship, ranging from parasitism to mutualism, usually through highly coordinated cellular processes (Bais et al. 2004; Pozo et al. 2005; see Chapter 2 by Provorov and Vorobyov). A tight control in the regulation of plant defence mechanisms appears to be a key aspect in AM fungal colonisation and compatibility with the host (Gianinazzi-Pearson 1996). Remarkably, a correlation between mycorrhiza-induced protection and the 'autoregulation of mycorrhization' has been proposed (Vierheilig et al. 2008). The autoregulation is manifested as a reduction in root colonisation by AMF once plants are already mycorrhizal. Mechanisms operating in such autoregulation may also impact plant interactions with pathogenic fungi (Vierheilig et al. 2008).

There is evidence for the accumulation of defensive plant compounds related to mycorrhization, although to a much lower extent than in plant–pathogen interactions (Gianinazzi-Pearson et al. 1996). Accumulation of reactive oxygen species, activation of phenylpropanoid metabolism and accumulation of specific isoforms of hydrolytic enzymes such as chitinases and glucanases has been reported in mycorrhizal roots. However, these reactions are generally localised, suggesting a role in AM establishment or control of the symbiosis (Dumas-Gaudot et al. 2000;

García-Garrido and Ocampo 2002; Pozo et al. 2002). It should be noted that the pattern of PR protein accumulation and the expression of defence-related genes varies during interactions with different AMF (Pozo et al. 1999; Pozo et al. 2002; Gao et al. 2004).

Concerning aboveground effects, transcriptional regulation of defence-related genes and accumulation of insect antifeedant compounds have been reported in shoots of mycorrhizal plants (Gange 2006; Liu et al. 2007). Liu and co-workers (2007) described a complex pattern of changes in gene expression in roots and shoots associated with mycorrhizal colonisation in *Medicago truncatula*. Defence-related genes were among those with altered expression levels, and the authors correlated that finding with increased resistance to shoot pathogens. Furthermore, the volatile blends released by AM plants can be more attractive to aphid parasitoids than those from non-mycorrhizal ones, as shown in tomato plants (Guerrieri et al. 2004). These results indicate that not only direct, but also indirect, plant defence mechanisms may be modulated in mycorrhizal plants. There is also evidence for systemic repression of plant defence associated with AM symbioses: a delay in the accumulation of PR proteins in response to some defence-related stimuli has been observed in mycorrhizal tobacco (Shaul et al. 1999; Dumas-Gaudot et al. 2000). Altogether, experimental evidence confirms the systemic modulation of plant defence in AM. This modulation may explain the pattern of enhanced resistance/susceptibility of mycorrhizal plants to diverse pests on the basis of the different signalling pathways involved in the plant response to particular attackers. In addition, it would explain the fact that AM can modulate the effectiveness of chemically-induced plant resistance (Sonnemann et al. 2005).

9.6 Priming of Defence Mechanisms in Mycorrhizal Plants

With the exception of the limited activation of plant defence discussed above, a direct activation of defences has not been observed in mycorrhizal plants. This contrasts with the systemic acquired resistance (SAR) triggered in plants after infection with necrotising pathogens. Indeed, systemic accumulation of PR proteins, salicylic acid or expression of marker genes associated with SAR has not been reported in tissues of mycorrhizal plants. This is also the case for systemic resistance achieved after colonisation by other beneficial organisms, such as certain rhizobacteria and other beneficial fungi (Van Wees et al. 2008). Despite the vital character of defence responses, constitutive expression of defence is too costly for the plant. Thus, beneficial micro-organisms have developed the ability of enhancing resistance not through a direct activation of defence, which would be too expensive for the plant in the absence of challenging attackers, but through priming of the defence mechanisms.

Molecular studies have confirmed that quantitative, rather than qualitative, differences in the defence mechanisms determine plant resistance or susceptibility to a pathogen (Nimchuk et al. 2003). Indeed, a rapid and strong activation of defence mechanisms is crucial for success in controlling attackers. Accordingly, preconditioning of plant tissues for a quick and more effective activation of defence upon attack has important ecological fitness benefits and seems to be a common feature of the plant's immune system (Conrath et al. 2006). This boost of basal defences, known as priming, seems to be successfully triggered by certain beneficial micro-organisms (Pozo et al. 2005; Van Wees et al. 2008), including AMF (Pozo and Azcón-Aguilar 2007).

Several studies point to priming as a main mechanism operating in MIR, indicated by stronger defence reactions triggered in the mycorrhizal plant upon attack. Mycorrhizal transformed carrot roots displayed stronger defence reactions at challenge sites by *Fusarium* (Benhamou et al. 1994). Mycorrhization also amplified the accumulation of the phytoalexins rishitin and solavetivone in *Rhizoctonia*-infected potato plantlets, while AMF themselves did not alter the levels of these compounds (Yao et al. 2003). Priming for callose deposition seems to be the mechanism involved in the protection achieved by *G. intraradices* against *Colletotrichum* in cucumber (Lee et al. 2005). Furthermore, colonisation by AMF systemically protects root systems. This was first illustrated for tomato plants against *Phytophthora parasitica* infection (Cordier et al. 1998; Pozo et al. 2002). Only mycorrhizal plants formed papilla-like structures around the sites of pathogen infection in non-mycorrhizal regions with deposition of non-esterified pectins and callose, preventing the pathogen from spreading further, and they accumulated significantly more PR-1a and basic β-1,3 glucanases than non-mycorrhizal plants upon *Phytophthora* infection (Cordier et al. 1998; Pozo et al. 1999; Pozo et al. 2002). Mycorrhizal protection of grapevine roots against *Meloidogyne incognita* has also been associated with primed expression, ubiquitously throughout the whole root system, of a chitinase gene, *VCH3*, in response to the nematode (Li et al. 2006). These different observations illustrate that primed responses are not restricted to AMF colonised areas, but they occur in the whole root system.

To investigate whether mycorrhizal colonisation leads to systemic priming of defence in aerial tissues, we compared the response of non-mycorrhizal plants or plants colonised by either *G. mosseae* or *G. intraradices* to the application in the shoots of different defence-related stimuli. Gene expression and enzymatic activities were monitored in a time course experiment after shoot treatment with jasmonic acid (JA), ethylene (ET) or salicylic acid (SA). Transcript profiling of leaves of mycorrhizal and non-mycorrhizal plants 24 h after treatment with JA revealed a stronger induction in mycorrhizal plants, particularly in *G. mosseae* colonised plants, of JA-regulated genes, including typical defence-related JA responsive genes such as those coding for proteinase inhibitors (Fig. 9.2; A. Verhage and M.J. Pozo, unpublished results). Our results show different defence-related gene regulation patterns in mycorrhizal plants, and point to a prominent role of priming for JA-dependent responses in AM-induced resistance.

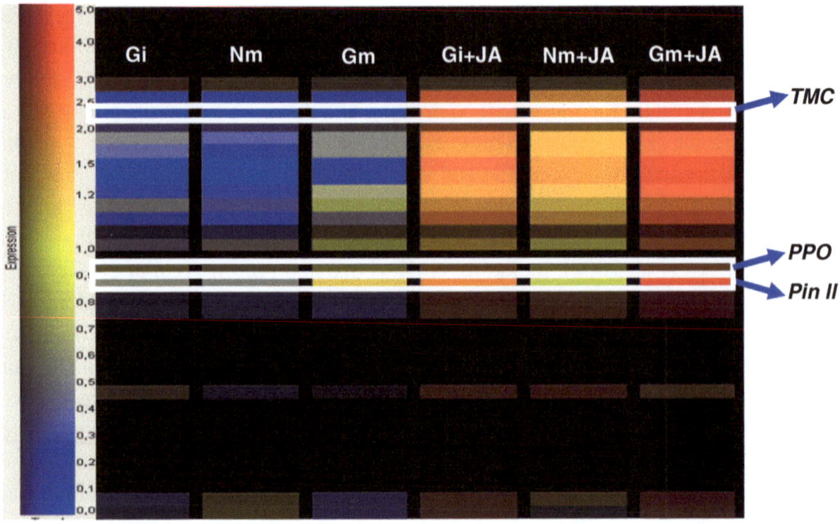

Fig. 9.2 Primed expression of JA-responsive genes in mycorrhizal plants. Induction of gene expression in tomato leaves upon treatment with methyl jasmonate (*+JA*) as a defence inducer was compared in non-mycorrhizal (*NM*) and mycorrhizal plants colonised by either *Glomus mosseae* (*Gm*) or *G. intraradices* (*Gi*). Transcriptomic analysis was performed using Affymetrix GeneChip® tomato Arrays, including probe sets for about 10,000 tomato genes. The heat map depicts a cluster of JA responsive genes with primed expression in mycorrhizal plants, including those coding for the defence-related tomato proteins multicystatin (*TMC*), polyphenol oxidase (*PPO*) and proteinase inhibitor II (*PIN II*). The expression values are represented following the colour scale on the left, ranging from blue (low expression level) to red (high expression level). Note that primed expression upon treatment was more pronounced in *G. mosseae* (*Gm+JA*) than in *G. intraradices* (*Gi+JA*) colonised plants

9.7 Signalling Pathways Involved in Mycorrhiza Priming of Defence

Plant defence mechanisms are tightly regulated through an interconnected network of signalling pathways in which JA, ET and SA play major roles. Priming is often manifested as a sensitisation of the tissues to one or some of the signalling molecules (Conrath et al. 2006). Salicylic acid coordinates defence mechanisms generally effective against biotrophic pathogens, whereas JA regulates wounding responses and resistance against necrotrophs (Ton et al. 2002; Glazebrook 2005). Nevertheless, there is some overlap in their spectrum of efficiency, especially concerning pathogens with intermediate lifestyles (Thaler et al. 2004; Pozo et al. 2005). Insect feeding guilds also determine the response they trigger in the plant. Generalist chewing insects, but not phloem-feeding ones, cause wounding and trigger JA-regulated responses (Heidel and Baldwin 2004; De Vos et al. 2005). These signalling pathways are not independent: intensive interactions ranging from synergism to antagonism shape a complex regulation network, in which trade-offs between SA and JA pathways are well documented (Bostock 2005; Beckers and Spoel 2006; Koornneef and Pieterse 2008).

As obligate biotrophs, AMF share similarities with biotrophic pathogens (Paszkowski 2006). Coherently, transcriptional profiling comparing plant responses during interactions with AM and pathogenic fungi showed that plant responses to AMF overlap more with those induced by the hemibiotroph *Magnaporthe grisea* than with those by the necrotroph *Fusarium moniliforme* (Güimil et al. 2005). Thus, AMF sensitivity to SA-regulated defences is likely. Indeed, exogenous SA application delays mycorrhizal colonisation, and plant mutants altered in endogenous SA levels point to a role of this pathway in the control of the AM symbiosis (García-Garrido and Ocampo 2002). It is plausible that AMF partially repress SA-dependent defence responses in the host in order to achieve a compatible interaction. A suppression of SA responses is also necessary for the establishment of the *Rhizobium*–legume symbiotic association (Stacey et al. 2006). In the case of mycorrhizal plants, such attenuation could explain the delay in systemic accumulation of PR proteins upon treatment with SA or analogs (Shaul et al. 1999; Dumas-Gaudot et al. 2000) and the reported enhanced susceptibility of mycorrhizal plants to viruses and certain biotrophic pathogens (Pozo and Azcón-Aguilar 2007). How would attenuation of plant defence fit with the widespread and mutualistic character of AM associations? A possible explanation may be by compensation via other defence signalling pathways.

A symbiotic program has to be activated in the host plant to achieve a successful mutualistic interaction upon recognition of the AM fungal partner (Reinhardt 2007). This program should allow a redistribution of nutrients and active physical accommodation of the fungal symbiont inside root cells (Genre et al. 2005). Both aspects may be regulated by jasmonates (Hause et al. 2007). Roots of mycorrhizal plants are associated with higher levels of endogenous JA as compared to non-mycorrhizal ones. The increase occurs after the onset of mycorrhization, and is probably related to fully established mycorrhizas (Hause et al. 2002). Elevated levels of basal JA production could be related to the increased resistance of mycorrhizal plants to certain pathogens and insects. In line with this hypothesis, experimental evidence linking the JA pathway to primed deposition of callose and enhanced resistance to oomycetes (Hamiduzzaman et al. 2005) argues for a role of JA signalling in the primed papillae formation observed in mycorrhizal tomato root systems upon infection with *Phytophthora* (Cordier et al. 1998). Recently, a central role for JA in systemic immunity has also been proposed in *Arabidopsis* (Truman et al. 2007). It is tempting to speculate that JA also serves as endogenous signal in MIR.

It is noteworthy that parallels exist between rhizobacteria and mycorrhiza-induced resistance. Like MIR, rhizobacteria-mediated ISR is mainly effective against necrotrophic pathogens and is based on priming of JA regulated responses (Verhagen et al. 2004). ISR by other beneficial organisms also seems to be associated with priming of JA-inducible responses (Pozo et al. 2005; Van Wees et al. 2008). Thus, modulation of plant defence mechanisms and conditioning of plant tissues for a more efficient activation of JA responses may be a common feature of beneficial interactions.

In summary, we propose that a functional mycorrhiza implies partial suppression of SA-dependent responses in the plant, compensated by an enhancement of those that are JA-regulated. This would result in priming of JA-dependent defence mechanisms (Pozo and Azcón-Aguilar 2007). Experimental evidence supports such a hypothesis: AM induced systemic protection against take-all disease is independent of systemic accumulation of salicylic acid (Khaosaad et al. 2007), and AM symbiosis primes tomato plants for a stronger activation of JA-dependent defence responses (Fig. 9.2). This defence regulation model is coherent with the spectrum of effectiveness described for MIR: increased susceptibility to biotrophs, and increased resistance to necrotrophs and generalist chewing insects (summarised in Fig. 9.1).

9.8 Conclusions

Mycorrhizal symbioses have an important impact on plant interactions with pathogens and insects. The association generally leads to reduction of damage caused by soil-borne pathogens, but effects on shoot-targeting organisms depend greatly on the attacker lifestyle. Mycorrhiza-induced resistance (MIR) in aboveground tissues seems effective against necrotrophic pathogens and generalist chewing insects, but not against biotrophs. Instead of constitutive activation of plant defence, MIR is associated with priming for an efficient activation of defence mechanisms upon attack. The spectrum of MIR efficiency correlates with a potentiation of JA-dependent plant defences. This low-cost type of induced resistance may be among the reasons to explain why root associations with AMF have been conserved during evolution and are widespread among plant species.

There is growing awareness about the importance of soil microbiota in natural and man-made ecosystems. Indeed, progresses in basic knowledge of plant interactions with mycorrhizal fungi, identification of markers associated with induced resistance, as well as the generation of predictive models for the outcome of particular interactions, may have important practical implications regarding the effectiveness of AMF in the biological control and integrated management of pests and diseases.

Acknowledgements Our research on MIR was financed by grant ERG 014770 from the Marie Curie Program (EU) and grant AGL2006-08029 from the Spanish Ministry of Science and Technology. A.V. and J.G-A. were supported by grants from the Socrates-Erasmus mobility programme (EU) and the Spanish National Research Council (CSIC), respectively.

References

Azcón-Aguilar C, Barea JM (1996) Arbuscular mycorrhizas and biological control of soil-borne plant pathogens - An overview of the mechanisms involved. Mycorrhiza 6:457–464

Bais HP, Park SW, Weir TL, Callaway RM, Vivanco JM (2004) How plants communicate using the underground information superhighway. Trends Plant Sci 9:26–32

Barea JM, Pozo MJ, Azcon R, Azcon-Aguilar C (2005) Microbial co-operation in the rhizosphere. J Exp Bot 56:1761–1778

Beckers GJM, Spoel SH (2006) Fine-tuning plant defence signalling: salicylate *versus* jasmonate. Plant Biol 8:1–10

Benhamou N, Fortin JA, Hamel C, St-Arnaud M, Shatilla A (1994) Resistance responses of mycorrhizal Ri T-DNA-transformed carrot roots to infection by *Fusarium oxysporum* f sp *chrysanthemi*. Phytopathology 84:958–968

Bostock RM (2005) Signal crosstalk and induced resistance: straddling the line between cost and benefit. Ann Rev Phytopathol 43:545–580

Chandanie WA, Kubota M, Hyakumachi M (2006) Interactions between plant growth promoting fungi and arbuscular mycorrhizal fungus *Glomus mosseae* and induction of systemic resistance to anthracnose disease in cucumber. Plant Soil 286:209–217

Conrath U, Beckers GJM, Flors V, Garcia-Agustin P, Jakab G, Mauch F, Newman MA, Pieterse CMJ, Poinssot B, Pozo MJ, Pugin A, Schaffrath U, Ton J, Wendehenne D, Zimmerli L, Mauch-Mani B (2006) Priming: getting ready for battle. Mol Plant Microbe Interact 19:1062–1071

Cordier C, Pozo MJ, Barea JM, Gianinazzi S, Gianinazzi-Pearson V (1998) Cell defense responses associated with localized and systemic resistance to *Phytophthora* induced in tomato by an arbuscular mycorrhizal fungus. Mol Plant Microbe Interact 11:1017–1028

de la Noval B, Perez E, Martinez B, Leon O, Martinez-Gallardo N, Delano-Frier J (2007) Exogenous systemin has a contrasting effect on disease resistance in mycorrhizal tomato (*Solanum lycopersicum*) plants infected with necrotrophic or hemibiotrophic pathogens. Mycorrhiza 17:449–460

de la Peña E, Rodriguez-Echeverria S, van der Putten WH, Freitas H, Moens M (2006) Mechanism of control of root-feeding nematodes by mycorrhizal fungi in the dune grass *Ammophila arenaria*. New Phytol 169:829–840

De Vos M, Van Oosten VR, Van Poecke RMP, Van Pelt JA, Pozo MJ, Mueller MJ, Buchala AJ, Metraux JP, Van Loon LC, Dicke M, Pieterse CMJ (2005) Signal signature and transcriptome changes of *Arabidopsis* during pathogen and insect attack. Mol Plant Microbe Interact 18:923–937

Dumas-Gaudot E, Gollotte A, Cordier C, Gianinazzi S, Gianinazzi-Pearson V (2000) Modulation of host defence systems. In: Kapulnik Y, Douds DDJ (eds) Arbuscular mycorrhizas: physiology and function. Kluwer, Dordrecht, pp 173–200

Elsen A, Gervacio D, Swennen R, De Waele D (2008) AMF-induced biocontrol against plant parasitic nematodes in *Musa sp.*: a systemic effect. Mycorrhiza 18:251–256

Fritz M, Jakobsen I, Lyngkjaer MF, Thordal-Christensen H, Pons-Kuehnemann J (2006) Arbuscular mycorrhiza reduces susceptibility of tomato to *Alternaria solani*. Mycorrhiza 16:413–419

Gange AC (2006) Insect-mycorrhizal interactions: patterns, processes, and consequences. In: Ohgushi T, Craig TP, Price PW (eds) Indirect interaction webs: nontrophic linkages through induced plant traits. Cambridge University Press, Cambridge, pp 124–144

Gao LL, Knogge W, Delp G, Smith FA, Smith SE (2004) Expression patterns of defense-related genes in different types of arbuscular mycorrhizal development in wild-type and mycorrhiza-defective mutant tomato. Mol Plant Microbe Interact 17:1103–1113

García-Garrido JM, Ocampo JA (2002) Regulation of the plant defence response in arbuscular mycorrhizal symbiosis. J Exp Bot 53:1377–1386

Genre A, Chabaud M, Timmers T, Bonfante P, Barker DG (2005) Arbuscular mycorrhizal fungi elicit a novel intracellular apparatus in *Medicago truncatula* root epidermal cells before infection. Plant Cell 17:3489–3499

Gernns H, von Alten H, Poehling HM (2001) Arbuscular mycorrhiza increased the activity of a biotrophic leaf pathogen - is a compensation possible? Mycorrhiza 11:237–243

Gianinazzi-Pearson V (1996) Plant cell responses to arbuscular mycorrhizal fungi: getting to the roots of the symbiosis. Plant Cell 8:1871–1883

Gianinazzi-Pearson V, Dumas-Gaudot E, Gollotte A, Tahiri-Alaoui A, Gianinazzi S (1996) Cellular and molecular defence-related root responses to invasion by arbuscular mycorrhizal fungi. New Phytol 133:45–57

Glazebrook (2005) Contrasting mechanisms of defense against biotrophic and necrotrophic pathogens. Ann Rev Phytopathol 43:205–227

Guerrieri E, Lingua G, Digilio MC, Massa N, Berta G (2004) Do interactions between plant roots and the rhizosphere affect parasitoid behaviour? Ecol Entomol 29:753–756

Güimil S, Chang HS, Zhu T, Sesma A, Osbourn A, Roux C, Ioannidis V, Oakeley EJ, Docquier M, Descombes P, Briggs SP, Paszkowski U (2005) Comparative transcriptomics of rice reveals an ancient pattern of response to microbial colonization. Proc Natl Acad Sci USA 102:8066–8070

Hamiduzzaman MM, Jakab G, Barnavon L, Neuhaus JM, Mauch-Mani B (2005) Beta-Aminobutyric acid-induced resistance against downy mildew in grapevine acts through the potentiation of callose formation and jasmonic acid signaling. Mol Plant Microbe Interact 18:819–829

Hause B, Maier W, Miersch O, Kramell R, Strack D (2002) Induction of jasmonate biosynthesis in arbuscular mycorrhizal barley roots. Plant Physiol 130:1213–1220

Hause B, Mrosk C, Isayenkov S, Strack D (2007) Jasmonates in arbuscular mycorrhizal interactions. Phytochemistry 68:101–110

Heidel AJ, Baldwin IT (2004) Microarray analysis of salicylic acid- and jasmonic acid-signalling in responses of *Nicotiana attenuata* to attack by insects from multiple feeding guilds. Plant Cell Environ 27:1362–1373

Khaosaad T, García-Garrido JM, Steinkellner S, Vierheilig H (2007) Take-all disease is systemically reduced in roots of mycorrhizal barley plants. Soil Biol Biochem 39:727–734

Koornneef A, Pieterse CMJ (2008) Cross-talk in defense signaling. Plant Physiol 146:839–844

Lee CS, Lee YJ, Jeun YC (2005) Observations of infection structures on the leaves of cucumber plants pre-treated with arbuscular mycorrhiza *Glomus intraradices* after challenge inoculation with *Colletotrichum orbiculare*. Plant Pathol J 21:237–243

Li HY, Yang GD, Shu HR, Yang YT, Ye BX, Nishida I, Zheng CC (2006) Colonization by the arbuscular mycorrhizal fungus *Glomus versiforme* induces a defense response against the root-knot nematode *Meloidogyne incognita* in the grapevine (*Vitis amurensis Rupr.*), which includes transcriptional activation of the class III chitinase gene VCH3. Plant Cell Physiol 47:154–163

Lingua G, D'Agostino G, Massa N, Antosiano M, Berta G (2002) Mycorrhiza-induced differential response to a yellows disease in tomato. Mycorrhiza 12:191–198

Liu J, Maldonado-Mendoza I, Lopez-Meyer M, Cheung F, Town CD, Harrison MJ (2007) Arbuscular mycorrhizal symbiosis is accompanied by local and systemic alterations in gene expression and an increase in disease resistance in the shoots. Plant J 50:529–544

Nimchuk Z, Eulgem T, Holt III BF, Dangl JL (2003) Recognition and response in the plant immune system. Ann Rev Gen 37:579–609

Paszkowski U (2006) Mutualism and parasitism: the yin and yang of plant symbioses. Curr Opin Plant Biol 9:364–370

Pozo MJ, Azcón-Aguilar C (2007) Unravelling mycorrhiza induced resistance. Curr Opin Plant Biol 10:393–398

Pozo MJ, Azcón-Aguilar C, Dumas-Gaudot E, Barea JM (1999) ß-1,3-glucanase activities in tomato roots inoculated with arbuscular mycorrhizal fungi and/or *Phytophthora parasitica* and their possible involvement in bioprotection. Plant Sci 141:149–157

Pozo MJ, Cordier C, Dumas-Gaudot E, Gianinazzi S, Barea JM, Azcón-Aguilar C (2002) Localized *vs* systemic effect of arbuscular mycorrhizal fungi on defence responses to *Phytophthora* infection in tomato plants. J Exp Bot 53:525–534

Pozo MJ, Van Loon LC, Pieterse CMJ (2005) Jasmonates - Signals in plant-microbe interactions. J Plant Growth Regul 23:211–222

Reinhardt D (2007) Programming good relations - development of the arbuscular mycorrhizal symbiosis. Curr Opin Plant Biol 10:98–105

Shaul O, Galili S, Volpin H, Ginzberg I, Elad Y, Chet I, Kapulnik Y (1999) Mycorrhiza-induced changes in disease severity and PR protein expression in tobacco leaves. Mol Plant Microbe Interact 12:1000–1007

Slezack S, Dumas-Gaudot E, Paynot M, Gianinazzi S (2000) Is a fully established arbuscular mycorrhizal symbiosis required for bioprotection of *Pisum sativum* roots against *Aphanomyces euteiches*? Mol Plant Microbe Interact 13:238–241

Sonnemann I, Streicher NM, Wolters V (2005) Root associated organisms modify the effectiveness of chemically induced resistance in barley. Soil Biol Biochem 37:1837–1842

St-Arnaud M, Hamel C, Vimard B, Caron M, Fortin JA (1997) Inhibition of *Fusarium oxysporum* f.sp. dianthi in the non-VAM species *Dianthus caryophyllus* by co-culture with *Tagetes patula* companion plants colonized by *Glomus intraradices*. Can J Bot 75:998–1005

Stacey G, McAlvin CB, Kim SY, Olivares J, Soto MJ (2006) Effects of endogenous salicylic acid on nodulation in the model legumes *Lotus japonicus* and *Medicago truncatula*. Plant Physiol 141:1473–1481

Thaler JS, Owen B, Higgins VJ (2004) The role of the jasmonate response in plant susceptibility to diverse pathogens with a range of lifestyles. Plant Physiol 135:530–538

Ton J, Van Pelt JA, Van Loon LC, Pieterse CMJ (2002) Differential effectiveness of salicylate-dependent and jasmonate/ethylene-dependent induced resistance in *Arabidopsis*. Mol Plant Microbe Interact 15:27–34

Truman W, Bennett MH, Kubigsteltig I, Turnbull C, Grant M (2007) Arabidopsis systemic immunity uses conserved defense signaling pathways and is mediated by jasmonates. Proc Natl Acad Sci USA 104:1075–1080

Van Wees SC, Van der Ent S, Pieterse CM (2008) Plant immune responses triggered by beneficial microbes. Curr Opin Plant Biol 11:443–448

Verhagen BWM, Glazebrook J, Zhu T, Chang HS, Van Loon LC, Pieterse CMJ (2004) The transcriptome of rhizobacteria-induced systemic resistance in *Arabidopsis*. Mol Plant Microbe Interact 17:895–908

Vierheilig H, Steinkellner S, Khaosaad T, García-Garrido JM (2008) The biocontrol effect of mycorrhization on soil-borne fungal pathogens and the autoregulation of the AM symbiosis: one mechanism, two effects? In: Varma A (ed) Mycorrhiza: biology, genetics, novel endophytes and biotechnology. Springer, Heidelberg, Germany

Whipps JM (2004) Prospects and limitations for mycorrhizas in biocontrol of root pathogens. Can J Bot 82:1198–1227

Yao MK, Desilets H, Charles MT, Boulanger R, Tweddell RJ (2003) Effect of mycorrhization on the accumulation of rishitin and solavetivone in potato plantlets challenged with *Rhizoctonia solani*. Mycorrhiza 13:333–336

Zhu HH, Yao Q (2004) Localized and systemic increase of phenols in tomato roots induced by *Glomus versiforme* inhibits *Ralstonia solanacearum*. J Phytopathol 152:537–542

Chapter 10
Reconciliation of Conflicting Phenotypic and rRNA Gene Phylogenies of Fungi in Glomeromycota Based on Underlying Patterns and Processes

Joseph B. Morton

Abstract A true species phylogeny requires congruence amongst all independent character data sets that inform on continuity of descent. A morphology-based phylogeny conflicts with rRNA phylogenies in basal nodes, but all are in greater accord at the family level and above. Incongruent patterns are attributed to population-level processes in gene evolution that create signals of homoplasy and distorted relationships widely recognized as problematic in other higher eukaryotic lineages. A species phylogeny can be misrepresented by two processes, either separately or together. Prerequisites for both are: (1) cladogenesis of a gene occurs prior to species cladogenesis, and (2) polymorphisms from gene cladogenesis are preserved in both diverging species lineages and then are sorted by selection or loss. Invoking the former process, polymorphisms arise in rRNA gene loci on different chromosomes and therefore are not homogenized by concerted evolution. Conflicts arise biologically, when extinction or sorting of these polymorphisms do not correspond with other characters indicative of speciation, or methodologically when intra-isolate clones are undersampled. With the latter process, polymorphisms arise at a single multicopy locus, after which sorting of lineages in subsequent cladogenic events can result in orthologous sequences in one lineage and paralogous sequences in another. A species phylogeny will be misleading when sequences from different fungal isolates of a species are undersampled. When either process is considered, the discord between rRNA gene and morphology-based trees can be explained. There is no definitive data to support either process, but credence is reflected in the complete lack of support for the proposed rRNA gene phylogeny from morphological, biochemical, and ecological characters correctly assessed as homologs. New gene trees can clarify species evolution at the molecular level, but they also can be in conflict. When this occurs, data from all classes of characters must be analyzed within the framework of a balanced multidisciplinary approach.

J.B. Morton
1090 Agricultural Sciences Building, West Virginia University
Morgantown, WV, 26506-6108, USA
e-mail: jbmorton@mail.wvu.edu

C. Azcón-Aguilar et al. (eds.), *Mycorrhizas - Functional Processes and Ecological Impact*,
DOI: 10/1007-978-3-540-87978-7_10,

10.1 Introduction

All of the fungal species in phylum Glomeromycota (Schüßler et al. 2001a) form mutualistic associations, most as arbuscular mycorrhizal (AM) associates with a majority of terrestrial plant species (Smith and Read 1997) and one with *Nostoc* cyanobacteria (Schüßler and Kluge 2001). AM fungi can colonize cortical roots of a wide range of hosts in almost all habitats. This absence of specificity, together with ancient origin, suggests co-evolution with land plants from their emergence on land (Morton 1990a; Redecker et al. 2000). AM fungi are totally dependent on their hosts for growth and reproduction, whereas some plant lineages have evolved completely away from the symbiosis (Trappe 1987). *Geosiphon pyriforme*, the only non-mycorrhizal fungus in the phylum, shares some parallel similarities in that the association with *Nostoc* is also an obligate mutualism. The extent these similarities complement the genetic linkage with AM fungi is evaluated later in this chapter.

All available evidence indicates AM fungi reproduce asexually, which has wide-ranging implications for genetic, and by extension, evolutionary processes (Pawlowska 2005). Polymorphisms within multicopy genes (e.g., those encoding 18S and 25S ribosomal RNA transcripts) from single spore samples may be low (Stukenbrock and Rosendahl 2005) or high (references in Rodriguez et al. 2004), so patterns (and causal mechanisms) remain enigmatic. Occurrence of sequence polymorphisms extends even to protein-encoding genes (Pawlowska and Taylor 2004). The presence of multiple nuclei adds another scale of process in which genetic mechanisms can influence population-level and species-level evolution. Much attention has been given to the question of whether these nuclei are homokaryotic (allelic variation is the same in all nuclei; Pawlowska and Taylor 2004) or heterokaryotic (allelic variation is distributed among nuclei; Kuhn et al. 2001; Hijri and Sanders 2005). Methodological differences and/or poorly understood population-level processes contribute to delays in resolution of this controversy. Regardless, either or both mechanisms can influence phylogenetic signal of genetic characters. Some evidence of low-level cryptic recombination has been reported by Gandolfi et al. (2003), and over geologic timescales these processes also could greatly influence evolution of genetic characters.

Morton (1990b) erected the first hypothesis of a taxic phylogeny of AM fungi based on morphological characters. A classification resulting from this analysis separated these fungi from the Zygomycota (Morton and Benny 1990). Some strong theoretical grounds existed to unite the group in its own phylum (Morton 1990a). However, no suitable outgroup unequivocably revealed monophyly with respect to other known phyla – a drawback of morphological characters. Caution, therefore, dictated that the fungi be classified as a monophyletic group at the order level, Glomales. Evidence for monophyly at a broader phylogenetic scale was obtained from partial sequences of the 18S rRNA gene (Schüßler et al. 2001a). A plethora of new families and genera have been erected since then based on morphology and partial sequences of the rRNA gene family (Morton and Redecker 2001; Walker et al. 2007), morphology alone (Oehl and Sieverding 2005; Spain et al.

2006), and rRNA sequences alone (Walker and Schüßler 2004). Three clades in *Glomus* resolved by the 18S rRNA gene tree are designated as "groups" (Schwartzott et al. 2001). Because of length limitations to this chapter, I refer the reader to Redecker and Raab (2006) for additional details on most of the currently defined taxa.

Ribosomal RNA genes comprise the most widely studied multicopy gene family shared by all eukaryotes. The 18S (SrRNA), 5.8S, and 25S (LrRNA) genes are arranged in a cluster with internal spacers (ITS1 and ITS2) between them. Together, they encode one large transcription unit that undergoes further processing. The SrRNA integrates with proteins in the small ribosomal subunit, and the 5.8S and LrRNA integrate into the large ribosomal subunit (Long and Dawid 1980). Each cluster is organized in arrays of tandem repeats. It is unclear whether an array is present at one locus or is present at multiple loci on different chromosomes (Trouvelot et al. 1999). The 5S RNA of the large subunit may be positioned with other rRNA genes (e.g., *Saccharomyces*) or as repeats interspersed throughout the genome (Rooney and Ward 2005). The conserved regions of the SrRNA and LrRNA can show strong homology, so they are used extensively in phylogenetic analyses encompassing evolutionarily distant taxa. The more highly variable ITS regions are used in conjunction with conserved regions to better characterize anagenesis between closely related species (e.g., Walker et al. 2007).

In the study of Glomeromycota systematics, most attention has been given to phylogeny of the SrRNA gene (Schüßler et al. 2001a; Schwartzott et al. 2001; Redecker and Raab 2006). A limited LrRNA gene tree was generated mostly from deposited Genbank sequences (de Silva et al. 2006), and even smaller trees were constructed from actin and transcription factor alpha genes (Helgason et al. 2003) as well as alpha and beta-tubulin genes (Corradi et al. 2004).

Most systematists acknowledge that great care must be taken in distinguishing "gene trees" (patterns indicative of evolution of a single gene) from "species trees" (patterns indicative of evolution of a species) and applying a multidisciplinary approach to examine both "molecular" and "non-molecular" characters (Pamilo and Nei 1988; Doyle 1992, 1997; Patterson et al. 1993). When all are in congruence, confidence that a true phylogeny is being depicted increases. When conflicts exist, then underlying processes at all levels of organization must be carefully appraised. At the present time, the SrRNA gene and flanking regions are presumed to represent a phylogeny that is isomorphic with a phylogeny of glomeromycotan species.

So far, the focus has been completely on analysis of pattern (generation of trees). But interpretation of those trees and then attaching names to branches (inferring a species phylogeny) is potentially misleading without an analysis of the processes that determine these patterns. I attempt such a goal in this chapter and, in so doing, hypothesize processes that offer a putative explanation of conflicts between morphology and SrRNA patterns at deeper nodes in Glomeromycota phylogeny. To do this, I commit heresy and stipulate a phylogeny of morphological and other phenotypic traits as an equally valid hypothesis of species phylogeny and, with that as a reference, appraise underlying processes of rRNA gene evolution. I re-examine characters by the scale of pattern and process where they are most informative and the phylogenetic prerequisites of independence and homology in character analysis.

I also examine patterns and processes of character evolution that explain current tree topologies. Some of the data and analyses communicated in this chapter are from research in my laboratory. This information has been stripped to essential details so as not to compromise publications in preparation.

10.2 Character Analysis

Characters most often are grouped into dichotomous classes described as either "molecular" and "non-molecular" or "molecular" and "morphological" (Nei 1987; Patterson et al. 1993). These dichotomies need to be discussed because each has equal probability of confusing homology (descent from a common ancestor) with homoplasy (descent from different ancestors), and each can contain equally important phylogenetic information (Sanderson and Donoghue 1989; Patterson et al. 1993). I revise the semantics only slightly to suit my purpose in the discussion that follows.

Phenotypic characters are those which possess only a few states because expressed variation is constrained by canalized ontogenetic, epigenetic, or biochemical pathways. The players in this class encompass morphological traits, macromolecules that are not the direct products of genes (e.g., carbohydrates, fatty acids), and possibly even conserved life history traits with discrete "reaction norms" (e.g., Schlichting and Pigliucci 1998). The underlying processes that lead to expression of these characters can be quite complex, but they are filtered and sorted so as to minimize perturbations detrimental to survival of the organism. On the positive side, the synapomorphies expressed by ontogeny and other canalized programs are more immune to change and thus possess a strong phylogenetic signal (Doyle 1992), especially in deeper nodes of a tree (Lanyon 1988). On the negative side, the finer details of genetic divergence that contribute to understanding mechanisms of species cohesion and anagenetic change are masked. With so few states, combinations of characters must be considered to generate a phylogenetic tree of any merit. Moreover, branches on the tree are limited to those taxa whose properties pass rigorous tests of homology (see below).

Genetic characters are those which contain many states with far fewer constraints and therefore behave more like populations (Nei 1987; Doyle 1992; Maddison 1997). The complexities of population-driven evolutionary processes can complicate, distort, or even mask phylogenetic signal. In AM fungi, these processes occur at two levels: the gene and the nucleus. The players in this class of characters are genes (with nucleotides as states), proteins (with amino acids as states), and possibly allozymes (with allelic arrays at loci as states). With so many states per character, information content is high enough that each character can be used to reconstruct a phylogenetic tree which contains taxa spanning large evolutionary distances and which is interpretable on its own merit. When genes are analyzed together (like morphological characters), then controversial issues such as character congruence ("total evidence") versus topological congruence become important (de Queiroz et al. 1996).

Both classes of characters, regardless of the scale at which they are measured, require two preconditions to be informative in phylogeny reconstruction. The first is independence from each other. In other words, the state of one character should not predict the state of another. The second is the stipulation of homology, in which similarity in character states reflect common rather than independent ancestry.

10.2.1 Character Independence

Within the phenotype, morphological characters require ontogenetic analysis to separate structures of unique origin from those which develop together as part of interrelated complexes. Consideration of this criterion was absent in the phylogeny constructed by Morton (1990b) and is being re-evaluated. Independence among genes, especially those with multiple copies in the genome (e.g., rRNA genes), depends not only on gene organization, but also on constraints that affect function such as secondary structure of rRNA transcripts. SrRNA and LrRNA genes clearly are interdependent at the level of the large transcript they encode, such that loss of sequence in any region (through gene conversion, insertion of mobile elements, etc.) relegates the entire repeat to pseudogene status. Each of these genes can generate its own phylogenetic tree, but those trees cannot be interpreted as behaving independently at the scale of species evolution. Independence among nucleotide sites also varies. In rRNA molecules, paired nucleotides in stem regions are preserved through selection so that sites in these regions are not independent and thus less informative than loop regions (Wheeler and Honeycutt 1988). Data from *Glomus coronatum* sequences support this conclusion (Clapp et al. 2001).

10.2.2 Character Homology

The occurrence of homoplasy often is assumed to be high in analysis of morphology. However, this perception is overstated if characters are carefully delineated by ontogenetic analysis and all criteria for homology are applied (Patterson 1982; Doyle 1992; Morton et al. 1994). Superficial similarity is used as a starting point and then ancillary criteria are employed. A priori tests include measures of topographic and compositional correspondence among structures as well as similarity in origins, ordering, and positioning of characters relative to each other as they undergo transformations in ontogeny. The most definitive test is an a posteriori analysis of phylogenetic congruence (Patterson 1982). These criteria have been applied exhaustively to AM fungal groups (reviewed in Morton et al. 1994), and results have borne fruit in clarifying independence of characters (Morton, unpublished results). While the number of characters now are fewer than those listed in Morton (1990b), interpretations of homology are much stronger. As I will show below, phenotypic characters used selectively to support the SrRNA phylogeny usually stop with interpretations of superficial similarity so that conclusions of homology can be in error.

Ignoring problems of homoplasy at nucleotide sites when analyzing alignments, I focus on a larger issue of misinterpretation of homology by paralogous sequences. Only orthologous (homologous) sequences identify cladogenetic events because they inform on speciation events. Paralogous (homoplasious) sequences arising from gene duplication or differential sorting of competing polymorphisms (e.g., Pamilo and Nei 1988; Maddison 1997) can misrepresent or distort evidence of species cladogenesis. These gene lineages can be difficult to separate when biological causation is not distinguishable from sampling error. Synteny, or topographic correspondence of one gene character relative to other gene characters by arrangement on a chromosome, offers new and more robust criteria for homology. However, this test can be applied only where more extensive genomic data is available (Rokas et al. 2003). Numerous examples of polymorphisms in spores of glomeromycotan fungi have been reported, especially within rRNA genes or spanning ITS sequences (see references in Rodriguez et al. 2004). It is unclear how many of these sequences are paralogous because of undersampling.

10.3 Trees and Their Interpretation

For all groups of organisms, regardless of their biology and ecology, only one true phylogenetic tree exists. This phylogeny is discovered by incorporating informative characters of any type which show continuity of lineage from the most remote ancestor to the most contemporary species.

Before discussing trees generated from species, I must address a fundamental question: What kind of change would occur in clonally reproducing members of a population such that two divergent lineages would emerge and thereafter evolve independently of each other as separate species? Reproductive isolation is not a viable mechanism because all members of an AM fungal population propagate clonally. But fungal organisms are unique in that, despite clonality, gene flow can still occur within a local population if vegetative compatibility allows hyphal anastomosis (Giovannetti et al. 1999; de la Providencia et al. 2005). Morphological innovations associated with mycorrhizal development and reproduction have been invoked in speciation events (Morton 1990a, b), but more realistically these are secondary manifestations of whatever isolating event led to the cladogenic event. The mechanism of isolation must be genetic, because organismal phenotypic properties are too plastic and morphological characters that define species have absolutely no ecological significance (that make sense anyway). Only one mechanism comes to mind that confers such genetic isolation, and that involves alteration in those genes conferring vegetative incompatibility so that one segment of a clonal population no longer interacts (or "sees") the remainder. The taxonomic level at which hyphae are able to anastomose with reasonable frequency still is unclear, but at the very least it likely is below the species level (Giovannetti et al. 2003). For the two species that now co-exist, each would evolve separately (represented by characters at all levels) based on the genetic background (and polymorphisms) present in its population *at the time of* the isolating (speciation) event. Reproductive strategy would not be a factor in subsequent evolution of each species. In effect, each species lineage becomes "containerized" (e.g., Maddison 1997) and cannot interact genetically with any other species.

To illustrate the most ordered gene lineage patterns that could emerge from a speciation event, I consider evolution of three hypothetical species from a common ancestor (Fig. 10.1) based on five gene trees (even though many more exist). In this example, the timing of gene cladogenesis is simultaneous with that of species cladogenesis. All gene lineages are orthologs because each gene that survives in species populations to terminal taxa recognize the same common ancestor. One gene lineage is lost because of extinction of the population(s) carrying it in each species, but other preserved gene lineages provide congruent evidence of a species phylogeny. Other population-level dynamics (gene duplication, gene loss/extinction, lineage sorting) discussed below would have no effect on phylogenetic signal *provided that* gene and species cladogenesis occurs within the same time frame.

In Fig. 10.2, I focus exclusively on speciation events that led to emergence of three major clades where morphology and rRNA gene-based phylogenetic hypotheses are discordant: Acaulosporaceae, Gigasporaceae, and Glomeraceae. I depict two overlaying rRNA gene phylogenies (dark lines) of identical topology, but which result from different causal processes at the genetic level.

The underlying tree in Fig. 10.2a was constructed using both SrRNA and LrRNA genes (concatenated and trimmed to 2.5 kB). Bootstrap values are only slightly lower than those published by SrRNA alone (not shown). A larger 82 taxon LrRNA tree (R Amarasinghe and J.B. Morton, unpublished results) maintains

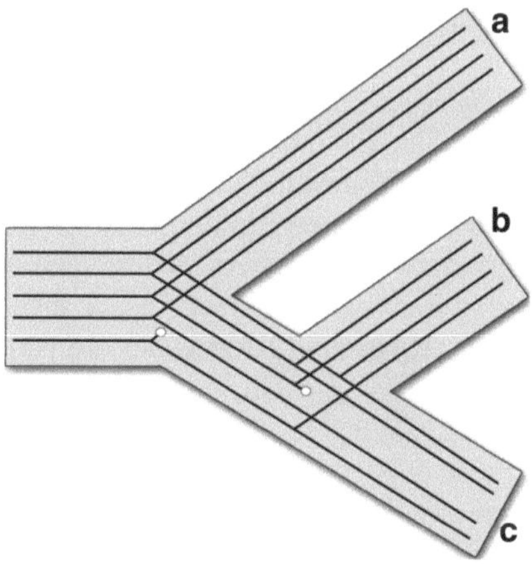

Fig. 10.1 Generalized illustration of the evolution of three fictional AM fungal species (a, b, c) based on five gene lineages. All sequences are orthologous because cladogenesis of genes and the species is isomorphic

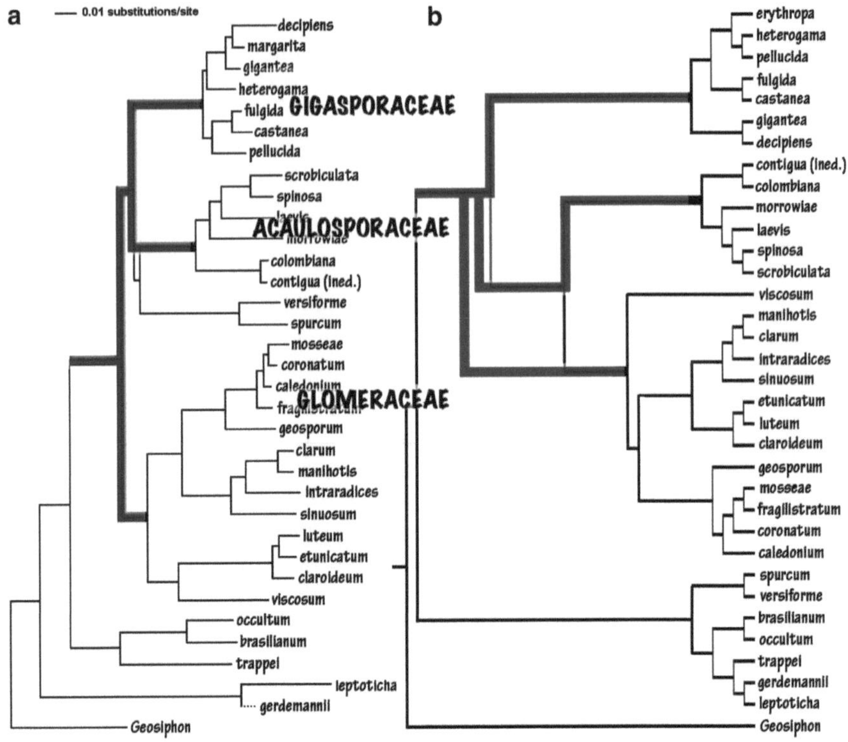

Fig. 10.2 Comparison of two phylogenies of glomeromycotan species that conflict at deeper nodes (interfamily relationships), with a hypothesis of topologically congruent gene evolution (*dark overlays*) that reconciles this conflict (see Sect. 10.3 for explanation). Taxa chosen were those used in Schüßler et al. (2001b). (**a**) Underlying neighbor-joining tree based on 2.5 kB of concatenated SrRNA and LrRNA sequences. (**b**) Underlying parsimony tree based an abbreviated set of 12 morphological characters, so only deeper nodes are well-supported

strong support and congruence with the one depicted here at the family level, but deeper nodes are more unstable. The overlay phylogeny represents the presumption of Schüßler et al. (2001b) and Schwartzott et al. (2001) that all of the sequences sampled are homogenized by concerted evolution and all are orthologous. In other words, gene evolution proceeds in the pattern as described in Fig. 10.1.

However, genes exhibit stochastic behavior (e.g., Lanyon 1988) channeled by population-like interactions. Ribosomal RNA genes are particularly susceptible to these processes because they exist as multiple copies clustered at one or more loci in the genome. Gene evolution at this level is analogous to population-level interactions that include birth (gene duplication), death (by inactivation or degradation into pseudogenes), fitness (rate at which genes are copied), and selection (lineage sorting at gene or nuclear levels). A well-documented example is the "birth" and "death" of small dispersed 5S gene repeats in fungi with fully-sequenced genomes (Rooney and Ward 2005). The proposed mechanism of retrotransposition of 5S gene copies would not apply to other rRNA genes because clusters are larger and gene truncation or degradation is not likely to impair overall function of all rRNA repeats at a locus.

Homogenization of gene variants is explained by concerted evolution via unequal crossing over and gene conversion (Liao 1999; Eickbush and Eickbush 2007). There can be little doubt that this process occurs in glomeromycotan genomes. The absence of meiosis in AM (and other) fungi is not a hindrance. Studies of yeast, in which all tandem repeats of the rRNA genes are arrayed at a locus on one chromosome, show that unequal crossing over between chromatids during mitosis occurs far more frequently than interchromosomal recombination during meiosis (references in Eickbush and Eickbush 2007). Even intra-gene or intra-repeat crossing over can occur in nucleolar-organizing regions when DNA strands are demethylated and undergoing transcription (Liao 1999). However, the extent that these processes are "relaxed" would affect how rapidly polymorphisms disappear or become differentially fixed in evolving populations. There is no clear evidence that this occurs.

In Fig. 10.2b, the underlying phylogeny is a parsimony analysis based on an abbreviated set of 12 homologous characters selected to resolve only deeper nodes and verify homologies through phylogenetic congruence. These include: mycobiont–phytobiont relationship, arbuscule architecture, arbuscule to cell topology (arum/paris), structure of intraradical and extraradical hyphae, frequency of hyphal coiling in cortex, mycorrhiza staining intensity, lipid storage structures, mode of spore formation, organization of spore and germinal walls, and mode of spore germination. These characters only discriminate deeper nodes, but that is the focus of this discussion. Relationships at the tips of the tree are ambiguous here because species-defining synapomorphies are not represented. Greater detail for nodes at all levels will be communicated separately (Morton et al., unpublished). This phylogeny of essentially morphological characters conflicts with the SrRNA gene phylogeny in that Acaulosporaceae and Glomaceae group as sister taxa (see also Morton 1990b). I consider this to be a fairly robust estimate of species phylogeny because (1) the tree is generated from multiple synapomorphies that have withstood direct selection for millions of years (Redecker et al. 2000; Lanyon 1988), and (2) they involve traits of such huge ecological and functional significance that evolutionary significance at the organismal level is obvious. Almost all new discoveries of relationships involving other *independent* phenotypic or life history traits are congruent with this hypothesis, such as differential infective propagules (Klironomos and Hart 2002), differential carbohydrate chemistry of cell walls (Gianinazzi-Pearson et al. 1994), differential colonization patterns in roots and soil (Hart and Reader 2002), and differential patterns of anastomosis (de la Providencia et al. 2005). These broad patterns likely extend to genome size (Hosny et al. 1998) and possibly even other genetic properties such as number of repetitive elements (Hijri and Sanders 2005). Absence of these characters in a computational phylogenetic analysis is not a function of their biological limitations, but rather inadequate sampling of taxonomic units. I do not presume this tree to be a *true* representation of the species phylogeny, because the depth of character data sets to be explored at all levels has yet to be plumbed.

However, incongruence between a gene tree and a species tree, whether the comparison is between genetic and phenotypic characters or just between genetic characters, can occur when (1) a gene lineage splits (producing polymorphisms) prior to a species cladogenic event, and (2) the time between speciation events following the gene split is short (Nei 1987; Pamilo and Nei 1988). In either event, polymorphisms are retained in the diverging species lineages and then are sorted by selection or gene loss within evolving populations. Two processes are widely accepted as pivotal to such incongruence, either separately or together: *gene duplication* (birth of putative paralogous gene lineages in the absence of concerted evolution) and *lineage sorting* (Doyle 1992; Lyons-Weiler and Milinkovitch 1997; Maddison 1997).

10.3.1 Gene Duplication

Two studies indicate the presence of multiple loci of LrDNA (Trouvelot et al. 1999) and ITS2 (Kuhn et al. 2001) genes based on in situ DNA-DNA hybridization using fluorescent probes in interphase nuclei of *Gigaspora margarita* and *Scutellospora castanea* spores, respectively. The methodology is problematic, so the results are not definitive. However, they raise the distinct possibility of independently evolving loci within individual nuclei. Concerted evolution most likely occurs within each locus, but if the nucleolar organizer regions within which transcription occurs remain separated, then polymorphic (and paralogous) sequences in the same lineage have the potential to evolve separately. These separated loci would have the same effect as a gene duplication event.

Polymorphic loci present in separate members of a fungal population each undergo concerted evolution but evolve independently of each other. If both loci are preserved in populations of both emergent species at cladogenesis, then stochastic or directed selection could lead to sorting of one polymorphism into lineages that evolve into Gigasporaceae and Acaulosporaceae while the other polymorphism is sorted into Glomaceae. Segregation of polymorphisms then would not reflect exact correspondence with a species phylogeny and cause the conflict we see. Alternatively, the orthologous loci might not have been completely lost and might coexist with paralogous loci. Discovery of both genes would require sequencing of many clones from spores of each fungal isolate of a species (e.g., Clapp et al. 2001). Without adequate sampling, the issue is not just whether the gene tree reflects the species tree, but whether the constructed gene tree reflects the true gene phylogeny.

The issue of unpredictable sorting of polymorphic loci applies to all multicopy genes, such as those encoding beta-tubulins (Corradi and Sanders 2006), elongation factor 1alpha (James et al. 2006), P-type II ATPases (Corradi and Sanders 2006), and even mitochondrial genes (Redecker and Raab 2006). The study by Corradi and Sanders (2006) suggests that gene duplication, as a process for evolutionary innovation, is not a rare event. The occurrence of multiple copies of the DNA polymerase

I gene (Pawlowska and Taylor 2004), regardless of whether they are house in "like" or "unlike" nuclei, is also suggestive of gene proliferation in AM fungal genomes.

10.3.2 Lineage Sorting

The events in this process that cause discordance are straightforward. Cladogenesis of a gene before a speciation event leads to polymorphisms distributed among members of a population that may be preserved in both emergent species lineages (Nei 1987; Pamilo and Nei 1988). These polymorphisms are differential sorted, either among members of populations that undergo subsequent speciation events or when populations carrying them become extinct at different times in different lineages. One possible result is gene lineages which incorrectly infer a close relationship between Gigasporaceae and Acaulosporaceae. If orthologous genes are present, undersampling of isolates of a species would also obscure recognition of the true gene phylogeny.

The phylogeny overlaying the phenotypic hypothesis of species evolution in Fig. 10.2b represents an outcome if either (or both) processes described above occurred during rRNA gene evolution. The topology of this phylogeny is congruent with phylogeny hypothesized in Fig. 10.2a and thus provides one possible explanation of the discord at deeper nodes.

There is little direct evidence to support the above-mentioned hypotheses of rRNA gene evolution in glomeromycotan fungi, mostly because the empirical focus has been on pattern rather than process. Regardless, it is these very processes which provide the first reasonable explanation of the discord with a morphological phylogeny. It is a first step away from blind acceptance of limited genetic data as somehow being "privileged" relative to other equally valid data sets. The phylogeny depicted in Fig. 10.2a has been formalized in a classification (Walker and Schüßler 2004) which likely will prove to be premature. I have never accepted this classification for one simple reason: the gene phylogeny divorces the historical record of evolution by natural selection and organismal development. I manage an international collection of AM fungi (INVAM) which contains fungi from habitats worldwide. I use the phenotypic phylogenetic hypothesis (Fig. 10.2b) in making decisions of how members of each clade should be manipulated and stored. If I treated fungi in Gigasporaceae and Acaulosporaceae equally as the gene phylogeny would suggest is appropriate (sister clades, after all), then all species of the former would be extinct from the culture collection. As it stands, most have been kept alive in good condition for almost 20 years!

The new order containing these homoplasious families is named Diversisporales (Walker and Schüßler 2004). The name conveys the true nature of this group: a "grab-bag" of taxa. Efforts have been made to try and include some (any!) phenotypic characters that would support sister-group relationship between Acaulosporaceae and Gigasporaceae, but none exist that are valid synapomorphies. Redecker and Raab (2006) suggest that shared flexible germinal walls might represent a synapomorphy uniting these two families, but this character fails all tests of homology except superficial similarity. There are at least six apomorphic structural characters that recognize Gigasporaceae as a unique clade. Mycorrhizal structures are so unusual in this clade that species (collectively, not individually) can be distinguished from members of any other clade when colonizing roots together (see also Merryweather and Fitter 1998).

10.3.3 Geosiphon pyriforme – the "Lone Ranger"

Geosiphon pyriforme is a single taxon in the family Geosiphonaceae of the order Archaeosporales based on the SrRNA gene tree (Schüßler et al. 2001a) and LrRNA. There is no dispute that the genetic data provides evidence of a phylogenetic connection between this unique fungus and AM fungi, and that this taxon is better suited in Glomeromycota than Zygomycota. But the view that this connection is such a close one that conclusions of paraphyly in basal AM fungal groups has little merit when phenotypic characters are appraised critically for phylogenetic signal (= homology). I analyze these interpretations to expose tenuous claims of organismal relatedness emboldened by the SrRNA gene linkage.

Geosiphon and AM fungi both form obligate mutualistic symbioses with a phytobiont (*Nostoc* in the former, higher terrestrial plants in the latter), but there is no topological congruence in structures associated with symbiotic interchange. *Geosiphon* is the "macrosymbiont" in that specialized bladders of the fungus contain the "endosymbiont" *Nostoc* colonies (Schüßler and Kluge 2001). Roles are reversed in the AM association, as specialized arbuscules of the fungi (endosymbiont) are contained within cortical cells of host plants (macrosymbiont). The morphological structures in the AM fungal thallus (arbuscules, intraradical hyphae, vesicles, auxiliary cells) show no congruence with those in the *Geosiphon* thallus (bladders from soil-borne hyphae, internal symbiosomes). Homology fails by two criteria.

Much is made of the "similarities" in the interface of symbiotic partners (Schüßler and Kluge 2001). The general organization and structure of this interface is clearly homoplasious, evolving in bacteria–bacteria symbioses and in mutualistic as well as pathogenic fungus–fungus symbioses with obvious parallel functionality (Voegele and Mendgen 2003). Again, there is no topological correspondence (membrane component arising from fungal cell vs plant cell), and no congruence with other components of ontogeny of each organism. The hypothesis that *Geosiphon* might produce arbuscules and thus form associations with plants is highly speculative. It implies that the fungus has evolved two specialized symbioses in which the hierarchy of units under the most pervasive selection pressures can be switched on or off. The restricted habitat and geographic range of *Geosiphon* attests indirectly to the absence of any such genetic and developmental plasticity.

Regarding other "similar" traits, *Geosiphon* and AM fungi reside in or on soil and grow via branching of coenocytic hyphae. These clearly are homoplasies because they are distributed in fungal groups of separate phyla. The similarity in spore wall structure is also a polyphyletic character. Within Glomeromycota, it is the SrRNA gene phylogeny which exposes this homoplasy. Three clades in Glomeraceae (groups A, B, and C) receive strong support as evolutionarily distinct clades, and members in each group form spores of similar organization to those of *Geosiphon*. This example is a clear case where patterns of descent at the genetic level expose homoplasy in an expressed phenotype that mistakenly passes tests of structural homology.

Genetic data also may be distorting phylogenetic interpretations because of one important issue: the absence of transitional taxa between *Geosiphon* and AM fungal clades. Genetic distance is partially inferred from an LrRNA tree of 82 species (100 sequences), in which *Geosiphon* could be interchanged with the Zygomycete fungus *Mortierella* as outgroup and patterns of ingroup (AM fungi) phylogeny were identical (R. Amarasinghe and J.B. Morton, unpublished results).

A further test was performed to ascertain the sensitivity of tree topology to omission of all but one taxon from a known clade. In this case, I used another enigmatic fungus, *Entrophospora infrequens*, as the focal species. LrRNA sequences from single spores of two isolates (one in pure culture and one in a mixed culture) position this species in *Glomus* group B. However, if all *Glomus* species in this clade are excluded, *E. infrequens* becomes a solitary basal clade, and Acaulosporaceae, *Glomus* group C, and Gigasporaceae form a monophyletic group distinct from the *Glomus* group A clade. These relationships clearly are incorrect and also bear no relationship to *Geosiphon* evolution. Without at least one or more transitional taxa and the absence of valid morphological homologs, the position of *G. pyriforme* in Glomeromycota is more tenuous than the current literature would suggest.

10.4 Congruence in the rDNA Phylogeny

The discords so evident in basal clades are less obvious and more infrequent among higher nodes of the rRNA gene phylogeny. Both rRNA genes provide strong support for the monophyly of most (not all) family level groups. Of all the clades that have undergone nomenclatorial revisions, those untouched may deserve it the most. *Glomus* always has been a problematic genus, with the greatest number of species and the greatest range in spore morphotypes, life history traits, and mycorrhizal behavior. Both rRNA genes discriminate three well-supported clades in *Glomus* discussed above which do not possess any clearly delineated phenotypic synapomorphies. This is not a case of conflict with morphology, but rather a situation where morphological characters are too simple, too plastic, or intergrade too much to provide a strong phylogenetic signal. It is likely that sorting of polymorphisms and gene lineages is alive and well at these levels. However, there may be greater congruence relative to the timing of species cladogenesis so that many of the sequences encompass orthologs and possible paralogs with low rates of anagenesis. Intergrading LrRNA sequences have been found within one species (*G. coronatum*) and related taxa (Clapp et al. 2001) and between two related species, *G. etunicatum* and *G. claroideum* (Rodriguez et al. 2005). Use of multiple spores to extract DNA confounds interpretation somewhat, but the evidence is suggestive of lineage sorting of gene polymorphisms that arose before species cladogenesis. It is also possible that the genetic processes underlying the evolutionary history of species in this clade may be distinctive either in type, intensity, or outcome relative to other clades. Alternatively, such processes may be similar in all clades but the data just have not been collected to make informative comparisons.

At the species level, the "containment" of a majority of sequences within the boundary conditions of a strong sampling of morphological species is astounding. We find few conflicts between morphospecies within our international culture collection and LrDNA sequences (R. Amarasinghe and J.B. Morton, unpublished results). Such strong congruence has been a major reason why we have prioritized development of an extensive database of LrRNA sequences. Mechanisms of cohesion at the molecular level are not well delineated or understood, and their correlation with morphological constraints may simply be a function of the "containerization" of initial genetic information that comes following a speciation event (see above).

I do not consider phenotypic criteria alone to be sufficient to reconstruct a species phylogeny for three interrelated reasons: (1) species are terminal units and so the homologs which define them are apomorphies, (2) simplicity of these apomorphies (the phenotype of one or more layers in the spore wall) presupposes involvement of only a few genes, and (3) morphological properties, at least at the species level, cannot possibly provide any causal mechanism for species cladogenesis and therefore are more likely correlative characters. The burden for true delineation of phylogenetic species then rests with analyses of gene lineages (trees) independent of each other followed by a taxonomic congruence approach that resolves grouping and ranking criteria (Taylor et al. 2000).

Evidence is mounting that processes affecting population dynamics of gene lineages differ significantly between species and possibly even at higher taxonomic levels. For example, species in the genus *Gigaspora* contain only one beta-tubulin gene, whereas species of the sister taxon, *Scutellospora*, contain two (Corradi et al. 2004; Z. Msiska and J.B. Morton, unpublished results). Genes encoding P-type II ATPases are duplicated in some *Glomus* species but not in others (Corradi and Sanders 2006). The solution will require analysis of truly independent single copy genes, and these are not always clearly recognizable.

10.5 Conclusions

The discussion in this chapter shows what can happen when one gene (or two interdependent genes) are given "privileged" status, and mechanisms of their evolution and possible outcomes from these processes are not considered. Biases prevail when other informative characters (phenotypic) are used selectively or outside a valid phylogenetic framework to support this gene phylogeny. New meaning is given to the phrase "competing phylogenies." But these hypotheses should not be competing with each other, because each character set provides valuable information; the true phylogeny must be concordant for *all* phylogenetically informative (homologous/orthologous) characters.

The analysis of rRNA gene evolution discussed above came about because of a fundamental belief that important elements of truth are contained in the distribution of phenotypic traits, especially in basal nodes, and so the focus had nowhere else to go. Most of the ideas presented in this chapter are not new to science, but they have not been discussed within the context of Glomeromycota systematics.

This chapter has been an epiphany of sorts for me, because for the first time I see both the strengths and limitations of both phenotypic and genetic characters with respect to pattern versus process. A balanced multidisciplinary approach which takes into account *all* synapomorphic characters, regardless of their origin, is widely practised by systematists of other groups, including fungi (Lutzoni et al. 2004). Researchers working on glomeromycotan fungi should not be the exception. If there is conflict, there has to be a causal basis for it. As "the deciders," we must approach resolution with "eyes wide open" lest we lose sight of the integration and uniqueness of evolutionary processes at all levels of organization.

Acknowledgments R. Amarasinghe, S. Bentivenga, J. Bever, M. Franke-Snyder, Z. Msiska and S. Stürmer working in my laboratory contributed data and ideas that led to this work. I also appreciate insightful thoughts from T. Pawlowska and D. van Tuinen. My laboratory and INVAM is totally dependent on the able technical assistance of W. Wheeler. Financial support comes from the WVU experiment station and NSF grants DBI-0650735 and DEB-0649341.

References

Clapp JP, Rodriguez A, Dodd JC (2001) Inter- and intra-isolate rRNA large subunit variation in *Glomus coronatum* spores. New Phytol 149:539–554

Corradi N, Sanders IR (2006) Evolution of the P-type II ATPase gene family in the fungi and presence of structural genomic changes among isolates of *Glomus intraradices*. BMC Evol Biol 6:21

Corradi N, Hijri M, Fumagalli L, Sanders IR (2004) Arbuscular mycorrhizal fungi (Glomeromycota) harbour ancient fungal tubulin genes that resemble those of the chytrids (Chytridiomycota). Fungal Genet Biol 41:1037–1045

de la Providencia IEF, de Souza D, Fernandez F, Séjalon-Delmas N, Declerck S (2005) Arbuscular mycorrhizal fungi reveal distinct patterns of anastomosis formation and hyphal healing mechanisms between different phylogenetic groups. New Phytol 165:261–271

de Queiroz A, Donoghue MJ, Kim J (1996) Separate versus combined analysis of phylogenetic evidence. Annu Rev Ecol Syst 26:657–681

de Silva GA, Lumini E, Maia LC, Bonfante P, Bianciotto V (2006) Phylogenetic analysis of Glomeromycota by partial LSU rDNA sequences. Mycorrhiza 16:183–189

Doyle JJ (1992) Gene trees and species trees: molecular systematics as one-character taxonomy. Syst Bot 17:144–163

Doyle JJ (1997) Trees within trees: genes and species, molecules and morphology. Syst Biol 46:537–553

Eickbush TH, Eickbush DG (2007) Finely orchestrated movements: evolution of the ribosomal RNA genes. Genetics 175:477–485

Gandolfi A, Sanders IR, Rossi V, Menozzi P (2003) Evidence of recombination in putative ancient asexuals. Mol Biol Evol 20:754–761

Gianinazzi-Pearson V, Lemaine MC, Arnould C, Gollotte A, Morton JB (1994) Localization of β(1–3)-Glucans in spore and hyphal walls of fungi in the Glomales. Mycologia 86:478–485
Giovannetti M, Azzolini D, Citernesi AS (1999) Anastomosis formation and nuclear and protoplasmic exchange in arbuscular mycorrhizal fungi. Appl Environ Microbiol 65:5571–5575
Giovannetti M, Sbrana E, Strani P, Agnolucci M, Rinaudo V, Avio L (2003) Genetic diversity of isolates of *Glomus mosseae* from different geographic areas detected by vegetative compatibility testing and biochemical and molecular analysis. Appl Environ Microbiol 69:616–624
Hart MM, Reader RJ (2002) Taxonomic basis for variation in the colonization strategy of arbuscular mycorrhizal fungi. New Phytol 153:335–344
Helgason T, Watson IJ, Young PW (2003) Phylogeny of the Glomerales and Diversisporales (Fungi: Glomeromycota) from actin and elongation factor 1-alpha sequences. FEMS Microbiol Lett 229:127–132
Hijri M, Sanders IR (2005) Low gene copy number shows that arbuscular mycorrhizal fungi inherit genetically different nuclei. Nature 433:160–163
Hosny M, Gianinazzi-Pearson V, Dulieu H (1998) Nuclear DNA content of 11 fungal species in Glomales. Genome 41:422–428
James TY, KauffF, Schoch CL et al. (2006) Reconstructing the early evolution of fungi using a six-gene phylogeny. Nature443:818–822
Klironomos JN, Hart MM (2002) Colonization of roots by arbuscular mycorrhizal fungi using different sources of inoculum. Mycorrhiza 12:181–184
Kuhn G, Hijri M, Sanders IR (2001) Evidence for the evolution of multiple genomes in arbuscular mycorrhizal fungi. Nature 414:745–748
Lanyon SM (1988) The stochastic mode of molecular evolution: what consequences for systematic investigations. Auk 105:565–573
Liao D (1999) Concerted evolution: molecular mechanism and biological implications. Am J Hum Genet 64:24–30
Long EO, IB Dawid (1980) Repeated genes in eukaryotes. Annu Rev Biochem 49:727–764
Lutzoni F, Kauff F, Cox CJ, McLaughlin D et al. (2004) Assembling the fungal tree of life: progress, classification, and evolution of subcellular traits. Am J Bot 91:1446–1480
Lyons-Weiler J, Milinkovitch MC (1997) A phylogenetic approach to the problem of differential lineage sorting. Mol Biol Evol 14:968–975
Maddison WP (1997) Gene trees in species trees. Syst Biol 46:523–536
Merryweather J, Fitter AH (1998) The arbuscular mycorrhizal fungi of *Hyacinthoides non-scripta* I. Diversity of fungal taxa. New Phytol 138:117–129
Morton JB (1990a) Species and clones of arbuscular mycorrhizal fungi (Glomales, Zygomycetes): Their role in macro- and microevolutionary processes. Mycotaxon 37:493–515
Morton JB (1990b) Evolutionary relationships among arbuscular mycorrhizal fungi in the Endogonaceae. Mycologia 82:192–207
Morton JB, Benny GL (1990) Revised classification of arbuscular mycorrhizal fungi (Zygomycetes): a new order, Glomales, two new suborders, Glomineae and Gigasporineae, and two new families, Acaulosporaceae and Gigasporaceae, with an emendation of Glomaceae. Mycotaxon 37:471–491
Morton JB, Redecker D (2001) Two new families of Glomales, Archaeosporaceae and Paraglomaceae, with two new genera *Archaeospora* and *Paraglomus*, based on concordant molecular and morphological characters. Mycologia 93:181–195
Morton J, Franke M, Bentivenga S (1994) Developmental foundations for morphological diversity among endomycorrhizal fungi in Glomales (Zygomycetes). In: Varma A, Hock B (eds) Mycorrhiza: structure, function, molecular biology, and biotechnology. Springer, Berlin, pp 669–683
Nei M (1987) Molecular evolutionary genetics. Columbia University Press, New York
Oehl F, Sieverding E (2005) Pacispora, a new vesicular arbuscular mycorrhizal fungal genus in the Glomeromycetes. J Appl Bot 78:72–82
Pamilo P, Nei M (1988) Relationships between gene trees and species trees. Mol Biol Evol 5:568–583

Patterson C (1982) Morphological characters and homology. In: Joysey KA, Friday AE (eds) Problems of phylogenetic reconstruction. Academic, New York, pp 21–74
Patterson C, Williams DM, Humphries CJ (1993) Congruence between molecular and morphological phylogenies. Annu Rev Ecol Syst 24:153–188
Pawlowska T (2005) Genetic processes in arbuscular mycorrhizal fungi. FEMS Microbiol Lett 251:185–192
Pawlowska TE, Taylor JW (2004) Organization of genetic variation in individuals of arbuscular mycorrhizal fungi. Nature 427:733–737
Redecker D, Raab P (2006) Phylogeny of the Glomeromycota (arbuscular mycorrhizal fungi): Recent developments and new gene markers. Mycologia 98:885–895
Redecker D, Kodner R, Graham LE (2000) Glomalean fungi from the Ordovician. Science 289:1920–1921
Rodriguez A, Clapp JP, Robinson L, Dodd JC (2004) Ribosomal RNA gene sequence diversity in arbuscular mycorrhizal fungi (Glomeromycota). J Ecol 92:986–989
Rodriguez A, Clapp JP, Robinson L, Dodd JC (2005) Studies on the diversity of the distinct phylogenetic lineage encompassing *Glomus claroideum* and *Glomus etunicatum*. Mycorrhiza 15:33–46
Rokas A, Williams BL, King N, Carroll SB (2003) Genome-scale approaches to resolving incongruence in molecular phylogenies. Nature 425:798–804.
Rooney AP, Ward TJ (2005) Evolution of a large ribosomal RNA multigene family in filamentous fungi: birth and death of a concerted evolution paradigm. Proc Natl Acad Sci USA 102:5084–5089
Sanderson MJ, Donoghue MJ (1989) Patterns of variation in levels of homoplasy. Evolution 43:1781–1795
Schlichting CD, Pigliucci M (1998) Phenotypic evolution: a reaction norm perspective. Sinauer Associates, Inc., Sunderland, MA
Schüßler A, Kluge M (2001) *Geosiphon pyriforme*, an endosymbiosis between fungus and cyanobacteria, and its meaning as a model system for arbuscular mycorrhizal research. In: Hock B (ed) The Mycota IX: fungal associations. Springer, London/Berlin, pp. 151–161
Schüßler A, Gehrig H, Schwartzott D, Walker C (2001a) A new fungal phylum, the Glomeromycota: phylogeny and evolution. Mycol Res 105:1413–1421
Schüßler A, Gehrig H, Schwartzott D, Walker C (2001b) Analysis of partial Glomales SSU rRNA gene sequences: implications for primer design and phylogeny. Mycol Res 105:5–15
Schwartzott D, Walker C, Schüßler A (2001) *Glomus*, the largest genus of the arbuscular mycorrhizal fungi (Glomales) is nonmonophyletic. Mol Syst Evol 21:190–107
Smith SE, Read DJ (1997) Mycorrhizal symbiosis, 2nd edition. Academic, London UK
Spain JL, Sieverding E, Oehl F (2006) *Appendicispora*: a new genus in the arbuscular mycorrhiza-forming Glomeromycetes, with a discussion of the genus *Archaeospora*. Mycotaxon 97:163–182
Stukenbrock EH, Rosendahl S (2005) Clonal diversity and population genetic structure of arbuscular mycorrhizal fungi (*Glomus* spp.) studied by multilocus genotyping of single spores. Mol Ecol 14:743–752
Taylor JW, Jacobson DJ, Kroken S, Kasuga T, Geiser DM, Hibbett DS, Fishe MC (2000) Phylogenetic species recognition and species concepts in fungi. Fungal Genet Biol 31:21–32
Trappe JM (1987) Phylogenetic and ecological aspects of mycotrophy in the angiosperms from an evolutionary standpoint. In: Safir GR (ed) Ecophysiology of VA mycorrhizal plants. CRC Press, Boca Raton, FL, pp. 5–25
Trouvelot S, van Tuinen D, Hijri M, Gianinazzi-Pearson V (1999) Visualization of ribosomal DNA loci in spore interphasic nuclei of glomalean fungi by fluorescence in situ hydridization. Mycorrhiza 8:203–206
Voegele RT, Mendgen K (2003) Rust haustoria: nutrient uptake and beyond. New Phytol 159:93–100
Walker C, Schüßler A (2004) Nomenclatural clarifications and new taxa in the Glomeromycota. Mycol Res 108:981–982
Walker C, Vestberg M, Demircik F, Stockinger H, Saito M, Sawaki H, Nishmura I, Schüßler A (2007) Molecular phylogeny and new taxa in the Archaeosporales (Glomeromycota):

Ambispora fennica gen. sp. nov., Ambisporaceae fam. nov., and emendation of *Archaeospora* and Archaeosporaceae. Mycol Res 111:137–153

Wheeler WC, Honeycutt RL (1988) Paired sequence difference in ribosomal RNAs: Evolutionary and phylogenetic implications. Mol Biol Evol 5:90–96

Chapter 11
Spatial Ecology of Ectomycorrhizas: Analytical Strategies

Brian J. Pickles, David Genney, Ian C. Anderson, and Ian J. Alexander

Abstract Spatial analysis techniques have recently become more common in ecological research. The application of these techniques to ecological problems has coincided with the development of molecular techniques for the study of ectomycorrhizal (ECM) fungal communities. Through determination of the spatial properties of ECM fungal species' distribution (i.e. patch size and shape, degree of clustering), it may be possible to derive much more information about these communities than can be obtained from simple species abundance and frequency data. For example, the use of spatial analysis may enable detection of species interactions and temporal changes in species distribution, as well as illustrating how environmental properties may relate to ECM fungal distribution. This review discusses the application of spatial ecology concepts to, and the issues surrounding, spatially explicit sampling of ECM fungal communities in relation to current trends in ECM community research, and suggests directions for future research.

11.1 Introduction

Since the introduction of molecular techniques to ECM fungal research, there has been a rapid increase in the number of studies aimed at understanding the below-ground ecology and structure of ECM communities (Horton and Bruns 2001; Taylor 2002; Lilleskov et al. 2004; Izzo et al. 2005; Koide et al. 2005a, b; see also Chapter 12 by Wolfe et al.). This interest in the ecology of ECM fungi has

B.J. Pickles(✉)
Ecosystem Science and Management, University of Northern British Columbia, 3333 University Way, Prince George, BC, V2N 4Z9, Canada.
e-mail: pickles@unbc.ca

D. Genney, I.J. Alexander
School of Biological Sciences, Plant and Soil Science, University of Aberdeen, St. Machar Drive, Aberdeen, AB24 3UU, UK

I.C. Anderson
The Macaulay Institute, Craigiebuckler, Aberdeen, AB15 8QH, UK

C. Azcón-Aguilar et al. (eds.), *Mycorrhizas - Functional Processes and Ecological Impact*, DOI: 10/1007-978-3-540-87978-7_11,

coincided with an upsurge of interest in the application of spatial analysis techniques to ecological problems (Tilman and Kareiva 1997; Dale et al. 2002; Ettema and Wardle 2002). Given the recent advances in species identification techniques (reviewed by Anderson and Cairney 2004), it is now possible to ask questions about spatial aspects of belowground ECM community ecology, such as how the distributions of individual species interact with each other, with other organisms, and with environmental factors such as pH, soil moisture content, soil depth and substrate availability. Furthermore, an understanding of the spatial structure of ECM fungi should allow researchers to disentangle the effects of spatial autocorrelation, the similarity of (a property of) neighbouring samples as a function of spatial separation distance (Legendre and Legendre 1998; Koenig 1999; Lennon 2000; Ettema and Wardle 2002; Taylor 2002), from the actual ecological factors influencing their distribution. With this in mind, we discuss the application of spatial ecology to studies of ECM fungi, examine issues related to the spatially explicit sampling of ECM fungal communities and make suggestions about the direction of future research in this field.

11.2 Spatial Ecology in Microbiology

Whilst ecologists have long recognised the importance of space (i.e. MacArthur and Levins 1964) it is still too often the case that studies of ecological communities do not explicitly consider spatial effects, perhaps due to difficulties in embracing the mathematical concepts involved or the sampling intensities required (Tilman and Kareiva 1997). However, interest in explicitly linking ecological studies with spatial properties has rapidly increased over the last decade, and the importance of spatial ecology has become widely acknowledged as a route to gain information that may have otherwise remained unobtainable (Legendre and Fortin 1989; Dale et al. 2002; Ettema and Wardle 2002). For example, the importance of space varies for different species, but may include access to resources, the potential to escape predation or competition, the ability to survive stochastic events (i.e. forest fires, climatic fluctuations, volcanism, etc.), dispersal ability, community stability and the potential for intraspecific interactions as well as anthropogenic impacts (Dale et al. 2002; Kneitel and Chase 2004; see also Chapter 13 by Simard).

The choice of analytical methods to deal with spatially explicit data has also greatly increased over the last decade (Dale et al. 2002). Many of these have been successfully applied to studies of plant and animal communities, for example spatial techniques have been used to reveal synchronicity of mast seeding in populations of North American Oak species (Liebhold et al. 2004), relationships between spatial structure and environmental factors in bird communities (Cushman and McGarigal 2004), and the spatial and temporal responses of predators to prey in insect communities (Winder et al. 2001). However, it is only relatively recently that attention has turned to micro-organisms (Paerl and Steppe 2003).

Green et al. (2004) used spatial scaling techniques to examine how the diversity of ascomycetous taxa changes with area. They found evidence of similar spatial structure to that observed for eukaryotes in marine and freshwater habitats.

Similarity between samples decreased with increasing distance, and whilst this pattern has been seen many times for other groups, the interesting point to note here is that the decline in similarity took place over significantly greater geographical scales than has been observed for larger organisms. This suggests that detection of structure in these communities may require much larger sampling areas than are necessary for studies of plants and animals. Additionally, Horner-Devine et al. (2004) studied taxa–area changes in bacterial communities. They reported a significant relationship between the number of taxa and area for bacteria, suggesting structure in these communities, although the rate of taxon turnover across space was much lower than that observed for any other previously studied taxa–area relationship. These studies also help to reject the idea of soil communities as a structure-less 'black-box' (Ettema and Wardle 2002). In their study of the distribution of two species of cyst nematodes in soil communities, Webster and Boag (1992) applied geostatistics by using a sampling strategy involving extraction of soil cores at a nested series of distances. This technique revealed that most of the spatial variation occurred over scales of 5–50 m, with both species displaying highly aggregated distributions. More recently, spatial autocorrelation in community structure of arbuscular mycorrhizal fungi (AMF) was studied (Whitcomb and Stutz 2007) with a variety of spatial techniques. The perceived diversity and species composition of these communities varied significantly with sampling effort, and spatial structure only became apparent at the smallest sampling distances. Another recent study investigated AMF abundance and community composition in a wetland habitat, in relation to mapped depth of the water table (Wolfe et al. 2007). In this analysis, a positive correlation was found between AMF abundance and depth of the water table. Basic spatial information of this type would be a useful step in the study of ECM communities.

Recently studies of ECM fungi have begun to incorporate spatial analysis techniques, or at least taken spatial properties into account (Horton and Bruns 2001; Taylor 2002; Koide et al. 2005b; Genney et al. 2006; Peay et al. 2007; Thiet and Boerner 2007). In order to advance our understanding of belowground ECM fungal ecology, research questions should begin to focus on how the distribution patterns of different species relate to each other, and to other biotic and abiotic variables, in time and space. A number of spatial analysis techniques exist which may be suitable for this purpose, and there is a large body of literature devoted to the discipline of spatial ecology (e.g. Tilman and Kareiva 1997; Legendre and Legendre 1998; Dale et al. 2002), although, so far, there have been few applications to ECM fungal research.

11.3 Methodological Approaches and Recent Spatial Studies of ECM Fungi

While molecular techniques have removed some of the barriers to belowground investigations of ECM communities (Anderson and Cairney 2004), there are still several issues relating to ECM detection and sampling that must be dealt with. When formulating research questions about ECM communities, especially if spatial and temporal properties are under investigation, the cryptic nature of some species, the relative importance of root tips and extramatrical mycelium (EM), and optimisation of the sampling regime should all be carefully considered.

11.3.1 Cryptic Species

Several ECM fungal species forming ectomycorrhizal associations either produce sporocarps which are cryptic (Kõljalg et al. 2000) or do not produce sporocarps at all. For example, *Cenococcum geophilum* is often one of the most abundant and frequently encountered species in studies of ECM fungal communities from many varied geographical locations (Tedersoo et al. 2003; Dickie and Reich 2005; Koide et al. 2005a, b), appearing to display little or no host specificity (Wurzburger and Bledsoe 2001), yet may not form sporocarps (but see Fernández-Toirán and Águeda 2007). Tomentelloid fungi are important and widespread ECM fungi which produce resupinate sporocarps that were often ignored in early studies of ECM communities but have recently received more attention (Kõljalg et al. 2000). An additional complication is the issue of intra-specific cryptic variation, as detected by molecular methods (Horton 2002). For example, some molecular methods can suggest the presence of multiple species when sequencing reveals that only one is present, as described earlier. This means that any study of a community of ECM fungi must carefully consider the contribution of the species present below ground, and determine if and how it will deal with cryptic species.

11.3.2 Sporocarps Versus Root Tips Versus Mycelia

Comparison of a colonised root tip approach to a detection of mycelia approach suggested that studies based on mycelia may lead to different conclusions about ECM fungal community structure from those based on root tips (Koide et al. 2005b). The frequency of occurrence of ECM fungi also appears to vary depending on whether root tips or mycelia are studied (Genney et al. 2006; Kjøller 2006). This again raises the question of what exactly is the most appropriate way of examining the spatial structure of an ECM fungal community, as sampling designs may detect different community structure and composition for sporocarps, root tips and hyphae. Knowledge of the techniques available for analysis of these different aspects of an ECM community, coupled with specific research goals, should allow the researcher to highlight the most appropriate methodological approach for the question/s of interest.

11.3.3 Random Sampling Techniques

Random sampling techniques have been used in studies of ECM fungal community structure, in an attempt to minimise the effects of spatial autocorrelation. Koide et al. (2005a, b) randomly located 80 sampling plots within a red pine (*Pinus resinosa*) plantation which they further randomly sub-sampled at various time intervals. They also selected root tips apparently at random from those collected within the sub-samples for molecular analysis. A random sampling design coupled with small transects has also been used in the study of ECM fungal communities to gather nested, yet randomly distributed, spatial data (Izzo et al. 2005). Given that the extent of spatial structure in belowground ECM fungal communities is unknown, it may be preferable to use an approach that combines both structured and random elements in an attempt to quantify spatial autocorrelation (Fortin et al. 1989).

11.3.4 Sampling Grids

Lian et al. (2006) examined the distribution of *Tricholoma matsutake* above- and belowground in *P. densiflora* forests. Of particular interest was their use of a sampling grid to investigate the distribution of ECM species present both inside and outside *T. matsutake* 'fairy-rings' as well as those species that were found directly below sporocarps. It appeared that the ECM community below *T. matsutake* sporocarps was species poor and mainly composed of *T. matsutake*, with a sharp boundary separating it from the very different, species-rich community found outside the fairy rings. Hirose et al. (2004) used a structured sampling grid to investigate the spread of different genets of a single species (*Suillus pictus*) within an ECM fungal community. The spatial distribution of *S. pictus* sporocarps as a whole was found to overlap with itself from year to year, yet was very different to the distribution of sporocarps from all other species, whilst the distribution of individual genets of *S. pictus* did not overlap with each other. The novelty of this approach was to examine both sporocarp and root tip distributions at a variety of scales, and show the potential for spatially structured studies to elucidate ecological processes, such as spatio-temporal turnover of ECM fungal species and/or individuals from year to year.

A geostatistical approach to ECM fungal distribution has been attempted through the use of sampling grids (Boerner et al. 1996; Thiet and Boerner 2007). Boerner et al. (1996) examined the overall colonisation of roots by ECM fungi on six sites representing successional change between vegetation communities. ECM fungi were distributed over an increasingly large area as time since disturbance increased, and basic morphotyping suggested a concurrent increase in species richness. This study used relatively few sampling points, which reduces the power of the geostatistical approach. Thiet and Boerner (2007) used similar methods to assess the colonisation patterns of ECM fungi across an ecotone from forest into open prairie. Spread of ECM into prairie appeared to follow two forms: the contagion model (where inoculum spreads out as a gradient from an existing infected area) and the centres of infection model (in which inoculum is found in patches surrounded by non-infected areas). Using a structured sampling grid approach, and extracting an appropriate number of samples, it should be possible to go beyond these studies and use geostatistics to examine and compare the belowground spatial distribution of individual ECM species.

Micromapping (Agerer et al. 2002) represents another form of sampling grid for the study of ECM root tips, albeit at an extremely fine scale. This procedure involves carefully noting and recording the positions of all root tips within a 1 cm × 25 cm^2 slice of soil, followed by identification of the ECM species colonising each tip. Species and their individual tip positions are then converted into pixels, which are used to estimate the degree of 'spatial relation' (Agerer et al. 2002) as a preliminary indicator of associations and dissociations between root tips at a very fine scale.

11.3.5 Patch Size Estimation

Studies of the distribution of individual genets of ECM fungi have been carried out for a few species and have suggested that patch sizes could vary from a few tens of centimetres to many tens of metres (Sawyer et al. 1999; Guidot et al. 2002; Hirose et al. 2004; Riviere et al. 2006). It seems likely that host root tip abundance itself is patchy and spatiotemporally variable as a result of interactions with biotic (e.g. inter- and intra-specific competition, herbivory, senescence, etc.) and abiotic (e.g. distribution of nutrients, depth of the water table, fire intensity, substrate type, etc.) factors. This in turn leads to a patchy distribution of ECM fungal root tips with additional spatial variability brought about by differences in ECM fungal exploration type, biology and function as well as predation by fungivores. Below this scale, EM distribution is also patchy (Genney et al. 2006) and likely related to the distribution of substrates throughout the soil profile. Above these scales, sporocarp distribution also appears to be patchy (Hirose et al. 2004; Riviere et al. 2006), and is not necessarily related to the belowground distribution of root tips. Finally, the species composition of ECM communities is likely to also be patchily distributed, at a scale above that of individual species as root tips. Given that there are differences observed between the distribution patterns of sporocarps, root tips and EM, the questions of how to define the patch size of individual species of ECM fungi and 'individuals' within these species appears complex.

11.3.6 Temporal Change

The effects of temporal change on spatial properties of ECM fungal communities have also been investigated. Temporal changes were observed within the ECM fungal community of old-growth mixed conifer forest on the West coast of the USA, in which differences in the year-to-year abundance of species took place, with changes apparently more pronounced at smaller scales (Izzo et al. 2005). However, this study did not take account of the scale of different species' distributions within their study site, so it is difficult to say whether the differences observed from year to year are simply due to a failure to take the spatial autocorrelation of species distributions into account (Koenig 1999). A recent study by Koide et al. (2007) examined changes in the composition of ECM fungal communities between seasons.

Fungal species were present at different relative frequencies between seasons, with root tips and hyphae observed to follow different patterns in some cases. This is the first time that a trajectory-based model had been applied to ECM fungi, in this case in an attempt to group species by their pattern of change in relative frequency over time. If these results are confirmed, this would provide support for the theory that the high diversity of fungal species observed in many communities is due to temporal partitioning amongst groups of ECM fungi.

11.4 Future Directions for ECM Community Research

One of the most important questions still to be addressed in ECM fungal community research is how to determine the appropriate sampling scale to address the ecological question being asked (Horton and Bruns 2001; Taylor 2002; Lilleskov et al. 2004). Whilst individual studies will obviously have different research goals, it is important to understand the scales at which the communities under investigation operate (see Chapter 12 by Wolfe et al.). Lilleskov et al. (2004) searched the literature for studies of ECM fungal communities and examined these pre-existing datasets of ECM abundance for evidence of spatial structure. Several statistical methods were used, including Mantel tests and correlograms, to look for evidence of spatial autocorrelation in the abundance of ECM fungi. Often the indicated distance at which spatial autocorrelation occurred was at, or below, the minimum core separation distance, which suggests that the sampling regime was not optimal for determining the patch sizes of ECM species in any of the studies that they examined. High levels of inter-core dissimilarity were found at all spatial scales, and this may partly result from the studies not being specifically set up to consider spatial structure. Spatial structure was found only at scales of approximately 3 m (with a range from 0 to ≥17 m) in almost all cases. Spatial autocorrelation could not be detected in several of the datasets, indicating that patch sizes were smaller than the minimum sampling distance used in these studies. This may be reassuring for ecological studies that are not concerned with uncovering the patchiness of ECM fungal species, because the data can be analysed without fear of errors caused by autocorrelation (Legendre and Legendre 1998). However, it is possible that this scale changes depending on the community being studied, and the quality of ecological data will be questionable unless further attempts are made to estimate the most relevant scale of analysis for these organisms (see Chapter 12 by Wolfe et al.).

11.4.1 Key questions to be answered using spatial approaches

One way of tackling these issues is through the application of appropriate spatial sampling and analysis techniques. When deciding on the correct analysis techniques it is important to consider what has been used in the past and what questions are being asked, as this will enable an informed decision about the best approach. So there are several key questions which the application of spatial analysis techniques should help to answer:

(1) What constitutes a belowground community of ECM fungi, i.e. which species are present, what is their relative abundance, and is it possible to determine the limits of a given 'community'?
(2) How does the distribution of fine roots themselves influence the structure and heterogeneity of the ECM fungal community?
(3) What do the distribution patterns of individual ECM fungal species tell us about patch size, relative dispersal ability and colonisation strategy?
(4) Are there optimum sampling regimes for belowground studies of ECM fungi which might allow better comparison between investigations?
(5) Do significant interactions take place between ECM fungal species belowground?
(6) How do ECM fungi interact with other organisms and abiotic variables?
(7) Are ECM fungal communities stable over time, and if not, are the fluctuations in community composition and structure stochastic or determined by competition, facilitation, environmental change, etc.?

11.5 Conclusions

Whilst methodological advances are continually being made, it seems that the molecular and statistical tools currently available to mycorrhizal ecologists have reached the point at which detailed and accurate community data can be retrieved. Additionally, it may be possible to determine the factors which drive species co-existence and generate the observed diversity of ECM fungal species below ground. How similar are ECM communities between sites? The T-RFLP method appears to be a useful tool for examining large-scale changes in ECM communities, and should help to answer this question. In fact, such techniques of community analysis should be applied to a variety of ECM fungal communities in order to discover whether there are generally applicable patterns in the distribution of both individual species and community composition. At what scale does change from one to another distinct ECM community occur? This question requires scaling up the sampling across landscapes, and may best be served through a series of sampling grids spread out throughout the area of interest. This would help to establish whether belowground ECM communities behave like those of other soil microorganisms on geographical scales (i.e. Green et al. 2004; Horner-Devine et al. 2004). A more thorough investigation of the seasonal changes in ECM fungal community composition is also required, and future studies should consider taking multiple sampling events across seasons using T-RFLP techniques to look for changes in the frequency of detection throughout the year (e.g. Koide et al. 2007). Similar and simultaneous investigations of the ECM fungal community deeper in the soil profile should also be conducted, to see whether community composition and turnover changes with depth.

Species co-existence in ECM communities also requires further investigation in order to establish the processes by which so many species co-exist (i.e. niche differentiation, disturbance hypotheses, environmental fluctuations, temporal partitioning, relative rarity), and how less abundant species persist. Further investigation is also required to see whether there are common traits amongst dominant fungal species, such as colonisation strategy.

One of the major challenges for ECM research is to develop a fast throughput molecular method that can produce abundance data on ECM fungi. T-RFLP may be able to do this if it can be converted into a quantitative technique. This would allow rapid advances to be made in ECM community ecology by enabling more sophisticated spatial and temporal questions to be tackled.

Acknowledgements The authors would like to thank an anonymous referee for constructive feedback on this manuscript.

References

Agerer R, Grote R, Raidl S (2002) The new method of 'micromapping', a means to study species-specific associations and exclusions of ectomycorrhizae. Mycol Prog 1:155–166

Anderson IC, Cairney JWG (2004) Diversity and ecology of soil fungal communities: increased understanding through the application of molecular techniques. Environ Microbiol 6:769–779

Boerner REJ, DeMars BG, Leicht PN (1996) Spatial patterns of mycorrhizal infectiveness of soils along a successional chronosequence. Mycorrhiza 6:79–90.

Cushman SA, McGarigal K (2004) Patterns in the species-environment relationship depend on both scale and choice of response variables. Oikos 105:117–124

Dale MRT, Dixon P, Fortin M-J, Legendre P, Myers DE, Rosenberg MS (2002) Conceptual and mathematical relationships among methods for spatial analysis. Ecography 25:558–577

Dickie IA, Reich PB (2005) Ectomycorrhizal communities at forest edges. J Ecol 93:244–255

Ettema CH, Wardle DA (2002) Spatial soil ecology. Trends Ecol Evol 17:177–183

Fernández-Toirán LM, Águeda B (2007) Fruitbodies of *Cenococcum geophilum*. Mycotaxon 100:109–114

Fortin M-J, Drapeau P, Legendre P (1989) Spatial autocorrelation and sampling design in plant ecology. Vegetatio 83:209–222

Genney DR, Anderson IC, Alexander IJ (2006) Fine-scale distribution of pine ectomycorrhizas and their extramatrical mycelium. New Phytol 170:381–390

Green JL, Holmes AJ, Westoby M, Oliver I, Briscoe D, Dangerfield M, Gillings M, Beattie AJ (2004) Spatial scaling of microbial eukaryote diversity. Nature 432:747–750

Guidot A, Debaud J-C, Marmeisse R (2002) Spatial distribution of the below-ground mycelia of an ectomycorrhizal fungus inferred from specific quantification of its DNA in soil samples. FEMS Microb Ecol 42:477–486

Hirose D, Kikuchi J, Kanzaki N, Futai K (2004) Genet distribution of sporocarps and ectomycorrhizas of *Suillus pictus* in a Japanese white pine plantation. New Phytol 164:527–541

Horner-Devine MC, Lage M, Hughes JB, Bohannan JM (2004) A taxa-area relationship for bacteria. Nature 432:750–753

Horton TR (2002) Molecular approaches to ectomycorrhizal diversity studies: variation in ITS at a local scale. Plant Soil 244:29–39

Horton TR, Bruns TD (2001) The molecular revolution in ectomycorrhizal ecology: peeking into the black-box. Mol Ecol 10:1855–1871

Izzo A, Agbowo J, Bruns TD (2005) Detection of plot-level changes in ectomycorrhizal communities across years in an old-growth mixed-conifer forest. New Phytol 166:619–630

Kjøller R (2006) Disproportionate abundance between ectomycorrhizal root tips and their associated mycelia. FEMS Microb Ecol 58:214–224

Kneitel JM, Chase JM (2004) Trade-offs in community ecology: linking spatial scales and species coexistence. Ecol Lett 7:69–80

Koenig WD (1999) Spatial autocorrelation of ecological phenomena. Trends Ecol Evol 14:22–26

Koide RT, Xu B, Sharda J, Lekberg Y, Ostiguy N (2005a) Evidence of species interactions within an ectomycorrhizal fungal community. New Phytol 165:305–316

Koide RT, Xu B, Sharda J (2005b) Contrasting below-ground views of an ectomycorrhizal fungal community. New Phytol 166:251–262

Koide RT, Shumway DL, Xu B, Sharda JN (2007) On temporal partitioning of a community of ectomycorrhizal fungi. New Phytol 174:420–429

Kõljalg U, Dahlberg A, Taylor AFS, Larsson E, Hallenberg N, Stenlid J, Larsson K-H, Fransson PM, Kårén O, Jonsson L (2000) Diversity and abundance of resupinate thelephoroid fungi as ectomycorrhizal symbionts in Swedish boreal forests. Mol Ecol 9:1985–1996

Legendre P, Fortin M-J (1989) Spatial pattern and ecological analysis. Vegetatio 80:107–138

Legendre P, Legendre L (eds) (1998) Numerical Ecology, 2nd Edition. Elsevier, Amsterdam

Lennon J (2000) Red-shifts and red herrings in geographical ecology. Ecography 23:101–113

Lian C, Narimatsu M, Nara K, Hogetsu T (2006) *Tricholoma matsutake* in a natural *Pinus densiflora* forest: correspondence between above- and below-ground genets, association with multiple host trees and alteration of existing ectomycorrhizal communities. New Phytol 171:825–836

Liebhold A, Sork V, Peltonen M, Koenig W, Bjørnstand ON, Westfall R, Elkinton J, Knops JMH (2004) Within-population spatial synchrony in mast seeding of North American oaks. Oikos 104:156–164

Lilleskov EA, Bruns TD, Horton TR, Taylor DL, Grogan P (2004) Detection of forest stand-level spatial structure in ectomycorrhizal fungal communities. FEMS Microb Ecol 49:319–332

MacArthur RH, Levins R (1964) Competition, habitat selection, and character displacement in a patchy environment. Proc Natl Acad Sci USA 51:1207–1210

Paerl HW, Steppe TF (2003) Scaling up: the next challenge in environmental microbiology. Environ Microbiol 5:1025–1038

Peay KG, Bruns TD, Kennedy PG, Bergemann SE, Garbelotto M (2007) A strong species-area relationship for eukaryotic soil microbes: island size matters for ectomycorrhizal fungi. Ecol Lett 10:470–480

Riviere T, Natarajan K, Dreyfus B (2006) Spatial distribution of ectomycorrhizal Basidiomycete *Russula* subsect. Foetentinae populations in a primary dipterocarp rainforest. Mycorrhiza 16:143–148.

Sawyer NA, Chambers SM, Cairney JWG (1999) Molecular investigation of genet distribution and genetic variation of *Cortinarius rotundisporus* in eastern Australian sclerophyll forests. New Phytol 142:561–568

Taylor AFS (2002) Fungal diversity in ectomycorrhizal communities: sampling effort and species detection. Plant Soil 244:19–28

Tedersoo L, Kõljalg U, Hallenberg N, Larsson K-H (2003) Fine scale distribution of ectomycorrhizal fungi and root across substrate layers including coarse woody debris in a mixed forest. New Phytol 159:153–165

Thiet RK, Boerner REJ (2007) Spatial patterns of ectomycorrhizal fungal inoculum in arbuscular mycorrhizal barrens communities: implications for controlling invasion by *Pinus virginiana*. Mycorrhiza 17:507–517

Tilman D, Kareiva P (1997) Spatial Ecology, Princeton University Press, Chichester, UK

Webster R, Boag B (1992) Geostatistical analysis of cyst nematodes in the soil. J Soil Sci 43:583–595
Whitcomb S, Stutz JC (2007) Assessing diversity of arbuscular mycorrhizal fungi in a local community: role of sampling effort and spatial heterogeneity. Mycorrhiza 17:429–437
Winder L, Alexander CJ, Holland JM, Woolley C, Perry JN (2001) Modelling the dynamic spatio-temporal response of predators to transient prey patches in the field. Ecol Lett 4:568–576
Wolfe BE, Mummey DL, Rillig MC, Klironomos JN (2007) Small-scale spatial heterogeneity of arbuscular mycorrhizal fungal abundance and community composition in a wetland plant community. Mycorrhiza 17:175–183
Wurzburger N, Bledsoe CS (2001) Comparison of ericoid and ectomycorrhizal colonization and ectomycorrhizal morphotypes in mixed conifer and pygmy forests on the northern California coast. Can J Bot 79:1202–1210

Chapter 12
Spatial Heterogeneity in Mycorrhizal Populations and Communities: Scales and Mechanisms

Benjamin E. Wolfe, Jeri L. Parrent, Alexander M. Koch, Benjamin A. Sikes, Monique Gardes, and John N. Klironomos

Abstract The importance of a spatial context in understanding the ecology and evolution of organisms has become increasingly clear. Although there is a growing awareness of the importance of mycorrhizal fungi in many communities and ecosystems, much of this understanding is based on a spatially homogenized view of these soil fungi. This homogenized approach may limit our understanding of how these organisms interact with plants and other biota in the field. As an attempt to advance a spatial framework for understanding mycorrhizal ecology, we review our current understanding of the spatial structure of communities and populations of ectomycorrhizal and arbuscular mycorrhizal fungi at the scale of landscapes, communities, and individual host root systems. A variety of potential mechanisms such as disturbance, abiotic and biotic dispersal of mycorrhizal propagules, and biotic interactions may be responsible for generating and maintaining this spatial variation of populations and communities, but the links between observed spatial patterns and mechanisms have yet to be formed. Future work assessing the potential functional significance of spatial variation of mycorrhizal fungi for plant communities and ecosystem function, as well as measuring spatial variation in mycorrhizal function, will continue to advance our understanding of the spatial template for mycorrhizal–plant interactions in the field.

B.E. Wolfe
Department of Organismic and Evolutionary Biology, Harvard University, 16 Divinity Avenue, Cambridge, MA 02138, USA

J.L. Parrent, A.M. Koch, B.A. Sikes, and J.N. Klironomos (✉)
Department of Integrative Biology, University of Guelph, Guelph, ON, N1G 2W1, Canada
e-mail: jklirono@uoguelph.ca

M. Gardes
Laboratoire Evolution et Diversité Biologique, Université Paul Sabatier, 118 Route de Narbonne, 31062 Toulouse Cedex 4, France

C. Azcón-Aguilar et al. (eds.), *Mycorrhizas - Functional Processes and Ecological Impact*,
DOI: 10/1007-978-3-540-87978-7_12,

12.1 Introduction

Recent empirical and theoretical research on a variety of organisms has illustrated that spatial scale is important when researchers investigate the dynamics and structure of populations (e.g., Campbell and Dooley 1992; Schweizer et al. 2007), the composition of communities (Turnbull et al. 2007), and the functioning of ecosystems (Maestre et al. 2005). When studying large organisms in the aboveground, much of this spatial structure is intuitive to the investigator through simple observations in the field. But for small organisms that inhabit the belowground, the spatial dynamics of populations and communities are elusive. An understanding of how soil organisms are structured over space can have practical implications for sampling design in studies of soil populations and communities (Klironomos et al. 1999; Lilleskov et al. 2004; see also Chapter 11 by Pickles et al.), but can also inform basic questions about modes of dispersal, how species diversity is maintained at local and regional scales, and how aboveground communities interact with belowground communities (Ettema and Wardle 2002).

Mycorrhizal fungi are one group of soil organisms that should display clear spatial patterns and processes. Mycorrhizal fungi are symbionts that obtain part or all of their carbon from living plant hosts. Furthermore, plant populations and communities that associate with such fungi have their own spatial structure and distributions. We should expect that this extrinsic force of aboveground spatial structure, in addition to spatial variation of soil properties (e.g., organic matter, moisture, pH, nutrient availability) that directly influence fungal growth, should result in distinct spatial organization of mycorrhizal fungi. Other intrinsic properties of mycorrhizal populations and communities such as varying modes of dispersal, differences in rates of growth and types of mycelia, and interactions among different individuals and species could also lead to spatial patterns of mycorrhizal fungi. Thus, soils represent heterogeneous environments where any differences in fundamental or realized niches among genetically different mycorrhizal fungi are expected to result in complex interactions among genetically different mycorrhizal fungi, their host species as well as intrinsic and extrinsic factors. Such interactions are expected to alter growth and local fitness of distinct mycorrhizal fungi thereby creating spatially structured populations and communities.

Unfortunately, many of our studies of mycorrhizal populations and communities (intentionally or not) ignore spatial structure. In field studies, spatial variability is often considered a sampling inconvenience when assessing responses of mycorrhizal fungi to experimental treatments (see also Chapter 11 by Pickles et al.). In greenhouse studies or laboratory studies where the functioning of mycorrhizal fungi is assessed, soils or inoculum are often thoroughly homogenized by mixing to limit spatial variation. Clearly, these steps are necessary to obtain clean data on basic properties and functions of mycorrhizal communities. But to achieve biological relevance in the field, we need to understand the spatial template on which mycorrhizal populations and communities interact with each other, their hosts, and their environment.

In this paper, we review our current understanding of the spatial ecology of mycorrhizal fungi, with a focus on the groups of fungi that form arbuscular mycorrhizal (AM) and ectomycorrhizal (EM) associations. We first establish the scales relevant to the functioning of mycorrhizal fungi. We then review potential mechanisms that could determine spatial structure of EM and AM fungal populations and communities, and highlight studies that have found spatial variation at various scales. We conclude with a proposal of promising future research directions, including how a spatial context for mycorrhizal associations could be useful in ecological restoration. This review is not meant to be comprehensive – there is still a considerable amount of research to be done before we can truly understand spatial dynamics of mycorrhizal fungal populations and communities. Our goal is to provide a spatial framework for considering the ecology and evolution of mycorrhizal populations and communities that may provoke thought, discussion, and further research – and that will push the study of mycorrhizal ecology and evolution towards greater relevance to the natural world.

12.2 Defining Relevant Spatial Scales for Mycorrhizal Fungi

With any discussion of spatial pattern and process, it is important to identify the spatial scales relevant to the organisms of interest. We believe four scales are most relevant for our discussion of spatial pattern and process of mycorrhizal populations and communities: (1) across landscapes, (2) within plant communities/ecosystems, (3) within an individual host root system, and (4) within an individual mycelium. Kilometers would be the physical distance for landscape scale patterns and processes, meters for the plant community or ecosystem scale, millimeters to centimeters for the host root scale, micrometers to centimeters for individual mycelia. In addition to identifying the appropriate scales, it is important to consider the potential mechanisms that might lead to spatial structure of mycorrhizal fungal populations and communities. Both intrinsic and extrinsic factors can lead to spatial patterns and processes in mycorrhizal communities. Intrinsic factors are inherent characteristics of the biology of the organisms (e.g., mode of dispersal), whereas extrinsic factors are properties of the environment inhabited by the organisms (e.g., soil moisture or pH).

At what spatial scale studies should be conducted will depend on the research questions. Any microevolutionary change occurs at the population level and will generally be investigated at smaller scales than studies aiming to describe and understand the mechanisms that shape communities or species distributions among geographic locations. Clearly, genetic change within any species results as the action of the evolutionary forces, mutation (and recombination), selection, genetic drift, and gene flow (migration among populations). These basic forces represent the underlying mechanisms that are responsible for the make-up of all populations. Their relative importance and role in structuring the spatial distribution of mycorrhizal fungi, however, is still poorly understood.

12.3 Spatial Ecology of Mycorrhizal Populations

12.3.1 Size and Spatial Distribution of Genetic Individuals

A basic tenet of the spatial ecology of any taxonomic group is an understanding of the size of individuals and their distribution over space. For mycorrhizal fungi, we could consider genetic individuals (genets) or physiological individuals (ramets). Because EM fungi often produce conspicuous fruiting bodies that are easy to map over space, many studies have employed molecular markers to determine the size and orientation of EM fungal genets in forests. Studies across a range of ecosystems and a variety of EM fungal species have shown that sizes of genets can range from 1 (Baar et al. 1994; Gryta et al. 1997; Gherbi et al. 1999) to 300 m^2 (Bonello et al. 1998). Intrinsic factors that influence genet size and spatial distribution include dispersal mode and frequency (e.g., asexual via hyphae or sexual via basidiospores) (Redecker et al. 2001), mycelial persistence (Guidot et al. 2004; Bergemann et al. 2006), and life history strategy (e.g., early versus late successional stage colonizers) (Deacon et al. 1983; Dahlberg and Stenlid 1990). Extrinsic factors such as small-scale disturbances and land-use history can also affect genet size and persistence, as was shown for *Hebeloma cylindrosporum*, where forest disturbance due to human activity negatively correlated with genet size (Guidot et al. 2002). For those EM fungi that do not produce conspicuous fruit bodies or do so infrequently, other sampling methods are necessary to determine the spatial distribution of genetic individuals. For example, Kretzer et al. (2004) sampled tuberculate mycorrhiza in addition to fruiting bodies for two co-occurring *Rhizopogon* species to determine their genet sizes and distributions.

In contrast, there are few studies to date that have measured the size of genetic individuals of AM fungi. In fact, the coenocytic hyphae and multinucleate spores of AM fungi complicate the study of "individuals" as findings suggest co-occurrence of genetically different nuclei within the same cytoplasm (Kuhn et al. 2001; Hijri and Sanders 2005). Investigation of spatial structure at this very lowest level within spores and mycelia is extremely challenging as it necessitates the development of nuclei-specific genetic markers. Working in an undisturbed coastal perennial grassland in Denmark, Rosendahl and Stukenbrock (2004) found 11 *Glomus* phylotypes that had a patchy distribution. Based on the distribution of phylotypes, the authors suggested the occurrence of a single *Glomus* individual with a mycelium covering at least 10 m in length, while the other less dominant phylotypes observed were smaller individual mycelia.

12.3.2 Spatial Distribution of Genetic Variation with Mycorrhizal Populations

In addition to understanding the distribution of individuals over space, it is useful to understand how genetic variation is spatially structured within mycorrhizal populations. Gene flow is an important intrinsic factor that affects how genetic variation is distributed over local spatial scales in mycorrhizal populations (Redecker et al. 2001). For mycorrhizal fungi, gene flow between individuals is influenced by modes and rates of propagule dispersal. Whether a species fruits aboveground (epigeous) or belowground (hypogeous) has a profound effect on the maximum distance between individuals at which gene flow can still occur. The epigeous species *Russula brevipes* exhibited strong genetic differentiation between populations located in the west coast of California and the Rocky mountains of Colorado 1,5002002 km apart (Bergemann and Miller), but not at smaller spatial scales (approximately 1 km apart) (Bergemann et al. 2006). In contrast, studies of hypogeous EM fungal species in the genera *Rhizopogon*, *Cenococcum*, and *Tuber* have shown much greater population differentiation at smaller spatial scales (LoBuglio and Taylor 2002; Murat et al. 2004; Grubisha et al. 2007). That difference between population genetic structure of epigeous and hypogeous fungal species reflects a contrast in dispersal modes, whereby epigeous fruiters producing airborne spores can disperse far greater distances than hypogeous species whose dispersal agents are often small mammals with much smaller home ranges (Maser et al. 1978; Meyer et al. 2005), but see Carriconde et al. (2008) for a recent example of strong genetic differentiation at fine-scales (<140m) in a wind-dispersed epigeous species.

In addition to effects on genet size, the degree to which species disperse as spores or conidia versus vegetative spread also influences the spatial distribution of genetic diversity within a population. For example, Kretzer et al. (2005) studied the population structure of two co-occurring congeneric EM fungi, *Rhizopogon vinicolor* and *R. vesiculosus*, in three 50 × 100 m forest plots in the pacific northwestern United States. Genets of *R. vesiculosus* were much more clustered within a plot than *R. vinicolor* genets, suggesting a much greater degree of clonal expansion in *R. vesiculosus*. Similarly, Gryta et al. (2006) studied co-occurring *Tricholoma populinum* and *T. scalpturatum* populations and also found differences in the spatial distribution and size of genets between taxa, revealing that one species, *T. scalpturatum*, relied more heavily upon spore dispersal for genet establishment than *T. populinum*, which exhibited greater vegetative expansion.

All mycorrhizal fungi "disperse," at least locally, through vegetative growth of hyphae, and this can result in the formation of large mycelial networks (Giovannetti et al. 2004). While many EM fungi do produce fruiting bodies, dispersal of AM fungi is likely to be more limited. Quantification of dispersal of different AM genotypes has not been achieved to date. Extrinsic factors such as variation in local environmental conditions over space, either biotic or abiotic, may create spatial structure in mycorrhizal populations by creating spatial variation in selection pressures on AM fungi, although data for this are lacking. Strong gradients in nutrient availability or other abiotic factors as well as the distribution of plant species might select for certain genotypes of mycorrhizal species in different soil patches thereby creating patchy mycorrhizal populations (Koch 2006; Croll et al. 2008).

Several studies have assessed the influence of extrinsic factors on the spatial structure of genetic variation in EM fungi populations. In a study of four EM fungal species along a zinc pollution gradient in Belgium, Colpaert et al. (2004) showed a greater number of zinc tolerant genotypes in plots less than 2 km from the pollution source than in plots situated 7–15 km away for *Suillus luteus*, *S. bovinus*, and *R. luteolus*. However, subsequent work by Muller et al. (2004) found high genetic diversity within polluted sites and high levels of gene flow between zinc polluted and unpolluted sites. These results suggest that zinc tolerant subpopulations in polluted plots were not established by a single founder event, and that admixture between zinc tolerant and zinc sensitive individuals between subpopulations separated by 5–15 km is common, a result further supported by the presence of zinc tolerant genotypes in non-polluted plots (Colpaert et al. 2004). Comparing genetic diversity of *T. populinum* and *T. scalpturatum* populations in 20 × 10 m forest plots subjected to frequent flooding or not, Gryta et al. (2006) discovered that the genetic diversity of both species was greater within the regularly flooded plot than in the undisturbed plot 50 km away, despite the intrinsic differences between these species in dispersal mode discussed in the previous section.

The studies that have measured genetic variation within populations of AM fungi have mostly focused on the scale of plant communities. In a 90 × 110 m agricultural plot in Switzerland, Koch et al. (2004) showed genetic and phenotypic variation in the common AM fungal species *Glomus intraradices*. There was no link between the genetic variation observed and tillage treatments in this field, but variation was observed between different plots distributed over space. Subsequent work showed that the different genotypes had varying effects on the growth of host plants (Koch et al. 2006), which suggests that there could be spatial variation in AM fungal function in the field associated with the spatial variation in the different genotypes. In another agricultural system, Stukenbrock and Rosendahl (2005) also found spatial structuring within fields of organic and conventional agriculture of several genotypes of *Glomus* spp. The factors that influence co-existence and the spatial distribution of different genotypes in natural communities have yet to be identified. It is possible that genotypes differ in their growth requirements and that complex genotype × environment interactions promote genetically diverse populations. Strong competition among closely related genotypes or species may also be a mechanism creating patchy distributions. AM fungal hyphae from the same spore or isolate frequently fuse to form new cytoplasmatic connections (Giovannetti et al. 1999, 2004; Avio et al. 2006), which may enhance local fitness and persistence of genotypes and species.

12.4 Scaling Spatial Ecology of Mycorrhizal Communities

Just as individual species of mycorrhizal fungi can be structured over space, multispecies communities of mycorrhizal fungi can also have a distinct spatial structure. We have organized our discussion of the spatial ecology of mycorrhizal communities by scale, from across landscapes to within communities to within host root systems.

12.4.1 Spatial Variation in Mycorrhizal Communities Across Landscapes

A variety of mechanisms could contribute to landscape-level spatial patterns of mycorrhizal communities. Clearly, there are many gradients in both abiotic and biotic factors that could influence the composition of mycorrhizal communities, including soil type and the type of plant community. In North American temperate forests with similar vegetation and climate types, the species composition of AM fungi was similar but relative abundance of different species was different, perhaps due to differences in local soil conditions (Klironomos et al. 1993). EM fungal communities have also been shown to be related to soil factors and dominant tree composition (Kernaghan et al. 2003).

Changes in biotic and abiotic factors as ecosystems develop may result in a heterogeneous distribution of AM fungi across sites of different successional ages. AM fungal spore data have given conflicting results with either no changes in AM fungal composition, changes only in the abundance of species, or differences in species diversity (Benjamin et al. 1989; Johnson et al. 1991; Koske and Gemma 1997). Along an old-field to forest chronosequence, AM fungal species abundances became more even, due to the reduction of spores of a single species in older sites (Johnson et al. 1991). In developing sand dune plantings, AM fungal species changed in both their presence and abundance (Koske and Gemma 1997); however, these differences may be due to the import of AM fungi on planting stock or differential sporulation. In both cases, certain AM fungal species appeared to be early successional while others dominated later in succession, and this variation was tied to host plants and soil nutrients. Recent molecular methods indicate that spore data may not give a complete picture of variation in AM fungal diversity over sites of different age; however, spatial differences were still present in AM fungal communities across a primary volcanic succession at Mt. Fuji (Wu et al. 2007). Simultaneous changes in plants, environment, and AM fungi during ecosystem succession make it difficult to isolate cause and effect from these surveys. Human changes in land use for agriculture, industry, and residential development provide “natural” experiments that shift specific factors such as plant and habitat diversity which determine local AM fungal community composition (Cousins et al. 2003; Li et al. 2007).

Large-scale anthropogenic or natural disturbances also play a role in the spatial structure of mycorrhizal communities across landscapes. Many studies of EM fungi have shown that logging activities can alter abundance and community composition (Jones et al. 2003). For example, Twieg et al. (2007) sampled the EM communities from 5-, 26-, 65- and 100-year-old Douglas fir (*Pseudotsuga menziesii*) and paper birch (*Betula papyrifera*) forest stands recovering from either stand replacing forest fire, or clearcut logging (see also Chapter 13 by Simard). EM fungal diversity was significantly lower in all 5-year-old Douglas fir stands than in the older age class stands, but no differences in diversity were detected between different aged birch stands, likely due to the ability of paper birch to stump resprout. EM fungal community composition also differed among different aged stands. For example, the frequencies of *Russula* and *Piloderma* species increased in Douglas fir plots with time since disturbance, while other species, such as *Rhizopogon* species were more prevalent in the 5-year-old burned or logged plots. Since replicate stands were a minimum of 800 m apart, these results illustrate how disturbances such as clearcutting and fire can generate a mosaic of spatially-structured EM fungal communities across the landscape, over which additional factors, such as the distribution of host tree species or distance from remaining adult trees (Cline et al. 2005), can create additional, fine-scale spatial patterns.

In tropical forests in Panama, Mangan et al. (2004) suggested that the diversity of AM fungi in these systems could be influenced by forest fragmentation. Fragmentation of these forest systems occurred after major flooding around hilltop forests after the construction of the Panama Canal. AM fungal communities on the mainland sites in this study were more similar to each other over a distance of greater than 5 km than to sites on a nearby island that was less than 1 km away despite minimal differences across sites in topography and soil characteristics.

The composition of the regional species pool is an intrinsic property of mycorrhizal communities that could also explain landscape level spatial variation in abundance and composition. Lekberg et al. (2007) recently addressed this process with a study of 10 fields located 1 to 25 Km from each other in Zimbabwe. They found that although local soil characteristics explain some of the regional variation in species composition of AM fungi in the different fields, distance between sites also explained a significant portion of dissimilarity between sites, potentially due to dispersal limitation.

12.4.2 Spatial Variation in Mycorrhizal Communities Within Plant Communities

Nested within larger landscape-scale variation in mycorrhizal communities is variation within plant communities. Most of the work on spatial heterogeneity of mycorrhizal communities has focused on this spatial scale. One extrinsic factor that could lead to spatial structure in mycorrhizal communities is the spatial distribution of compatible plant hosts within a plant community. Numerous studies have shown that both EM and AM fungi show varying levels of host-specificity or host preference for different plant species (Molina and Trappe 1982; Bever et al. 1996; Massicotte et al. 1999; Vandenkoornhuyse et al. 2003). As the structure of aboveground plant communities can be spatially structured (Miller et al. 2002; Seabloom et al. 2005), the spatial distribution of individual plants aboveground could lead to small-scale spatial structure of mycorrhizal communities belowground.

At a coarser scale of resolution, the local diversity of the plant community may influence the composition of the mycorrhizal communities. Experimental studies have shown that plant community composition can affect the composition of mycorrhizal communities at a m^2 scale (Burrows and Pfleger 2002), so variation in plant community composition within a site could drive composition of the mycorrhizal community. In an old-field ecosystem in North Carolina, Schultz (1996) found that distinct morphospecies of AM fungi had different patch sizes throughout the site, and the species richness of the AM fungal community was positively associated with plant species richness. Landis et al. (2004) also found that AM fungal richness was positively correlated with plant species richness and soil N content in oak savannas, while Pringle and Bever (2002) found high spatial heterogeneity in the community composition of AM fungi at a scale of several meters at this same site. Boerner et al. (1996) found that the abundance of AM fungal propagules in a series of plant communities was spatially heterogeneous within several meters, and that patches of low abundance of AM fungi at some sites were associated with high densities of non-mycorrhizal plant species.

Lilleskov et al. (2004) examined EM fungal communities from eight forest stands and found that the EM fungal communities exhibited a high degree of fine-scale spatial heterogeneity, with the greatest degree of spatial autocorrelation detected at distances 2.6 m apart. In an old-growth conifer forest in California, Izzo et al. (2005) examined how the spatial structure of EM fungal communities changed over time. Soil samples collected over three successive years also showed higher EM fungal community similarity at spatial scales <4 m apart. At even finer spatial scales (<20 cm), the EM fungal community was temporally dynamic and similarity indices at this spatial scale were lower than that exhibited at the overall plot level, suggesting a high degree of species turnover possibly due to root senescence. In a study of EM fungal communities associated with eight co-occurring tree species in Japanese mixed hardwood–conifer forests, Ishida et al. (2007) found that many EM fungal species exhibited a high degree of host specificity, and that the degree of host overlap was positively correlated with phylogenetic relatedness between tree species. Although this study did not examine these results in a spatially-explicit context, the spatial distribution of potential host tree species will likely determine the spatial distribution of host specific EM fungi, as well as help to maintain high levels of EM fungal community diversity.

In addition to spatial variation in the local plant community, the soil environment can also be spatially variable in abiotic properties within a plant community at relatively small spatial scales (Robertson et al. 1997; Boerner et al. 1998). This small-scale environmental heterogeneity might also lead to the spatial structuring of mycorrhizal communities. In a study of maquis and salt marsh plant communities in Portugal, the spatial distribution of spores of AM fungi was closely linked with environmental variables and proximity to individual plants (Carvalho et al. 2003). Klironomos et al. (1999) also found similar patchiness of AM fungal spores in relation to the dominant shrub present in a southern California chaparral ecosystem. Few other studies have attempted to link spatial patterns of mycorrhizal fungi in the field with local environmental conditions, making it difficult to know the relative contribution of environment or plant on the spatial structure of mycorrhizal communities. A study by Toljander et al. (2006) examined the EM fungal community along a 90-m natural nutrient concentration gradient in northern Sweden, and found both EM fungal species richness increased and community structure varied along the gradient. Although host tree species was not responsible for this pattern, the high degree of spatial autocorrelation between soil N content and base cation concentration along this gradient makes it difficult to discern which of the soil chemical characteristics contributes most significantly to the observed community changes. In addition to differences in soil chemical properties, preference for specific substrates within the soil can also lead to spatial partitioning in EM fungal communities. Tedersoo et al. (2003) examined the spatial distribution of EM fungal species in soil and woody debris, and discovered that EM fungal species in the Thelephoroid clade, Athelioid clade, and the Sebacinales, showed a greater affinity for coarse woody substrates, whereas other EM fungi in the Agaricales and the Ascomycota were more abundant in the mineral soil.

Another potential mechanism leading to patchy distributions of mycorrhizal communities over space is the dispersal of mycorrhizal propagules. A variety of organisms have been shown to move viable spores of mycorrhizal fungi at scales ranging from cm to km (e.g., Maser et al. 1978; Meyer et al. 2005; Ashkannejhad and Horton 2006). Differential dispersal of mycorrhizal species by dispersal agents may lead to patchiness of species distributions in the field as dispersed propagules establish and spread. The relative abundance of dispersal agents should also affect the spatial heterogeneity of mycorrhizal fungi within an ecosystem. For example, in soils with a high abundance of earthworms and other macrofauna that may consume propagules of AM fungi, increased dispersal are expected to reduce patchiness within AM fungal communities. Although many dispersal agents have been identified, their exact role in contributing to the spatial structure of mycorrhizal communities remains unclear.

12.4.3 Spatial Variation in Mycorrhizal Communities Within Root Systems

Just as there is variation in mycorrhizal community structure between neighboring plant root systems, there can be spatial variation in the abundance and composition of mycorrhizal communities within the root systems of individual plants. This variation can occur in the rhizosphere, where hyphae and spores extend into the soil immediately adjacent to root systems, or can occur where the fungi colonize the roots of the host plants.

At the scale of individual roots, it would be expected that one of the major driving factors influencing the spatial distribution of mycorrhizal fungi would be interactions within and among mycorrhizal species. For both EM and AM fungi, there is evidence that, when the fungi are colonizing host plant root systems, competitive (Pearson et al. 1993; Kennedy et al. 2007) or facilitative (van Tuinen et al. 1998) interactions can occur between different mycorrhizal species. Just as interactions between species can create spatial structure in other communities (Seabloom et al. 2005), these interactions might lead to the formation of spatial patterns within roots. Spatial structure of mycorrhizal fungi within a root system may increase the total diversity of mycorrhizal fungal species within a root as species are spatially segregated and competition is minimized. To date, the role of this mechanism in creating within root spatial patterns of mycorrhizal fungi has not been explicitly tested. Recent work by Maherali and Klironomos (2007) showed that the number of co-existing AM fungal species grown on individual *Plantago lanceolata* also depends on the initial phylogenetic composition of the fungi. Their results suggest that conservation of similar traits among closely related AM species acts as a mechanism that promotes co-existence of phylogenetically diverse communities due to functional complementarity among taxa.

Alteration of mycorrhizal communities by neighboring plants could be an extrinsic factor that could lead to spatial variation in mycorrhizal abundance and composition within a root system. Using terminal restriction fragment length polymorphism (T-RFLP), Mummey et al. (2005) showed that AM fungal communities within the roots of the grass species *Dactylis glomerata* that had been growing next to the invasive exotic plant *Centaurea maculosa* were different in composition compared to roots growing without neighboring *C. maculosa* roots. Although the exact mechanisms to explain this observation remain untested, alteration of the local rooting environment through resource uptake or exudation of secondary compound by the roots of one species could change the environment surrounding a root and could select for certain AM fungal species.

Differences in the physiology or function of different types of roots within a root system may also play a role in fine-scale spatial structure of mycorrhizal communities. This phenomenon has been recognized for several decades in EM fungal communities, where late-stage fungi are found on roots closest to the trunk of the tree and early stage fungi are found on roots furthest from the base of the tree (Ford et al. 1980; Mason et al. 1982). This pattern has also been observed in AM fungal systems. In a study examining the molecular diversity of AMF in three nitrogen-fixing forbs, different AM fungal communities were present in root nodules than in roots without root nodules (Scheublin et al. 2004). The authors suggested that different AM fungi may have preferences for these spatially segregated locations within the root system because of variation in the availability of nitrogen or interactions between AM fungi and *Rhizobium* spp. within the root system.

12.4.4 Vertical Spatial Variation in Mycorrhizal Communities

Although the focus of this review so far has been on horizontal spatial variation in mycorrhizal communities, some studies have documented vertical spatial variation in both AM and EM fungal communities through soil profiles. Oehl et al. (2005) described the vertical distribution of AM fungi in various managed systems in the Upper Rhine Valley. In addition to finding a general decline in the species richness of AM fungi moving down the soil profile, these authors found that different AM fungi had different rates of sporulation at the various depths within the soil profiles of the sampled sites, with some species being only found in lower portions of the profiles. Similar patterns have been observed for EM fungal communities. Dickie et al. (2002) found that different EM fungal species showed preferences for different soil profile layers and that EM fungal species richness decreased with increasing soil depth. Mycorrhizas and their associated extraradical mycelium are not always found in the same soil profile (Genney et al. 2006). Furthermore, Moyersoen et al. (1998) did not find a vertical separation of AM and EM in a rainforest ecosystem, although such a vertical trend was found in the roots of the dual mycorrhizal tree, *Populus tremuloides* (Neville et al. 2002). Lindahl et al. (2007) also found a similar vertical partitioning between EM and saprotrophic fungi in the soil horizons. Vertical niche partitioning is thought to be one way by which the high species diversity of mycorrhizal fungi can be maintained at small spatial scales (Bruns 1995).

12.5 Knowledge Gaps and Future Directions

12.5.1 Linking Mechanisms with Patterns

At a variety of spatial scales, many mechanisms have been proposed as determinants of spatial structure in mycorrhizal populations and communities, but few direct links have been made between observed spatial patterns in the field and specific mechanisms. Probably the best way to attempt to make these links would be to complement observations in the field with experimental manipulations. For example, some authors have suggested that dispersal limitation of mycorrhizal fungi might lead to the spatial structure of mycorrhizal communities in the field (Mangan et al. 2004). Following the approach that plant community ecologists have used (Foster 2001), one could first map out the mycorrhizal community composition of a particular site, locate areas of low mycorrhizal fungal species richness, and add spores of various species to these sites. Follow-up studies could determine if the addition of mycorrhizal propagules increases the richness of the mycorrhizal community, or if other local factors such as local environment or host plant availability influence mycorrhizal community composition in low richness patches.

Other mechanistic explanations for creation of spatial structure of mycorrhizal fungi in the field rely on differential dispersal and spread of mycorrhizal fungi in the field. Although many dispersal agents of mycorrhizal propagules have been identified, few studies to date have successfully tracked the establishment and spread of mycorrhizal propagules within a site in relation to dispersal agents. Several studies have monitored the spread and establishment of EM fungi that have been intentionally inoculated into a site (Schwartz et al. 2006), but few studies have examined the process of natural establishment of mycorrhizal propagules. By continuing to develop molecular markers to track specific isolates of mycorrhizal fungi in the field, especially for AM fungi, we might be able to better understand the movement, growth, and interactions of mycorrhizal in natural settings. This approach might be especially important considering the widespread and unregulated transport of mycorrhizal inoculum throughout the world for commercial purposes. Because such practices have potentially positive or negative consequences (Schwartz et al. 2006), any intentional movement of non-native microbes should be considered with more care than is currently the case.

As with many aspects of the mycorrhizal symbiosis, it remains unclear whether the plants or the fungi are in control of the observed spatial structure of plant–mycorrhizal interactions or if there are processes that lead to feedbacks between the plants and fungi. Can the spatial patterns of mycorrhizal fungi influence the spatial patterns and processes in aboveground plant communities, or are the plant communities driving the spatial patterns in mycorrhizal communities? Do the processes that influence the spatial structure of plant communities happen at the same scale as processes that influence the spatial structure of mycorrhizal communities?

For the most part in our discussion, mycorrhizal "species" have been treated as if they are clearly defined entities. For EM fungi, their biology and genetics are generally better understood than for AM fungi. The comparatively low number of known AM fungal "species" (relative to that of known EM fungal species) is likely a gross underestimation (see also Chapter 10 by Morton). Even though most AM fungal species appear to be generalists, it is still unclear whether AM fungal individuals are generalists, as opposed to assemblies of more specialized genotypes. How are intraspecific functional and genetic diversity (Koch et al. 2004; Munkvold et al. 2004) linked and at what spatial scale? More work is needed to obtain a better understanding of the genetics of AM fungi, and to address at what spatial scale genetic diversity or species/taxa identity is relevant for the survival of AM individuals and their function and distribution across terrestrial ecosystems. In plant ecology, there are studies that have assessed the effect of individual plant species on the local community diversity (e.g., Wiegand et al. 2007). This approach may also be applicable for mycorrhizal fungi and could provide an alternative way to studying spatial patterns in natural communities.

12.5.2 Linking Spatial Patterns with Functional Significance

Although many studies have measured some form of spatial heterogeneity of mycorrhizal fungi in the field, there have been few attempts to assess whether spatial variation in composition of mycorrhizal communities translates into spatial variation in the function of mycorrhizal fungi. Spatial structure in mycorrhizal populations and communities could have important functional significance from a variety of perspectives. From a fungal biology/ecology perspective, understanding the patterns of spatial segregation between mycorrhizal species could help understand interactions between mycorrhizal fungi and controls of mycorrhizal diversity at all the spatial scales considered in this review. From a plant and ecosystem ecology perspective, considering the spatial structure of mycorrhizal fungi in the field could help further clarify the potential role of mycorrhizal fungi in aboveground and belowground properties and processes. Spatial variation in the function of mycorrhizal fungi in the field has been repeatedly proposed to have important community and ecosystem effects (Streitwolf-Engel et al. 2001; Lovelock and Miller 2002; Thiet and Boerner 2007), yet attempts to measure this functional variation are lacking potentially as a result of the methodological limitations of measuring mycorrhizal function in the field (Read 2002).

One potential method for measuring functional variation of mycorrhizal communities in the field is through rotated cores (Johnson et al. 2001). These cores provide a method of severing experimental plants from the hyphal network of mycorrhizal fungi within a plant community from which the potential function of the mycorrhizal community can be inferred by comparing a rotated core to a non-rotated core. By placing rotated cores into the field in a spatially explicit manner, it might be possible to estimate small-scale spatial variation in AM fungal function for individual plants or for ecosystem processes such as nutrient uptake and decomposition.

Another emerging method for addressing functional variation in mycorrhizal communities over space involves using molecular approaches to assess spatial patterns of expression of functional genes. For example, Luis et al. (2005) used primers that targeted laccase genes in basidiomycetes and showed that most dominant laccase sequences were probably associated with extraradical hyphae of EM fungi in small patches. At an even smaller scale of within a mycelium, work in mesocosms has also found spatial variation in gene expression in *Paxillus involutus* using cDNA microarrays (Wright et al. 2005).

It is also possible to investigate spatial patterns by making use of frequently occurring natural gradients in nature. Establishing pure fungal cultures of specific target species or taxa along such gradients and the analysis of their functional and genetic diversity might provide exciting insight into how species and communities are formed. Transplant experiments and subsequent genetic analyses of mycorrhizal communities in such systems may be a way to investigate to what degree mycorrhizal fungi are locally adapted.

12.5.3 Development of Spatially-explicit Models

Recent efforts have been made to model the interactions between mycorrhizal fungi and plants using theoretical approaches. For example, recent studies have modeled mechanisms leading to the coexistence of both plants and fungi (Johnson et al. 2006). While these models have provided useful frameworks for considering the dynamics of mycorrhizal interactions, they have generally not explicitly incorporated spatial structure. Given that a spatially-explicit approach can influence the outcome of models (Tilman and Kareiva 1997), attempts to parameterize these models with the empirical work discussed above could potentially improve their ability to predict outcomes of ecological interactions of mycorrhizal fungi. Coupling empirical estimates of spatial variation of plants and mycorrhizal fungi with theoretical models exploring the significance of this spatial heterogeneity may further advance our understanding of the ecology of mycorrhizal–plant interactions.

12.6 Conclusion

Mycorrhizal associations are, by their nature, multispecies assemblages. In addition, the associated fungi are biotrophs (many are obligate biotrophs), and they inhabit the cryptic soil environment, making it very challenging to study their ecology, even when we ignore the effects of space. Nonetheless, it will not be possible to understand the structure and dynamics of individual species and of multispecies communities unless we do take a spatial approach. This review illustrates that mycorrhizal ecologists have recognized the importance of understanding the effects of spatially-explicit processes on mycorrhizal populations and communities. However, it is also clear that this area of ecology is at its infancy, and there is much work yet to be done.

References

Ashkannejhad S, Horton TR (2006) Ectomycorrhizal ecology under primary succession on coastal sand dunes: interactions involving *Pinus contorta*, suilloid fungi and deer. New Phytol 169:345–354

Avio L, Pellegrino E, Bonari E, Giovannetti M (2006) Functional diversity of arbuscular mycorrhizal fungal isolates in relation to extraradical mycelial networks. New Phytol 172:347–357

Baar J, Ozinga WA, Kuyper TW (1994) Spatial distribution of *Laccaria bicolor* genets reflected by sporocarps after removal of litter and humus layers in a *Pinus sylvestris* forest. Mycol Res 98:726–728

Benjamin PK, Anderson RC, Liberta AE (1989) Vesicular arbuscular mycorrhizal ecology of little bluestem across a prairie forest gradient. Can J Bot 67:2678–2685

Bergemann SE, Miller SL (2002) Size, distribution, and persistence of genets in local populations of the late-stage ectomycorrhizal basidiomycete, *Russula brevipes*. New Phytol 156:313–320

Bergemann SE, Douhan GW, Garbelotto M, Miller SL (2006) No evidence of population structure across three isolated subpopulations of *Russula brevipes* in an oak/pine woodland. New Phytol 170:177–184

Bever JD, Morton JB, Antonovics J, Schultz PA (1996) Host-dependent sporulation and species diversity of arbuscular mycorrhizal fungi in a mown grassland. J Ecol 84:71–82

Boerner REJ, DeMars BG, Leicht PN (1996) Spatial patterns of mycorrhizal infectiveness of soils long a successional chronosequence. Mycorrhiza 6:79–90

Boerner REJ, Scherzer AJ, Brinkman JA (1998) Spatial patterns of inorganic N, P availability, and organic C in relation to soil disturbance: a chronosequence analysis. Appl Soil Ecol 7:159–177

Bonello P, Bruns TD, Gardes M (1998) Genetic structure of a natural population of the ectomycorrhizal fungus *Suillus pungens*. New Phytol 138:533–542

Bruns TD (1995) Thoughts on the processes that maintain local species-diversity of ectomycorrhizal fungi. Plant Soil 170:63–73

Burrows RL, Pfleger FL (2002) Arbuscular mycorrhizal fungi respond to increasing plant diversity. Can J Bot 80:120–130

Campbell DR, Dooley JL (1992) The spatial scale of genetic differentiation in a hummingbird-pollinated plant – Comparison with models of isolation by distance. Am Nat 139:735–748

Carriconde F, Gryta H, Jargeat P, Mouhamadou B, Gardes M (2008) High sexual reproduction and limited contemporary dispersal in the ectomycorrhizal fungus *Tricholoma scalpturatum*: new insights from population genetics and spatial autocorrelation analysis. Mol Ecol 17: 4433–4445

Carvalho LM, Correia PM, Ryel RJ, Martins-Loução MA (2003) Spatial variability of arbuscular mycorrhizal fungal spores in two natural plant communities. Plant Soil 251:227–236

Cline ET, Ammirati JF, Edmonds RL (2005) Does proximity to mature trees influence ectomycorrhizal fungus communities of Douglas-fir seedlings. New Phytol 166:993–1009

Colpaert JV, Muller LAH, Lambaerts M, Adriaensen K, Vangronsveld J (2004) Evolutionary adaptation to Zn toxicity in populations of Suilloid fungi. New Phytol 162:549–559

Cousins JR, Hope D, Gries C, Stutz JC (2003) Preliminary assessment of arbuscular mycorrhizal fungal diversity and community structure in an urban ecosystem. Mycorrhiza 13:319–326

Croll D, Wille L, Gamper HA, Mathimaran N, Lammers PJ, Corradi N, Sanders IR (2008) Genetic diversity and host plant preferences revealed by simple sequence repeat and mitochondrial

markers in a population of the arbuscular mycorrhizal fungus *Glomus intraradices*. New Phytol 178: 672–687.
Dahlberg A, Stenlid J (1990) Population-structure and dynamics in *Suillus bovinus* as indicated by spatial distribution of fungal clones. New Phytol 115:487–493
Deacon JW, Donaldson SJ, Last FT (1983) Sequences and interactions of mycorrhizal fungi on birch. Plant and Soil 71: 257–262.
Dickie IA, Xu B, Koide RT (2002) Vertical niche differentiation of ectomycorrhizal hyphae in soil as shown by T-RFLP analysis. New Phytol 156:527–535
Ettema CH, Wardle DA (2002) Spatial soil ecology. Trends Ecol Evol 17:177–183
Ford ED, Mason PA, Pelham J (1980) Spatial patterns of sporophore distribution around a young birch tree in three successive years. Trans Br Mycol Soc 75:287–296
Foster BL (2001) Constraints on colonization and species richness along a grassland productivity gradient: the role of propagule availability. Ecol Lett 4:530–535
Genney DR, Anderson IC, Alexander IJ (2006) Fine-scale distribution of pine ectomycorrhizas and their extramatrical mycelium. New Phytol 170:381–390
Gherbi H, Delaruelle C, Selosse MA, Martin F (1999) High genetic diversity in a population of the ectomycorrhizal basidiomycete *Laccaria amethystina* in a 150-year-old beech forest. Mol Ecol 8:2003–2013
Giovannetti M, Azzolini D, Citernesi AS (1999) Anastomosis formation and nuclear and protoplasmic exchange in arbuscular mycorrhizal fungi. App Environ Microbiol 65:5571–5575
Giovannetti M, Sbrana C, Avio L, Strani P (2004) Patterns of below-ground plant interconnections established by means of arbuscular mycorrhizal networks. New Phytol 164:175–181
Grubisha LC, Bergemann SE, Bruns TD (2007) Host islands within the California Northern Channel Islands create fine-scale genetic structure in two sympatric species of the symbiotic ectomycorrhizal fungus *Rhizopogon*. Mol Ecol 16:1811–1822
Gryta H, Debaud JC, Effosse A, Gay G, Marmeisse R (1997) Fine-scale structure of populations of the ectomycorrhizal fungus *Hebeloma cylindrosporum* in coastal sand dune forest ecosystems. Mol Ecol 6:353–364
Gryta H, Carriconde F, Charcosset JY, Jargeat P, Gardes M (2006) Population dynamics of the ectomycorrhizal fungal species *Tricholoma populinum* and *Tricholoma scalpturatum* associated with black poplar under differing environmental conditions. Environ Microbiol 8:773–786
Guidot A, Gryta H, Gourbiere F, Debaud JC, Marmeisse R (2002) Forest habitat characteristics affect balance between sexual reproduction and clonal propagation of the ectomycorrhizal mushroom *Hebeloma cylindrosporum*. Oikos 99:25–36
Guidot A, Debaud JC, Effosse A, Marmeisse R (2004) Below-ground distribution and persistence of an ectomycorrhizal fungus. New Phytol 161:539–547
Hijri M, Sanders IR (2005) Low gene copy number shows that arbuscular mycorrhizal fungi inherit genetically different nuclei. Nature 433:160–163
Ishida TA, Nara K, Hogetsu T (2007) Host effects on ectomycorrhizal fungal communities: insight from eight host species in mixed conifer-broadleaf forests. New Phytol 174:430–440
Izzo A, Agbowo J, Bruns TD (2005) Detection of plot-level changes in ectomycorrhizal communities across years in an old-growth mixed-conifer forest. New Phytol 166:619–630
Johnson D, Leake JR, Read DJ (2001) Novel in-growth core system enables functional studies of grassland mycorrhizal mycelial networks. New Phytol 152:555–562
Johnson NC, Zak DR, Tilman D, Pfleger FL (1991) Dynamics of vesicular-arbuscular mycorrhizae during old-field succession. Oecologia 86:349–358
Johnson NC, Hoeksema JD, Bever JD, et-al. (2006) From Lilliput to Brobdingnag: extending models of mycorrhizal function across scales. BioScience 56:889–900
Jones MD, Durall DM, Cairney JWG (2003) Ectomycorrhizal fungal communities in young forest stands regenerating after clearcut logging. New Phytol 157:399–422
Kennedy PG, Hortal S, Bergemann SE, Bruns TD (2007) Competitive interactions among three ectomycorrhizal fungi and their relation to host plant performance. J Ecol 95:1338–1345
Kernaghan G, Sigler L, Khasa D (2003) Mycorrhizal and root endophytic fungi of containerized *Picea glauca* seedlings assessed by rDNA sequence analysis. Microb Ecol 45:128–136

Klironomos JN, Moutoglis P, Kendrick B, Widden P (1993) A comparison of spatial heterogeneity of vesicular-arbuscular mycorrhizal fungi in two maple-forest soils. Can J Bot 71:1472–1480
Klironomos JN, Rillig MC, Allen MF (1999) Designing below-ground field experiments with the help of semi-variance and power analyses. Appl Soil Ecol 12:227–238
Koch AM (2006) Linking the population genetics and ecology of arbuscular mycorrhizal fungi. PhD thesis, Lausanne, Switzerland
Koch AM, Kuhn G, Fontanillas P, Fumagalli L, Goudet J, Sanders IR (2004) High genetic variability and low local diversity in a population of arbuscular mycorrhizal fungi. Proc Natl Acad Sci USA 101:2369–2374
Koch AM, Croll D, Sanders IR (2006) Genetic variability in a population of arbuscular mycorrhizal fungi causes variation in plant growth. Ecol Lett 9:103–110
Koske RE, Gemma JN (1997) Mycorrhizae and succession in plantings of beachgrass in sand dunes. Am J Bot 84:118–130
Kretzer AM, Dunham S, Molina R, Spatafora JW (2004) Microsatellite markers reveal the below ground distribution of genets in two species of *Rhizopogon* forming tuberculate ectomycorrhizas on Douglas fir. New Phytol 161:313–320
Kretzer AM, Dunham S, Molina R, Spatafora JW (2005) Patterns of vegetative growth and gene flow in *Rhizopogon vinicolor* and *R. vesiculosus* (Boletales, Basidiomycota). Mol Ecol 14:2259–2268
Kuhn G, Hijri M, Sanders IR (2001) Evidence for the evolution of multiple genomes in arbuscular mycorrhizal fungi. Nature 414:745–748
Landis FC, Gargas A, Givnish TJ (2004) Relationships among arbuscular mycorrhizal fungi, vascular plants and environmental conditions in oak savannas. New Phytol 164:493–504
Lekberg Y, Koide RT, Rohr JR, Aldrich-Wolfe L, Morton JB (2007) Role of niche restrictions and dispersal in the composition of arbuscular mycorrhizal fungal communities. J Ecol 95:95–105
Li LF, Li T, Zhao ZW (2007) Differences of arbuscular mycorrhizal fungal diversity and community between a cultivated land, an old field, and a never-cultivated field in a hot and arid ecosystem of southwest China. Mycorrhiza 17:655–665
Lilleskov EA, Bruns TD, Horton TR, Taylor DL, Grogan P (2004) Detection of forest stand-level spatial structure in ectomycorrhizal fungal communities. FEMS Microb Ecol 49:319–332
Lindahl BD, Ihrmark K, Boberg J, Trumbore SE, Hogberg P, Stenlid J, Finlay RD (2007) Spatial separation of litter decomposition and mycorrhizal nitrogen uptake in a boreal forest. New Phytol 173:611–620
LoBuglio KF, Taylor JW (2002) Recombination and genetic differentiation in the mycorrhizal fungus *Cenococcum geophilum* Fr. Mycologia 94:772–780
Lovelock CE, Miller R (2002) Heterogeneity in inoculum potential and effectiveness of arbuscular mycorrhizal fungi. Ecology 83:823–832
Luis P, Kellner H, Zimdars B, Langer U, Martin F, Buscot F (2005) Patchiness and spatial distribution of laccase genes of ectomycorrhizal, saprotrophic, and unknown basidiomycetes in the upper horizons of a mixed forest cambisol. Microb Ecol 50:570–579
Maestre FT, Escudero A, Martinez I, Guerrero C, Rubio A (2005) Does spatial pattern matter to ecosystem functioning? Insights from biological soil crusts. Funct Ecol 19:566–573
Maherali H, Klironomos JN (2007) Influence of phylogeny on fungal community assembly and ecosystem functioning. Science 316:1746–1748
Mangan SA, Eom AH, Adler GH, Yavitt JB, Herre EA (2004) Diversity of arbuscular mycorrhizal fungi across a fragmented forest in Panama: insular spore communities differ from mainland communities. Oecologia 141:687–700
Maser C, Trappe JM, Nussbaum RA (1978) Fungal-small mammal interrelationships with emphasis on Oregon coniferous forests. Ecology 59:799–809
Mason PA, Last FT, Pelham J, Ingleby K (1982) Ecology of some fungi associated with an ageing stand of birches (*Betula pendula* and *B. pubescens*). For Ecol Manage 4:19–39
Massicotte HB, Molina R, Tackaberry LE, Smith JE, Amaranthus MP (1999) Diversity and host specificity of ectomycorrhizal fungi retrieved from three adjacent forest sites by five host species. Can J Bot 77:1053–1076

Meyer MD, North MP, Kelt DA (2005) Fungi in the diets of northern flying squirrels and lodgepole chipmunks in the Sierra Nevada. Can J Zool 83:1581–1589
Miller TF, Mladenoff DJ, Clayton MK (2002) Old-growth northern hardwood forests: spatial autocorrelation and patterns of understory vegetation. Ecol Monogr 72:487–503
Molina R, Trappe JM (1982) Lack of mycorrhizal specificity by the ericaceous host *Arbutus menziesii* and *Arctostaphylos uva-ursi*. New Phytol 90:495–509
Moyersoen B, Fitter AH, Alexander IJ (1998) Spatial distribution of ectomycorrhizas and arbuscular mycorrhizas in Korup National Park rain forest, Cameroon, in relation to edaphic parameters. New Phytol 139:311–320
Muller LAH, Lambaerts M, Vangronsveld J, Colpaert JV (2004) AFLP-based assessment of the effects of environmental heavy metal pollution on the genetic structure of pioneer populations of *Suillus luteus*. New Phytol 164:297–303
Mummey DL, Rillig MC, Holben WE (2005) Neighboring plant influences on arbuscular mycorrhizal fungal community composition as assessed by T-RFLP analysis. Plant Soil 271:83–90
Munkvold L, Kjoller R, Vestberg M, Rosendahl S, Jakobsen I (2004) High functional diversity within species of arbuscular mycorrhizal fungi. New Phytol 164:357–364
Murat C, Diez J, Luis P, Delaruelle C, Dupre C, Chevalier G, Bonfante P, Martin F (2004) Polymorphism at the ribosomal DNA ITS and its relation to postglacial re-colonization routes of the Perigord truffle *Tuber melanosporum*. New Phytol 164:401–411
Neville J, Tessier JL, Morrison I, Scarratt J, Canning B, Klironomos JN (2002) Soil depth distribution of ecto- and arbuscular mycorrhizal fungi associated with *Populus tremuloides* within a 3-year-old boreal forest clear-cut. Appl Soil Ecol 19:209–216
Oehl F, Sieverding E, Ineichen K, Ris E-A, Boller T, Wiemken A (2005) Community structure of arbuscular mycorrhizal fungi at different soil depths in extensively and intensively managed agroecosystems. New Phytol 165:273–283
Pearson JN, Abbott LK, Jasper DA (1993) Mediation of competition between 2 colonizing VA mycorrhizal fungi by the host plant. New Phytol 123:93–98
Pringle A, Bever JD (2002) Divergent phenologies may facilitate the coexistence of arbuscular mycorrhizal fungi in a North Carolina grassland. Am J Bot 89:1439–1446
Read (2002) Towards ecological relevance – progress and pitfalls in the path towards an understanding of mycorrhizal functions in nature. In: van der Heijden MGA, Sanders I (Eds.) Mycorrhizal ecology. Ecological studies, Vol. 157. Springer, Berlin, pp 3–29
Redecker D, Szaro TM, Bowman RJ, Bruns TD (2001) Small genets of *Lactarius xanthogalactus*, *Russula cremoricolor* and *Amanita francheti* in late-stage ectomycorrhizal successions. Mol Ecol 10:1025–1034
Robertson GP, Klingensmith KM, Klug MJ, Paul EA, Crum JR, Ellis BG (1997) Soil resources, microbial activity, and primary production across an agricultural ecosystem. Ecol Applic 7:158–170
Rosendahl S, Stukenbrock EH (2004) Community structure of arbuscular mycorrhizal fungi in undisturbed vegetation revealed by analyses of LSU rDNA sequences. Mol Ecol 13:3179–3186
Scheublin TR, Ridgway KP, Young JPW, van der Heijden MGA (2004) Nonlegumes, legumes, and root nodules harbor different arbuscular mycorrhizal fungal communities. App Env Microbiol 70:6240–6246
Schultz PA (1996) "Arbuscular mycorrhizal species diversity and distribution in an old field community" (PhD thesis), Duke University, Durham
Schwartz MW, Hoeksema JD, Gehring CA, Johnson NC, Klironomos JN, Abbott LK, Pringle A (2006) The promise and potential consequences of the global transport of mycorrhizal fungal inoculum. Ecol Lett 9:501–515
Schweizer M, Excoffier L, Heckel G (2007) Fine-scale genetic structure and dispersal in the common vole (*Microtus arvalis*). Mol Ecol 16:2463–2473
Seabloom EW, Bjornstad ON, Bolker BM, Reichman OJ (2005) Spatial signature of environmental heterogeneity, dispersal, and competition in successional grasslands. Ecol Monogr 75:199–214
Streitwolf-Engel R, van der Heijden MGA, Wiemken A, Sanders IR (2001) The ecological significance of arbuscular mycorrhizal fungal effects on clonal reproduction in plants. Ecology 82:2846–2859

Stukenbrock EH, Rosendahl S (2005) Clonal diversity and population genetic structure of arbuscular mycorrhizal fungi (*Glomus* spp.) studied by multilocus genotyping of single spores. Mol Ecol 14:743–752

Tedersoo L, Koljalg U, Hallenberg N, Larsson KH (2003) Fine scale distribution of ectomycorrhizal fungi and roots across substrate layers including coarse woody debris in a mixed forest. New Phytol 159:153–165

Thiet RK, Boerner REJ (2007) Spatial patterns of ectomycorrhizal fungal inoculum in arbuscular mycorrhizal barrens communities: implications for controlling invasion by *Pinus virginiana*. Mycorrhiza 17:507–517

Tilman D, Kareiva P (1997) Spatial ecology: the role of space on population dynamics and interspecific interactions. Princeton University Press, Princeton

Toljander JF, Eberhardt U, Toljander YK, Paul LR, Taylor AFS (2006) Species composition of an ectomycorrhizal fungal community along a local nutrient gradient in a boreal forest. New Phytol 170:873–883

Turnbull LA, Coomes DA, Purves DW, Rees M (2007) How spatial structure alters population and community dynamics in a natural plant community. J Ecol 95:79–89

Twieg T, Durall DM, Simard SW (2007) Ectomycorrhizal fungal succession in mixed temperate forests. New Phytol 176:437–447

Vandenkoornhuyse P, Ridgway KP, Watson IJ, Fitter AH, Young JPW (2003) Co-existing grass species have distinctive arbuscular mycorrhizal communities. Mol Ecol 12:3085–3095

van Tuinen D, Jacquot E, Zhao B, Gollotte A, Gianinazzi-Pearson V (1998) Characterization of root colonization profiles by a microcosm community of arbuscular mycorrhizal fungi using 25S rDNA-targeted nested PCR. Mol Ecol 7:879–887.

Wiegand T, Gunatilleke CVS, Gunatilleke IAUN, Huth A (2007) How individual species structure diversity in tropical forests. Proc Natl Acad Sci USA 104:19029–19033

Wright DP, Johansson T, Le Quere A, Söderström B, Tunlid A (2005) Spatial patterns of gene expression in the extramatrical mycelium and mycorrhizal root tips formed by the ectomycorrhizal fungus *Paxillus involutus* in association with birch (*Betula pendula*) seedlings in soil microcosms. New Phytol 167:579–596

Wu BY, Hogetsu T, Isobe K, Ishii R (2007) Community structure of arbuscular mycorrhizal fungi in a primary successional volcanic desert on the southeast slopc of Mount Fuji. Mycorrhiza 17:495–506

Chapter 13
Response Diversity of Ectomycorrhizas in Forest Succession Following Disturbance

Suzanne W. Simard

Abstract A diversity of responses to disturbance among ectomycorrhizal fungal (EMF) species contributing to similar ecosystem functions is important to forest successional recovery. Response diversity is particularly important in ecosystems subject to frequent or catastrophic disturbance, providing insurance that succession proceeds within a historical stability domain. This chapter examines patterns of EMF species diversity and composition in relation to succession following wildfire and clearcutting in Douglas-fir dominated ecosystems. Both disturbance types alter EMF communities, but succession proceeds predictably and similarly. However, severe disturbances, such as clearcutting followed by site preparation that removes living trees, root systems, or forest floor, can reduce EMF diversity on regenerating seedlings, often corresponding with reduced survival, growth rates, and foliar nutrients. Concurrent increases in invasive plants augment these changes. These results suggest a reduction in ecological function with reduced EMF diversity, or reduced response diversity. By contrast, retaining forest floor and key functional groups of plants, such as broadleaf trees, help maintain diverse EMF that persist through succession. Many of these fungi are host-generalists, linking broadleaves with conifers, and facilitating seedling survival and improving their physiological condition. This dynamic interplay among broadleaves, soils, and EMF may allow understory conifers to eventually dominate, in congruence with historic forest succession patterns.

13.1 Introduction

The functioning and stability of forest ecosystems are determined by disturbance regimes, plant succession and biodiversity. Fundamental problems currently facing western North American forests are the loss of old-growth after a long history of exploitation (Ludwig et al. 1993), loss of species (Botkin et al. 2007), and changes in

S.W. Simard
Department of Forest Sciences, University of British Columbia,
Vancouver, British Columbia, Canada V6T 1Z4
e-mail: suzanne.simard@ubc.ca

C. Azcón-Aguilar et al. (eds.), *Mycorrhizas - Functional Processes and Ecological Impact*, 187
DOI: 10/1007-978-3-540-87978-7_13,

natural disturbance regimes that appear to include more severe and extensive wildfires and insect infestations resulting from climate change (Chapin et al. 2000; Carroll et al. 2003; McKenzie et al. 2004). Forest management practices that include fire suppression, species conversions, or removal of soil organic matter can intensify these effects (Perry et al. 1989; British Columbia Ministry of Forests and Range 2006). Combined with this is the increasing risk of biological invasion of forests by exotic weeds as global climate changes and road networks increase (Dukes and Mooney 1999). Unusual disturbance regimes and invasion of new plant species can dramatically alter native plant communities, but we know very little about what happens to associated soil biota, particularly mycorrhizal fungi (Callaway et al. 2003). Further, we know little about the reciprocal effects of the mycorrhizal community on plant diversity, succession, and ecosystem stability (van der Heijden et al. 1998; Borowicz 2001; Hartnett and Wilson 2002; Kardol et al. 2006; McHugh and Gehring 2006).

The stability of a forest community can be determined by examining features of its responses to disturbances (Halpern 1988). Plant succession is an important and easily measured response and has been characterized following wildfire and clearcutting in many North American forest ecosystems (Oliver and Larson 1997). Many different forest successional pathways are possible, and depend on the disturbance regime (agent, severity, duration, intensity, and extent), initial community composition, local climate, herbivory, and chance (Connell et al. 1987; Chapin et al. 2002). Gap-phase disturbance is common in many of these forests, usually causing increases in species and structural diversity (Fahey and Puettman 2007), but stand-replacing disturbances are also commonly followed by predictable stages of stand development (Oliver and Larson 1997). The stability of the system depends on whether development along these various successional pathways returns the ecosystem to its former state within a reasonable timeframe (Halpern 1988). Ecological resilience depends on the speed at which this happens (Suding et al. 2004); it is a measure of the amount of disturbance the system can absorb, then self-organize and stay within the same stability state or domain of attraction (Holling 1996). Degraded ecosystems may result in positive feedback systems between abiotic and biotic factors, however, resulting in a persistent alternate state resistant to successional recovery (Fig. 13.1) (Perry et al. 1989; Suding et al. 2004). To avoid this possibility, most reforestation efforts have focused on ways to conserve or quickly restore historical abiotic and biotic conditions, promoting reasonably rapid return to the same stability state (Hobbs and Harris 2001). Maintaining ecological resilience appears essential for sustaining productivity and ecosystem services in forests, which naturally are complex, non-linear, and subject to stochastic disturbances (Gunderson and Holling 2002).

Biological diversity is thought to enhance ecosystem resilience and play an important role in sustaining natural forest ecosystems in the face of change (Peterson et al. 1998). Maintaining diversity in species, populations, and functional groups in ecosystems undergoing change provides redundancy and insurance against losing function. Moreover, in systems undergoing change, or which are subject to new disturbance regimes, response diversity is particularly important; response diversity is the diversity of responses (e.g., several species with the ability to take up available P or N in response to soil disturbance) contributing to the same

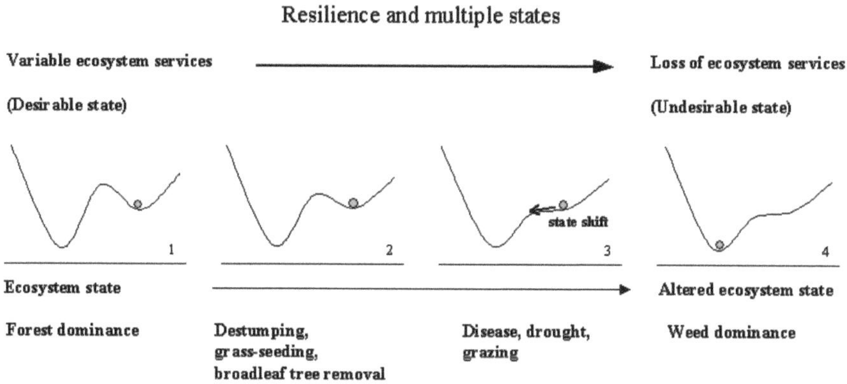

Fig. 13.1 Shifts in forested ecosystems from a desirable state as a consequence of forest management and drought. The *curves* represent stability landscapes, with basins representing domains of attraction. The shift from forest dominance to weed dominance is caused by disturbance and drought, which resilient ecosystems would otherwise absorb through re-organization supported by response diversity (Elmqvist et al. 2003)

ecosystem function (e.g., biomass accumulation, maintenance of slope stability) (Elmqvist et al. 2003). In temperate forest communities, for example, there is usually a mix of light-seeded, ruderale herbaceous species that rapidly occupy disturbed sites and maintain slope stability. The dominant species is fireweed (*Epilobium angustifolium* L.); should this species decline dramatically, a few other subdominants (e.g., *Circaea alpina* L., *Urtica dioica* L.) could fill the same functional role, thus providing response diversity to disturbance. At the same time, these species, along with other herbs in the community, are functionally dissimilar in certain traits (e.g., take up water or nutrients from different soil depths), providing functional diversity (i.e., niche complementarity).

Mycorrhizal fungi form symbiotic associations with most of these r-selected plants, fulfilling critical nutrient and water uptake roles that enable plant invasion and occupancy of disturbed sites (Callaway et al. 2003). Considering that over 200 species of arbuscular mycorrhizal fungi (AMF) form associations with most vascular plant species, the combined response diversity of the plant–fungal symbiosis is likely high (Hartnett and Wilson 2002). For example, multiple AMF taxa will likely colonize an invading plant community in a complex network for nutrient uptake following fire (Giovannetti et al. 2006). Response diversity is potentially even higher in ectomycorrhizal associations, with over 5,000 species of EMF (Molina et al. 1992) and even greater variation in cryptic EMF species (Douhan and Rizzo 2005; Gryta et al. 2006; Tedersoo et al. 2006), which more or less vary in functional attributes affecting plant productivity. These include the ability to use organic nutrient sources, secrete extracellular enzymes, take up water or nutrients from different soil patches, or protect against fungal pathogens (Smith and Read 1997; Jones et al. 2003), sometimes at a cost to plant growth (Klironomos 2003; Jones and Smith 2004). To ensure that forest ecosystems can re-organize and succeed to historical stability domains following disturbance, it appears that functional groups of plant and mycorrhizal fungal species with high response diversity must remain available. Indeed, the role of high response diversity in successional re-organization of complex systems, requiring high spatial and temporal niche packing, may favor the large diversity of mycorrhizal taxa that exists (Selosse and Duplessis 2006; see also Chapters 11 and 12 by Pickles et al. and Wolfe et al., respectively). This is becoming more critical as climate change alters disturbance regimes and imposes chronic stress on forest ecosystems (Elmqvist et al. 2003).

In this paper, EMF succession, diversity and, where possible, response diversity among different types and severities of disturbance are compared in mixed temperate forests of western North America. Two groups of studies are reviewed in the mixed Douglas-fir forests of southern British Columbia: one where disturbances appear to fall within the range of natural variability for this forest type (Arsenault and Klenner 2005), and a second one where they do not (Fleming et al. 2006). Three questions are asked: (1) Does the EMF community recover to the same stability domain following standard reforestation practices as it does following natural, mixed wildfire regimes? (2) Are EMF communities and seedling performance negatively affected by reforestation practices that remove critical legacies from the original forest? (3) Does replacement of these legacies help restore the EMF community to its original stability domain? In the first section, the experimental forests, including their climate, natural disturbance regime, composition, and typical management pathway are described. Then, a chronosequence study is described where EMF succession following clearcutting; this is compared to a study of one of the sites that was degraded to a persistent alternate state, and then partially restored using soil transfer. Conclusions are drawn as to management recommendations that can be used to maintain or restore resilience in altered forest ecosystems.

13.2 Disturbance and Management Regimes in Mixed Temperate Forests

The mixed temperate forests of the Interior Cedar Hemlock (ICH) biogeoclimatic zone of interior British Columbia grow in a favorable continental climate generally characterized by warm, moist summers and mild, snowy winters (Pojar et al. 1987). The forests are species-rich and productive, and in the wettest subzones, they are considered rainforests or, where disturbance is infrequent, antique forests (Goward and Arsenault 2000). Owing to climate variability across the vast distribution of the ICH zone, the historical natural disturbance regime has been characterized as diverse, complex, and episodic at multiple spatial and temporal scales (Arsenault and Klenner 2005), with fires varying widely in severity and median return intervals that range between 30 and 1,200 years (Johnson et al. 1990; Parminter 1998; Daniels et al. 2006; Sanborn et al. 2006) (Fig. 13.2: low and high severity natural disturbance pathways). The fire regime is considered mixed, but many fires are stand-replacing, and the complexity of developing stands is further increased by

Fig. 13.2 Examples of the complexity of outcomes of natural and managed disturbance regimes. Disturbance to pristine, primeval forest is shown to proceed along four archetypal pathways. (**First column from left**) A low severity fire as part of a mixed fire regime in interior Douglas-fir forests. Large trees are scorched but many survive, and forest floor is partially consumed. Legacies survive and regeneration is facilitated. (**Second column from left**) A high severity fire as part of a mixed fire regime in interior Douglas-fir forests. Trees are killed but the snags remain, retaining carbon stores on site. Large patches of forest floor are consumed by the fire, but sufficient legacies and neighboring seed sources facilitate regeneration of a diverse stand. (**Third column from left**) Standard reforestation pathway with low severity harvesting, where green trees and the forest floor are retained. This is followed by weeding of broadleaf trees and spacing to remove unwanted conifers in order to reduce competition to the preferred conifer crop species. The resulting forest is tree-species poor, but remains within the EMF stability domain characteristic of the natural disturbance regime. (**Fourth column from left**) Intensive reforestation pathway, where disease infected stumps are removed from the soil followed by seeding with domestic grasses for cattle forage and planting. The decadal drought contributes to the shift to a weed-dominated meadow, creating a positive feedback where the grasses compete with conifers and reduce EMF diversity, resulting in a new, persistent stable state resilient to restoration (Photographs courtesy of J Barker, F Teste, J Mather, S Simard and B Zimonick)

gap-phase disturbances caused by the pathogenic fungi *Armillaria ostoyae* (Romagn.) Herink and *Phellinus weirii* (Murr.) Gilb. Broadleaf tree species are resistant to these root diseases, allowing them to persist in older forests in gaps created by conifer mortality (Morrison et al. 2000). Other small-scale disturbances result from windthrow and insects, particularly host-specific bark beetles and spruce budworm (*Choristoneura occidentalis* Freeman).

The ICH forests are comprised of several conifer and broadleaf species, forming stratified, single to multi-cohort, mixed species stands (Cameron 1996). Up to ten species can regenerate on a single site following stand-replacing wildfire or clear-cutting, with shade intolerant species such as interior Douglas-fir (*Pseudotsuga menziesii* var. *glauca* (Beissn.) Franco), western larch (*Larix occidentalis* Nutt.) and lodgepole pine (*Pinus contorta* Dougl. ex Loud. var. *latifolia* Engelm.) growing alongside shade tolerant western redcedar (*Thuja plicata* D.) and western hemlock (*Tsuga heterophylla* (Raf.) Sarg.), as well as shade intolerant paper birch (*Betula papyrifera* Marsh.) and trembling aspen (*Populus tremuloides* Michx.). Species variation in height growth and shade tolerance results in early stand dominance by broadleaves and intolerant conifers, but with no further disturbance, intense vertical stratification quickly gives rise to mixed conifer–broadleaf stands, with eventual dominance by shade tolerant cedar and hemlock (Simard and Vyse 2006).

The typical management pathway in these forests aims to create free-growing, high-yield conifer stands. The stands are clearcut (30–50 ha in size; Sachs et al. 1998), the sites usually prepared by extraction of disease-infected stumps, leaving forest floor present but redistributed (i.e., destumping; Morrison et al. 2000), and broadleaf trees are removed (i.e., weeding; Simard and Vyse 2006). This is rapidly followed by planting of large (30–70 cm tall) 1- to 2-year-old Douglas-fir or lodgepole pine container seedlings and, commonly, seeding with domestic grasses to increase cattle forage production. Once the conifers are established, the plantations are again weeded of broadleaf trees and other encroaching shrub species using herbicides or cutting, and then manually thinned to reduce conifer densities, favoring the early successional conifer species. Recently, a few green trees have been retained on the sites for visual quality or as wildlife trees (Fig. 13.2: low severity harvest pathway; Mitchell and Beese 2002). This management pathway follows the basic management principles for high yield conifer production practised elsewhere in the world (Wagner et al. 2005).

13.3 Does the EMF Community Recover Following Standard Reforestation Practices?

Succession of the EMF community was examined to determine whether standard harvesting and reforestation practices resulted in reasonable EMF community recovery to the stability domain represented by fire-origin stands. Succession of EMF communities is poorly understood (Smith et al. 2002; Kranabetter et al. 2005; Durall et al. 2006), whereas several plant successional pathways following natural and managed disturbances have previously been described for the complex ICH forests (Cameron 1996; Parminter 1998; DeLong et al. 2005). An attempt to answer the question posed is made by reviewing several EMF studies in mixed Douglas-fir – paper birch forests conducted on (1) root tips of 1 to 2-year-old seedlings establishing on slash-burned clearcuts (Jones et al. 1997; Simard et al. 1997a; Philip 2006), (2) epigeous fruiting bodies in 5- and 120-year-old fire-origin stands (Durall et al. 2006) and, in particular, (3) root tips from soil cores sampled across a chronosequence of fire-origin (5, 26, 65, and 100 years old) and clearcut stands (5 and 26 years old only) (Twieg et al. 2007) (Fig. 13.3). All EMF communities of Douglas-fir and paper birch were characterized using ectomycorrhizal root tip morphology and/or ITS region DNA sequences. Burned sites had experienced a mixed fire severity regime leaving behind forest floor and root-stocks, whereas the clearcut sites contained the forest floor and stumps from the original stands (note that this is less intensive than the typical pathway described above). The ages studied represent important stand development stages in these forests: stand initiation, crown closure-to-stem exclusion, and stand re-initiation at the beginning and near the end of birch natural senescence, respectively (Oliver and Larson 1997; Simard et al. 2004).

These studies showed that EMF community composition and diversity on root tips of both Douglas-fir and paper birch changed rapidly from 1 to 26 years following disturbance, after which they stabilized (Fig. 13.3). The EMF community had stabilized near the time of canopy crown closure, congruent with stability patterns in plant community composition and crown cover (Simard et al. 2004). There was little difference in Douglas-fir or paper birch EMF diversity or community structure between the young clearcut and wildfire origin sites (Twieg et al. 2007), suggesting that clearcutting with forest floor and root retention did well at emulating the natural fire disturbance (Fig. 13.3). The EMF composition on root tips from the older forests (Twieg et al. 2007) overlapped with that of sporocarps sampled from the 120-year-old fire-origin forests by Durall et al. (2006), suggesting the EMF communities were stable across multiple sites in the ICH and stayed within the same stability domain following mixed fire disturbances.

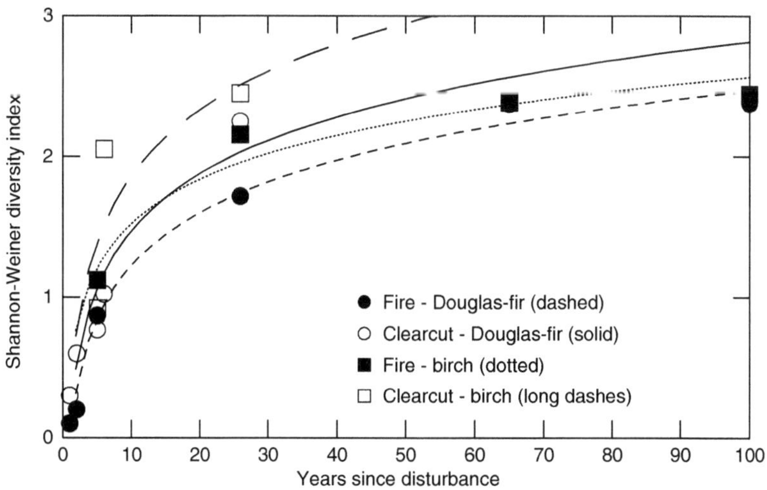

Fig. 13.3 Shannon-Weiner diversity of EMF fungal taxa on Douglas-fir and paper birch roots sampled from a chronosequence of stands representing early forest succession. Data sources are Twieg et al. (2007), Simard et al. (1997a, b), Jones et al. (2007), Philip (2006), and Barker (unpublished dataset). The data from Twieg et al. (2007) was derived from soil cores, where as all other data was derived from seedling root tips. *Curves* for clearcuts have been extrapolated beyond 26 years

Diversity of the EMF community was higher on regenerating paper birch than Douglas-fir in most of the studies, and this was attributed to its larger root system in the field or greenhouse studies (Jones et al. 1997; Simard et al. 1997a; Philip 2006), or to the persistence of inocula and mycorrhizal networks on sprouting paper birch stumps and root-stocks that had survived the previous disturbance in the chronosequence study (Twieg et al. 2007). Once linked into the legacy mycorrhizal network of the original root systems, EMF diversity on regenerating paper birch was little affected by stand age (Twieg et al. 2007; Fig. 13.3). Here, the old birch root-stocks served to link the new forest with the old through time and space, stabilizing the community. Douglas-fir does not regenerate by coppice (Klinka et al. 2000), but nursery seedlings are planted, which may account for its lower EMF diversity than on naturally regenerating paper birch (Twieg et al. 2007). Douglas-fir seedlings also appeared to benefit from linking into the mycorrhizal network with older birch stumps and trees because it shared in common many of the generalist EMF species (Fig. 13.4; Simard et al. 1997b). Retaining forest legacies, such as forest floor, old root systems, or live trees, appeared to provide important links to regenerating seedlings, aiding recovery following disturbance and stabilizing the ecosystem.

Host-generalist fungi were common on paper birch and Douglas-fir roots across all stand ages, indicating high potential for the two species to link together in a mycorrhizal network throughout succession (Fig. 13.4). Our results support the concept that generalist fungi are common in ecosystems with complex, long-lived plant communities, or that are subject to frequent and variable disturbances; here, they would play a role in facilitating regeneration under uncertain conditions (Perry et al. 1989; Simard and Durall 2004; Selosse et al. 2006). Resource transfer between

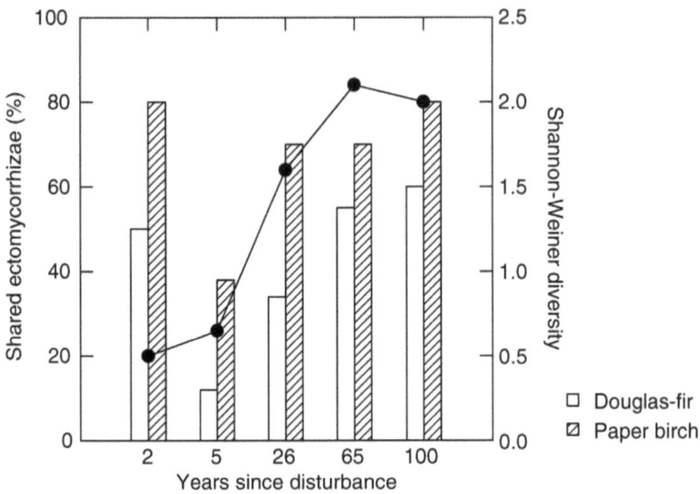

Fig. 13.4 Proportion of root tips colonized by EMF taxa shared in common by Douglas-fir and paper birch (*histograms*), and Shannon-Weiner diversity of EMF in the mycorrhizal network (*line with filled circles*)

paper birch and Douglas-fir through the mycorrhizal network might also reduce competitive stress and improve individual tree resistance against the ubiquitous root pathogens or drought stress, thus improving plantation survival and ensuring future carbon sources for the mycorrhizal fungi (Simard et al. 1997c; Woods 2003; Philip 2006; Selosse et al. 2006). In support, linking into a mycorrhizal network with older trees has facilitated establishment of Douglas-fir seedlings in nearby Douglas-fir forests (Simard et al. 1997b; Teste and Simard 2008). Between 2 and 5 years following seedling establishment, there was a brief shift from host-generalist to host-specific EMF dominance, followed by increasing proportions of host-generalists (Fig. 13.4); here, the diversity of host generalists also increased with stand age. This brief hiatus in the proportion of host-generalists may have resulted from loss of a few r-selected nursery EMF species (e.g., *Wilcoxina*) as the stands developed; the nursery fungi were probably weak competitors against the common, K-selected, native EMF, such as Douglas-fir-specific *Rhizopogon vinicolor*-type or paper birch-specific *Lactarius pubescens*, which also accumulated with stand age (Twieg et al. 2007). Host generalist EMF continued to increase in diversity and abundance with age, hence providing a complex mycorrhizal network, which could improve regeneration potential during the increasingly frequent gap-phase disturbances. *Russula* and *Piloderma* were notably more abundant in the older stands, where birch had matured and was in decline (Twieg et al. 2007). These patterns in EMF succession are tightly linked to secondary plant succession and fungal life history traits, and do not fit the "early-stage, late-stage" model ascribed to afforested European forests, which more closely follows a primary successional pathway (Deacon and Fleming 1992).

The persistence of the mycorrhizal network but its shift in EMF species composition with stand age (Fig. 13.4) suggests response diversity exists in these forests. The composition of the mycorrhizal network changed from dominance by r-selected to K-selected host-generalist EMF species following stand replacing disturbance, but its role in facilitating seedling establishment likely remained functional throughout stand development. Indeed, the mycorrhizal network in Douglas-fir forests has been shown to facilitate establishment following stand-replacing (Teste and Simard 2008; Teste et al. 2008) or gap-phase disturbance (Simard et al. 1997b). Whether disturbances create large openings or small gaps in a mature forest, there appears sufficient functional similarity in the redundancy in the EMF community to ensure that a mycorrhizal network exists to facilitate seedling establishment. The increasing EMF response diversity for maintaining a functional mycorrhizal network with stand age provides greater flexibility for responding to the increasingly variable gap phase disturbances that occur as ICH forest succession proceeds. Response diversity was also evident in EMF extracellular enzyme activity in these forests. Although activities of chitinase, phosphatase, and glucosidase varied among EMF species, the total activity of each enzyme summed over all EMF species did not vary with stand age (Twieg et al. 2008). Hence, sufficient response diversity exists among EMF species so that enzyme activity for organic nutrient transformation and translocation was maintained at consistent levels throughout forest succession, even though EMF composition changed.

13.4 Does Removal of Critical Legacies Shift the EMF Community?

Ecosystems normally stable under natural disturbance regimes can suddenly collapse resulting from changes in climate, human exploitation, and fragmentation (Pauly et al. 2002; Bellwood et al. 2004; Suding et al. 2004). Climate stress, sometimes combined with poor reforestation practices, have been associated with native forest decline in the Mediterranean basin of California (Gómez-Aparicio et al. 2004), the pinyon-juniper woodlands of northern Arizona (McHugh and Gehring 2006), and the upper montane mixed *Abies* forests of Oregon and California (Amaranthus and Perry 1989). The typical management pathway practised in the ICH (Fig. 13.2: low severity harvest pathway) commonly meets short-term conifer production goals, but its extensive application is raising concerns about the diversity, health, and resilience of the resulting forests (Woods et al. 2005; Lieffers et al. 2007). Recently, these conifer plantations have been subject to widespread insect and disease damage, such as from Dothistroma needle blight (*Dothistroma septosporum* (Dorog.) Morelet), mountain pine beetle (*Dendroctonus ponderosae* Scolytidae), and *A. ostoyae* root disease (Cruickshank et al. 1997; Carroll et al. 2003; Woods et al. 2005). Approximately 40% of older ICH plantations in the southern interior of British Columbia are no longer sufficiently stocked because of extensive mortality caused by *A. ostoyae* root disease (Woods 2006); pathogens may be accumulating on these sites over time due to replanting of the most disease-susceptible hosts, as in other temperate and tropical forests (Harms et al. 2000; Packer and Clay 2000). In some cases, multiple stressors, in combination with a decade of summer drought, have resulted in early plantation failure (Fig. 13.2: high severity harvest pathway). This kind of damage is expected to increase with changing climatic conditions (Hansen et al. 2005; Hamann and Wang 2006). Indeed, the climate of the ICH zone, as it is currently distributed in the southern interior of British Columbia, is projected to change to that of the Bunchgrass zone (dominated by bunchgrasses and sages; Pojar et al. 1987) over the next century (Hamann and Wang 2006; British Columbia Ministry of Forests and Range 2006).

Plantation collapse has occurred in the ICH following a loss of historic biotic legacies and an increase in positive feedbacks with damaging abiotic conditions. The collapse of one of these sites is described here. The original stand was comprised of western redcedar, Douglas fir, western redcedar and paper birch; it was clearcut in 1988 and planted to off-site interior spruce [*Picea engelmannii* (Moench) Voss subsp. *glauca* (Parry ex Engelm.)] in 1989. This plantation subsequently suffered extensive *A. ostoyae*-caused mortality as well as growth losses due to summer frost damage. The failed plantation was then replaced in 1992 with a tree species mixture comprised of the four native species; prior to this, the site was prepared in the fall of 1991 by destumping to remove the disease-infected stumps, roots, and forest floor. Unfortunately, there was a severe growing-season drought in 1992 (Durall et al. 2006), and this stress was compounded by immediate seeding of the site to a high-density arbuscular mycorrhizal grass mix (dominated by *Dactalys glomerata* L. and *Phleum pratense* L.)

followed by repeated cattle grazing by a local rancher. Not surprisingly, this was followed by extensive tree seedling mortality. Annual replanting of the same four native species and a reduction in grazing intensity was undertaken for 10 years in an effort to establish the plantation, but this coincided with a decade of summer drought, and hence these efforts were completely unsuccessful. The site appeared to remain in a hiatus indefinitely, and the arbuscular mycorrhizal-dominated community assumed an unusual succession pattern influenced by new exotic plant invasions. It appeared that the system had degraded into a persistent, alternative state comprised of an arbuscular mycorrhizal-dominated meadow that was resilient to our restoration efforts aimed at re-establishing the natural ectomycorrhizal-dominated forest (Fig. 13.2: high severity harvest pathway) (Suding et al. 2004; Didham et al. 2007). Feedbacks between the climate and the new arbuscular mycorrhizal-dominated plant community appeared to have broken the plant–soil link and shifted the soil biotic community composition, resulting in a new but stable state (Perry et al. 1989).

13.5 Does Replacement of Legacies Restore the EMF Community?

An experiment was established in 2002 to investigate the underlying causes of this 26-ha plantation failure. Using a 2 × 3 factorial experiment, effects of arbuscular mycorrhizal-grass competition removal and soil transfer from different ectomycorrhizal-forest tree species were examined on seedling growth, survival, foliar nutrients, and mycorrhization, as well as on soil chemical and microbiological characteristics. For the two competition removal treatments, either (1) all vegetation was completely removed (aboveground and belowground parts) from a 1 × 1 m plot centered on the seedling to a depth of 10 cm in the mineral soil, or (2) was left intact. In the three soil transfer treatments, Douglas-fir seedlings were planted into planting holes either with (1) 250 ml soil transferred from beneath mature paper birch trees, (2) 250 ml soil transferred from beneath mature Douglas-fir trees, or (3) untreated, un-transferred soil. The transferred soil was collected from under the crowns of four mature paper birch and Douglas-fir trees located one tree-length inside the forest adjacent to the clearcut, and included a mix of forest floor and the upper 10 cm of mineral soil. These six treatment combinations were randomly allocated to 1 × 1 m experimental units spaced 10 m apart in a grid pattern. Each treatment was replicated 12 times. One interior Douglas-fir seedling was introduced into the middle of each experimental unit.

In August 2004, seedlings were assessed for survival and living trees were measured for height and diameter. Mineral soils were collected to 15 cm depth adjacent to seedlings in the field for soil chemical analysis. All live seedlings were then harvested, subsampled for EMF (Goodman et al. 1996) and AMF (Brundrett et al. 1984) evaluations, foliar nutrient analysis (Kalra and Maynard 1991), and oven dry mass estimation. They were also examined for evidence of disease.

Grass removal increased Douglas-fir seedling survival, diameter growth, height growth, and biomass in all soil transfer treatments (Fig. 13.5a: only survival data

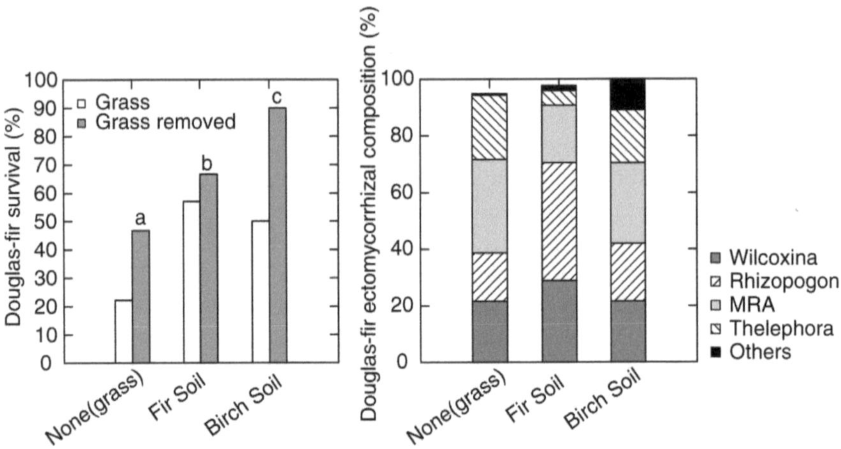

Fig. 13.5 Two-year Douglas-fir seedling survival (**left graph**), and EMF composition (**right graph**) among grass removal and soil transfer treatments. *Bars* or *proportions* of histograms represent means. At **left**, grass removal treatments were consistently different across all soil transfer treatments, and soil transfer treatments with different letters were significantly different from each other, as determined by two-way ANOVA and Bonferroni's mean separation test. At **right**, abundance of *Rhizopogon* and *Thelephora* differed among soil transfer treatments. Difference were considered significant at $P = 0.05$

shown). Soil transfer improved survival without affecting other growth parameters. A comparison of soil nutrients showed no difference among grass removal or soil transfer treatments, except there was greater boron (B) concentration in plots with grass and greater manganese in the Douglas-fir soil transfer treatment (data not shown). Foliar nutrients were significantly affected by soil transfer, with increases in foliar B, sulfur, and potassium (K) concentrations (data not shown). Grass removal appeared to reduce nitrogen (N)-induced deficiency of K, whereas birch-soil transfer improved K levels. In addition to grass competition, imbalances of foliar N relative to K, zinc or B may have contributed to seedling mortality (Perry 1994).

More than 92% of live root tips were colonized by EMF across all grass removal or soil transfer treatments (Fig. 13.5b). Where grass remained untreated, Douglas-fir root systems were also infected by AMF over 22–26% of their root length, and by the root endophytic pathogen *Olpidium*, indicating that arbuscular mycorrhizal grasses induced changes in the rhizosphere community of Douglas-fir. Arbuscular mycorrhizal colonization of Douglas-fir and other Pinaceae tree species growing in the presence of abundant herbs and grasses has been observed previously (Cázares and Trappe 1993; Horton et al. 1998; Smith et al. 1998), although it is thought not to be common. Neither vegetation nor soil transfer treatments had a significant effect on total mycorrhizal colonization, but more EMF taxa occurred in the grass-removal treatment (data not shown), probably because of greater seedling growth associated with reduced grass interference (Simard et al. 2003). Grass removal increased *Wilcoxina*, an abundant native of the woody understory plant community in Douglas-fir forests (Hagerman et al. 2001), and decreased *Thelephora*, a common nursery fungus that is easily out-competed by native fungi (Visser 1995). *Wilcoxina* has also been associated with stressed or disturbed sites in other studies (Visser 1995; Torres and Honrubia 1997; Baar et al. 1999; Grogan et al. 2000). The decade-long hiatus in the presence of healthy EMF hosts also probably contributed to loss of EMF inoculum potential from the site, and the partial infection of Douglas-fir roots with AMF suggests that AMF inoculum was high enough to either compete strongly for EMF host root colonization sites (Lodge and Wentworth 1990) or negatively interact with EMF hyphae (Baar and Stanton 2000). Compared with no soil transfer or transfer from beneath birch, soil transferred from beneath mature Douglas-fir trees increased seedling colonization by *Rhizopogon*-like by 2.5 times and decreased that by *Thelephora*-like by 4 times (Fig. 13.5b). *R. vinicolor*, an abundant native in these forests, is thought to readily inoculate healthy seedlings from a resident spore bank (Kjøller and Bruns 2003).

These results indicate that seeding with high densities of domestic arbuscular mycorrhizal grasses, combined with a decade of drought stress, caused substantial mortality of planted Douglas-fir; this was associated with grass competition for soil resources and changes in the rhizosphere microbial community (Hawkes et al. 2006). Changes in the EMF community composition reduced response diversity as expressed by reduced seedling nutrient uptake and survival. Douglas-fir survival increased with grass removal, and this was augmented by soil transfer from beneath mature trees. Transfer from beneath conspecific mature trees was particularly effective at restoring native EMF community on Douglas-fir seedlings. When Douglas-fir seedlings were released from competition, there ensued fuller expression of soil inoculations on the more robust root systems. Reductions in arbuscular mycorrhizal shrubs have similarly improved the EMF community composition and growth of ectomycorrhizal pinyon pines subject to chronic drought (McHugh and Gehring 2006).

13.6 Conclusions

The forests of the ICH zone are normally stable and resilient against shifts to alternative stability domains. EMF communities of harvested forests largely recover to pre-disturbance conditions by the time canopy crown closure occurs, provided standard reforestation practices are followed, where forest floor and some green trees are retained on site, and native tree species mixtures are planted or allowed to regenerate naturally. At least with respect to the ectomycorrhizal community of regenerating seedlings, these forestry practices appear to emulate the natural, mixed fire disturbance regime. Additionally, there appears to be sufficient response diversity in EMF species so that ecosystem functions, such as colonization rates, mycorrhizal network development, or enzyme production for nutrient uptake, all important to seedling survival and physiology, are maintained under natural disturbance patterns. Variation in EMF species composition of the mycorrhizal network is an example of response diversity; it varies with time since disturbance, but consistently facilitates regeneration through succession in the highly variable gap-phase disturbance regime.

The buffering capacity of long-lived forests is under threat, however, with climate change. Recent history of repeated drought and frequent changes in the common reforestation pathway, in which critical ecological legacies are removed, point to future vulnerabilities in these ecosystems, where they could degrade to alternative stable states resilient to restoration (Figs. 13.1 and 13.2). Poor management practices that remove forest floor, residual trees, stumps, and root stocks, can result in reduced EMF diversity, and when this is combined with reduced sporocarp production during drought (Durall et al. 2006), there may be limited colonization by fungi from neighboring sites, leading to local extinction of fungal species (McHugh and Gehring 2006). When this happens, EMF response diversity is weakened, which feeds back to the plant community, causing further changes in community structure. This can lead to regeneration failure following disturbance and shifts to a different ecosystem stability domain. Replacement of these legacies with unique methods may restore systems to their original stability domain (Suding et al. 2004), but even in the study reported here, Douglas-fir soil transfer and grass weeding increased survival to only 60%, which is still considered a plantation failure in these ecosystems (Simard and Vyse 2006). Given this, it is increasingly important that forest management practices include sufficient legacies from the primeval forests for EMF response diversity and networks to be maintained in order to facilitate forest establishment and succession within historical stability domains.

Acknowledgements I thank my family for their patience as I wrote this paper. I am very grateful to my husband, Don Sachs, who made all of graphs, and I thank Drs. Melanie Jones and Dan Durall for their contributions to the presentation we made at ICOM6 in Granada, not to mention the rich research we have conducted together over the past 15 years. I have benefited greatly from thoughtful discussions with Dreena Lindstrom, Sybille Haeussler, and Dave Perry on concepts of ecological resilience. I am particularly proud of my students, who have toiled in the trenches to make observations and conduct experiments, many of which contributed to this paper: Karen Baleshta, Jason Barker, Marcus Bingham, Trevor Blenner-Hassett, Julia Chandler, Deb Delong, Rhonda Delong, Julie Deslippe, Julia Dordel, Marty Kranabetter, Ameur Manceur, Federico Osorio, Leanne Philip, Francois Teste, Brendan Twieg, and Shannon Wright. There are many more who have helped me – thank-you. Writing of this paper was partially funded by my NSERC DG.

References

Amaranthus MP, Perry DA (1989) Rapid root tip and mycorrhiza formation and increased survival of Douglas-fir seedlings after soil transfer. *New Forests* 3:259–264

Arsenault A, Klenner W (2005) Fire regime in dry-belt forests of British Columbia: perspectives on historic disturbances and implications for management. In: Taylor L, Zelnik J, Cadwallander S, Hughes B (eds) Mixed severity fire regimes: ecology and management, Symposium Proceedings, November 17–19. Association of Fire Ecology MISC03, Washington State University, Pullman, WA, pp 105–121

Baar J, Stanton NL (2000) Ectomycorrhizal fungi challenged by saprotrophic basidiomycetes and soil microfungi under different ammonium regimes *in vitro*. *Mycol Res* 104:691–697

Baar J, Horton TR, Kretzer AM, Bruns TD (1999). Mycorrhizal colonization of *Pinus muricata* from resistant propagules after a stand-replacing wildfire. *New Phytol* 143:409–418

Bellwood DR, Hughes TP, Folke C, Nyström M (2004) Confronting the coral reef crisis. *Nature* 429:827–833

Borowicz VA (2001) Do arbuscular mycorrhizal fungi alter plant-pathogen relations. *Ecology* 82:3057–3068

Botkin DB, Saxe H, Araújo MB, Betts R, Bradshaw RHW, Cedhagan T, Chesson P, Dawson TP, Etterson JR, Faith DP, Ferrier S, Guisan A, Hansen AS, Hilbert DW, Loehle C, Margules C, New M, Matthew JS, Stockwell DRB (2007) Forecasting the effects of global warming on biodiversity. *BioScience* 57:227–236

British Columbia Ministry of Forests and Range (2006). The state of British Columbia's forests. Government of British Columbia, Victoria, Canada. ISBN 0-7726-5628-2. p 194

Brundrett MC, Piche Y, Peterson RL (1984) A new way for observing the morphology of vesicular arbuscular mycorrhizae. *Can J Bot* 62:2128–2134

Callaway RM, Mahall BE, Wicks C, Pankey J, Zabinski C (2003) Soil fungi and the effects of an invasive forb on grasses: neighbor identity matters. *Ecology* 84:129–135

Cameron IR (1996) Vertical stratification in some 50 year-old mixed species stands in the Interior Cedar-Hemlock zone. In: Comeau PG, Thomas KD (eds) Silviculture of temperate and boreal broadleaved-conifer species. BC Ministry of Forests, Victoria, Canada, Land Management Handbook 36:122–125

Carroll AL, Taylor SW, Régnière J, Safranyik L (2003) Effects of climate change on range expansion by the mountain pine beetle in British Columbia. In Shore TL, Brooks JE, Stone JE (eds) Mountain pine beetle Symposium: challenges and solutions, October 30–31, Kelowna, Canada. Natural Resources Canada, Canadian Forest Service, Pacific Forestry Centre, Information Report BC-X-399, pp 223–232

Cázares E, Trappe J (1993) Vesicular endophytes in roots of the Pinaceae. *Mycorrhiza* 2:153–156

Chapin III FS, Mcguire AD, Randerson J, Pielke Sr. R, Baldocchi DE, Hobbie SE, Roulet N, Eugster W, Kasischke E, Rastetter EB, Zimov SA, Running SW (2000) Arctic and boreal ecosystems of western North America as components of the climate system. *Global Change Biol* 6:211-223

ChapinIII FS, Matson PA, Mooney HA (2002) Principles of terrestrial ecosystem ecology. Springer, New York, p 436

Connell JH, Noble IR, Slatyer RO (1987) On the mechanisms producing successional change. *Oikos* 50:136-137

Cruickshank MG, Morrison DJ, Punja ZK (1997) Incidence of Armillaria species in precommercial thinning stumps and spread of Armillaria ostoyae to adjacent Douglas-fir trees. *Can J For Res* 27:481-490

Daniels LD, Cochrane J, Gray RW (2006) Refining mixed-severity fire regimes in the Rocky Mountain Forest District. Report to Tembec Inc., BC Division, Canadian Forest Products Ltd., Radium Hot Springs, Canada

Deacon JW, Fleming LV (1992) Specificity phenomena in mycorrhizal symbiosis: community-ecological consequences and practical implications. In: Allen MF (ed) Mycorrhizal functioning: an integrative plant-fungal process. Chapman & Hall, New York, NY, pp 249-300

DeLong DL, Simard SW, Comeau PG, Mitchell S (2005) Survival and growth responses of planted seedlings in root-disease infected partial cuts. *For Ecol Manage* 206:365-379

Didham RK, Tylianakis JM, Gemmell NJ, Rand TA, Ewers RM (2007) Interactive effects of habitat modification and species invasion on native species decline. *Trends Ecol Evol* 22:489-496

Douhan GW, Rizzo DM (2005) Phylogenetic divergence in a local population of the ectomycorrhizal fungus Cenococcum geophilum. *New Phytol* 166:263-271.

Dukes JS, Mooney HA (1999) Does global change increase the success of biological invaders. *Trends Ecol Evol* 14:135-139

Durall DM, Gamiet S, Simard SW, Kudrna L, Sakakibara SM (2006) Effects of clearcut logging and tree species composition on the diversity and community composition of epigeous fruit bodies formed by ectomycorrhizal fungi. *Can J Bot* 84:966-980

Elmqvist T, Folke C, Nystrom M, Peterson G, Bengtsson J, Walker B, Norberg J (2003) Response diversity, ecosystem change, and resilience. *Front Ecol Environ* 1:488-494

Fahey RT, Puettmann KJ (2007) Ground-layer disturbance and initial conditions influence gap partitioning of understorey vegetation. *J Ecol* 95:1098-1109

Fleming RL, Powers RF, Foster NW, Kranabetter JM, Scott DA, PonderJr. F, Berch S, Chapman WK, Kabzems RD, Ludovici KH, Morris DM, Page-Dumroese DS, Sanborn PT, Sanchez FG, Stone DM, Tiarks AE (2006) Effects of organic matter removal, soil compaction, and vegetation control on 5-year seedling performance: a regional comparison of long-term soil productivity sites. *Can J For Res* 36:529-550

Giovannetti M, Avio L, Fortuna P, Pellegrino E, Sbrana C, Strani P (2006) At the root of the wood wide web: self recognition and non-self incompatibility in mycorrhizal networks. *Plant Signal Behav* 1:1-5

Gómez-Aparicio L, Zamora R, Gómez JM, Hodar JA, Castro J, Baraza E (2004) Applying plant facilitation to forest restoration: a meta-analysis of the use of shrubs as nurse plants. *Ecol Appl* 14:1128-1138

Goodman DM, Durall DM, Trofymow JA, Berch SM (1996) Describing ectomycorrhizae. In: Goodman DM, Durall DM, Trofymow JA, Berch SM (eds) A manual of concise descriptions of North American ectomycorrhizae. Sydney, BC, Canada. Mycologue Publications 3A1-3A5

Goward T, Arsenault A (2000) Inland old-growth rain forests: safe havens for rare lichens? In: Darling LM (ed) Proceedings of a Conference on the Biology and Management of Species and Habitats at Risk, Kamloops, BC, 15-19 February, 1999. Vol. 2. BC Ministry of Environment, Lands and Parks, Victoria, BC, and University College of the Cariboo, Kamloops, BC, Canada, pp 759-766

Grogan P, Baar J, Bruns TD (2000) Below-ground ectomycorrhizal community structure in a recently burned bishop pine forest. *J Ecol* 88:1051-1062

Gryta H, Carriconde F, Charcosset J-Y, Jargeat P, Gardes M (2006) Population dynamics of the ectomycorrhizal fungal species Tricholoma populinum and Tricholoma scalpturatum associated with black poplar under differing environmental conditions. *Environ Microbiol* 8:773-778

Gunderson LH, Holling CS (2002) Panarchy: understanding transformations in human and natural systems. Island Press, Washington, DC

Hagerman SM, Sakibara SM, Durall DM (2001) The potential for woody understory plants to provide refuge for ectomycorrhizal inoculum at an interior Douglas-fir forest after clear-cut logging. *Can J For Res* 31:711-721

Halpern CB (1988) Early successional pathways and the resistance and resilience of forest communities. *Ecology* 69:1703-1715

Hamann A, Wang T (2006) Potential effects of climate change on ecosystem and tree species distribution in British Columbia. *Ecology* 87:2773-2786

Hansen J, Nazarenko L, Ruedy R, Sato M, Willis J, Del Genio A, Koch D, Lacis A, Lo K, Menon S, Novakov T, Perlwitz J, Russel G, Schmidt GA, Tausnev N (2005) Earth's energy imbalance: confirmations and implications. *Science* 308:1431-1435

Harms KE, Wright SJ, Calderón O, Hernandéz A, Herre EA (2000) Pervasive density-dependent recruitment enhances seedling diversity in a tropical forest. *Nature* 404:493-495

Hartnett DC, Wilson GWT (2002) The role of mycorrhizas in plant community structure and dynamics: lessons from grasslands. *Plant Soil* 244:319-331

Hawkes CV, Belnap J, D'Antonio C, Firestone MK (2006) Arbuscular mycorrhizal assemblages in native plant roots change in the presence of invasive exotic grasses. *Plant Soil* 281:369-380

Hobbs RJ, Harris JA (2001) Restoration ecology: repairing the earth's ecosystems in the new millennium. *Rest Ecol* 9:239-246

Holling CS (1996) Engineering resilience versus ecological resilience. In: Schulze P (ed) Engineering within ecological constraints. National Academy, Washington, DC, pp 31-44

Horton TR, Cázares E, Bruns TD (1998) Ectomycorrhizal, vesicular-arbuscular and dark septate fungal colonization of bishop pine (Pinus muicata) seedlings in the first 5 months of growth after wildfire. *Mycorrhiza* 8:11-18

Johnson EA, Fryer GJ, Heathcott MJ (1990) The influence of man and climate on frequency of fire in the interior wet belt forest, British Columbia. *J Ecol* 78:403-412

Jones MD, Smith SE (2004) Exploring functional definitions of mycorrhizas: Are mycorrhizas always mutualists. *Can J Bot* 82:1089-1109

Jones MD, Durall DM, Harniman SMK, Classen DC, Simard SW (1997) Ectomycorrhizal diversity on Betula papyrifera and Pseudotsuga menziesii seedlings grown in the greenhouse or outplanted in single-species and mixed plots in southern British Columbia. *Can J For Res* 27:1872-1889

Jones MD, Durall DM, Cairney JWG (2003) Ectomycorrhizal fungal communities in young forest stands regenerating after clearcut logging. *New Phytol* 257:399-422

Kalra YP, Maynard DG (1991) Methods manual for forest soil and plant analysis. For Can North. For Cent Inf Rep NOR-X-319

Kardol P, Bezemer TM, van der Putten WH (2006) Temporal variation in plant-soil feedback controls successions. *Ecol Lett* 9:1080-1088

Kjøller R, Bruns TD (2003) Rhizopogon spore bank communities within and among California pine forests. *Mycologia* 95:603-613

Klinka K, Worrall J, Skoda L, Varga P, Krajina VJ (2000) The distribution and synopsis of ecological and silvical characteristics of tree species of British Columbia. Canadian Cartographics Ltd., Coquitlam, Canada

Klironomos JN (2003) Variation in plant response to native and exotic arbuscular mycorrhizal fungi. *Ecology* 84:2292-2301

Kranabetter JM, Friesen J, Gamiet S, Kroeger P (2005) Ectomycorrhizal mushroom distribution by stand age in western hemlock - lodgepole pine forests of northwestern British Columbia. *Can J For Res* 35:1527-1539

Lieffers VJ, Stadt KJ, Feng Z (2007) Free-to-grow regeneration standards are poorly linked to growth of spruce in boreal mixedwoods. *For Chron* 83:818-824

Lodge DJ, Wentworth TR (1990) Negative associations among VA-mycorrhizal fungi and some mycorrhizal fungi inhabiting the same root system. *Oikos* 57:347-356

Ludwig D, Hilborn L, Walters C (1993) Uncertainty, resource exploitation and conservation: lessons from history. *Science* 260:17

McHugh TA, Gehring CA (2006) Belowground interactions with arbuscular mycorrhizal shrubs decrease with performance of pinyon pine and the abundance of its ectomycorrhizas. *New Phytol* 171:171-178

McKenzie D, Gedalof Z, Peterson DL, Mote P (2004) Climatic change, wildfire, and conservation. *Cons Biol* 18:890-902

Mitchell SJ, Beese WJ (2002) The retention system: reconciling variable retention with the principles of silvicultural systems. *For Chron* 78:397-403

Molina R, Massicotte H, Trappe JM (1992) Specificity phenomenon in mycorrhizal symbiosis: community-ecological consequences and practical implications. In: Allen MF (ed) Mycorrhizal functioning: an integrative plant-fungal process. Chapman & Hall, New York, pp 357-423

Morrison DJ, Pellow KW, Norris DJ, Nemec AFL (2000) Visible versus actual incidence of Armillaria root disease in juvenile coniferous stands in the southern interior of British Columbia. *Can J For Res* 30:405-414

Oliver C, Larson BC (1997) Forest stand dynamics: updated edition. Wiley, New York

Packer A, ClayK (2000)Soil pathogens and spatial patterns of seedling mortality in a temperate tree. *Nature* 404:278-281

Parminter J (1998) Natural disturbance ecology. In: Voller J, Harrison S (eds) Conservation biology principals of forested landscapes. University of British Columbia Press, Vancouver, Canada, pp 3-41

Pauly D, Christensen V, Guénette S, Pitcher TJ, Sumaila UR, Walters CJ, Watson R, Zeller D (2002) Towards sustainability in world fisheries. *Nature* 418, 689-695

Perry DA (1994) Forest Ecosystems. Johns Hopkins University Press, Baltimore, MD

Perry DA, Amaranthus MP, Borchers JG, Borchers SL, Brainerd RE (1989) Bootstrapping in ecosystems. *BioScience* 39:230-237

Peterson G, Allen CR, Holling CS (1998) Ecological resilience, biodiversity, and scale. *Ecosystems* 1:6-18

Philip LJ (2006) Carbon transfer between ectomycorrhizal paper birch (Betula papyrifera) and Douglas fir (Pseudotsuga menziesii). PhD thesis, University of British Columbia, Vancouver, Canada

Pojar J, Klinka K, Meidinger DV (1987) Biogeoclimatic ecosystem classification in British Columbia. *For Ecol Manage* 22:119-154

Sachs DL, Sollins P, Cohen WB (1998) Detecting landscape changes in the interior of British Columbia from 1975 to 1992 using satellite imagery. *Can J For Res* 28:23-36

Sanborn P, Geertsema M, Jull AJT, Hawkes B (2006) Soil and sedimentary charcoal evidence for Holocene forest fires in an inland temperate rainforest, east-central British Columbia, Canada. *Holocene* 16:415-427

Selosse M-A, Duplessis S (2006) More complexity in the mycorrhizal world. *New Phytol* 172:600-604

Selosse M-A, Richard F, He X, Simard SW (2006) Mycorrhizal networks: les liaisons dangeureuses. *Trends Ecol Evol* 21:621-628

Simard SW, Durall DM (2004) Mycorrhizal networks: a review of their extent, function and importance. *Can J Bot* 82:1140-1165

Simard SW, Vyse A (2006) Trade-offs between competition and facilitation: A case study of vegetation management in the interior cedar-hemlock forests of southern British Columbia. *Can J For Res* 36:2486-2496

Simard SW, Molina R, Smith JE, Perry DA, Jones MD (1997a) Shared compatibility of ectomycorrhizae on Pseudotsuga menziesii and Betula papyrifera seedlings grown in mixture in soils from southern British Columbia. *Can J For Res* 27:331-342

Simard SW, Perry DA, Smith JE, Molina R (1997b) Effects of soil trenching on occurrence of ectomycorrhizae on Pseudotsuga menziesii seedlings grown in mature forests of Betula papyrifera and Pseudotsuga menziesii. *New Phytol* 136:327-340

Simard SW, Perry DA, Jones MD, Myrold DD, Durall DM, Molina R (1997c) Net transfer of carbon between tree species with shared ectomycorrhizal fungi. *Nature* 388:579-582

Simard SW, Jones MD, Durall DM, Hope GD, Stathers RJ, Sorensen NS, Zimonick BJ (2003) Chemical and mechanical site preparation: effects on Pinus contorta growth, physiology, and microsite quality on steep forest sites in British Columbia. *Can J For Res* 33:1495-1515

Simard SW, Sachs DL, Vyse A, Blevins LL (2004) Paper birch competitive effects vary with conifer tree species and stand age in interior British Columbia forests: implications for reforestation policy and practice. *For Ecol Manage* 198:55-74

Smith JE, Johnston KA, Cázares E (1998) Vesicular mycorrhizal colonization of seedlings of Pinaceae and Betulaceae after spore inoculation with Glomus intraradices. *Mycorrhiza* 7:279-285

Smith JE, Molina R, Huso MMP, Luoma DL, McKay D, Castellano MA, Lebel T, Valachovic Y (2002) Species richness, abundance, and composition of hypogeous and epigeous ectomycorrhizal fungal sporocarps in young, rotation-age, and old-growth stands of Douglas-fir (Pseudotsuga menziesii) in the Cascade Range of Oregon, U.S.A. *Can J Bot* 80:186-204

Smith SE, Read DJ (1997) Mycorrhizal symbiosis, 2nd ed. Academic, London

Suding KN, Gross KL, Houseman GR (2004) Alternative states and positive feedbacks in restoration ecology. *Trends Ecol Evol* 19:46-53

Tedersoo L, Hansen K, Perry BA, Kjøller R (2006) Molecular and morphological diversity of pezisalen ectomycorrhiza. *New Phytol* 170:581-596

Teste FP, Simard SW (2008) Distance from mature trees and mycorrhizal networks alter patterns of competition and facilitation in dry Douglas-fir forests. *Oecologia* 158: 193–203.

Teste FP, Simard SW, Durall DM (2008) Role of mycorrhizal networks and tree proximity in ectomycorrhizal colonization of planted seedlings. Fungal Ecology, submitted.

Torres P, Honrubia M (1997) Changes and effects of a natural fire on ectomycorrhizal inoculum potential of soil in a Pinus halepensis forest. *For Ecol Manage* 96: 189-196

Twieg BD, Durall DM, Simard SW (2007) Ectomycorrhizal fungal succession in mixed temperate forests. *New Phytol* 176:437-447

Twieg BD, Durall DM, Simard SW, Jones MD (2008) Influence of stand age and soil properties on ectomycorrhizal communities in mixed temperate forests. Mycorrhiza, submitted.

van der Heijden MGA, Klironomos JN, Ursic M, Moutgoulis P, Streitwolf-Engel R, Boller T, Wiemken A, Sanders IR (1998) Mycorrhizal fungal diversity determines plant biodiversity, ecosystem variability and productivity. *Nature* 396:69-72

Visser S (1995) Ectomycorrhizal fungal succession in jack pine stands following wildfire. *New Phytol* 129:389-401

Wagner RG, Little KM, Richardson B, McNabb K (2005) The role of vegetation management for enhancing productivity of the world's forests. *Forestry* 79:57-79

Woods AJ (2003) Species diversity and forest health in northwest British Columbia. *For. Chron.* 79:892-897

Woods AJ (2006) Are free-growing stands meeting timber productivity expectations? Forest and Range Practices Act (FRPA) Effectiveness Evaluation. British Columbia Ministry of Forests and Range, Victoria, Canada, p 21

Woods AJ, Coates KD, Hamann A (2005) Is an unprecedented Dothistroma needle blight epidemic related to climate change. *BioScience* 55:761-769

Chapter 14
AMF Associated with Indigenous and Non-indigenous Plants at Urban and Desert Sites in Arizona

Robert J. Bills and Jean C. Stutz

Abstract Arbuscular mycorrhizal fungal (AMF) biodiversity from 30 sites throughout the Phoenix, USA, metropolitan area was compared to determine the impact of urbanization on AMF communities. Spores from pot cultures started with soil collected from non-indigenous and indigenous plants at urban sites and from indigenous plants at desert sites were identified. The total number of species detected, number of species per plant and species richness (no. of AMF species/sampling site) were fewer at the urban sites in comparison to desert sites, but were similar between urban sites with indigenous plants and those with non-indigenous plants. There was a significant overlap in the species composition between desert and urban sites with about 70% of the species detected at both urban and desert locations. The relative frequency of AMF species varied between desert and urban sites with several frequently detected species from desert areas (detected in ~40% of collected samples) detected in fewer than 10% of samples from urban areas. Although it appears that urbanization has an impact on AMF communities, this effect does not appear to be linked to the presence of non-indigenous plant species in these areas. Future studies will be needed to determine the drivers of alterations in AMF community structure in urban areas.

14.1 Introduction

Growing awareness of human impacts on ecosystems and biodiversity has resulted in increased research into how human impact biological communities (Austin 2004). For example, studies of arbuscular mycorrhizal fungal (AMF) communities have found that factors such as anthropogenic disturbance to plant communities (Egerton-Warburton and Allen 2000; Zhang et al. 2004), soil disturbance associated

R.J. Bills
School of Life Sciences, Arizona State University, Tempe, Arizona, USA

J.C. Stutz (✉)
Applied Biological Studies, Arizona State University Polytechnic, Mesa, Arizona, USA
e-mail: jstutz@asu.edu

C. Azcón-Aguilar et al. (eds.), *Mycorrhizas - Functional Processes and Ecological Impact*, 207
DOI: 10/1007-978-3-540-87978-7_14, © Springer-Verlag Berlin Heidelberg 2009

with agriculture (Abbott and Robson 1991; Abbott and Gazey 1994; Merryweather and Fitter 1998), and the intensity of that land-use (Oehl et al. 2003) influence and alter mycorrhizal-plant associations (see Chapter 13 by Simard).

Urbanization is one of the main demographic trends that results in land-transformation worldwide (Vitousek et al. 1997; Grimm et al. 2000). Roughly half of the world's population lives in urban areas with this number projected to rise to over 60% in the next 30 years (United Nations Population Division 1997). Developed countries are more highly urbanized with approximately 80% of the population of the United States residing in urban areas. Although urban areas currently occupy about 2% of the land surface of the earth, their ecological footprint extends beyond the city's boundaries with impacts on biogeochemical cycles, alterations of land use and cover, and impacts on biodiversity due to habitat fragmentation and introduction of non-indigenous species (Grimm et al. 2000).

Studies have shown that biological diversity and abundance may be dramatically altered in urban ecosystems when comparisons are made with biological communities in the surrounding areas. Urbanization is often associated with decreases in species richness of many taxa including mammals, birds, butterflies, and lizards (as reviewed by McKinney 2002). In contrast, species richness of plants (Hope et al. 2003) and some arthropod communities, including herbivorous arthropods, predatory arthropods, and detritivores (McIntyre 2000; Cook and Faeth 2006), often increase in urban environments due to the introduction of non-indigenous taxa and increases of available resources. Urbanization has also been shown to alter species composition and rank frequency. Studies of several groups, including birds, ground arthropods, plant-feeding arthropods (such as aphids and whiteflies), bees, and spiders, have found that the relative abundance of some indigenous and non-indigenous generalist species greatly increased with urbanization while the abundance of more specialized species decreased (Faeth et al. 2005). Exotic or non-indigenous bird (McKinney 2002) and earthworm (Szlavecz et al. 2006) species often dominate communities in urban areas.

Although the vast majority of terrestrial plants have been shown to form a symbiotic association in their roots with mycorrhizal fungi (Smith and Read 1997), little is known about the impact of urbanization on mycorrhizal fungal diversity and community structure. Physical factors associated with urbanization, such as soil disturbance associated with urban development and increases in atmospheric nitrogen deposition, air and soil pollutants, average ambient temperature (urban heat island effect), and soil compaction (Pickett et al. 2001; Grimm and Redman 2004), might be expected to impact mycorrhizal community structure. AMF communities might also be altered in urban areas by biotic factors such as the introduction of non-indigenous landscape plant materials, spatial variations in plant diversity (Hope et al. 2003), and human induced increases in primary productivity due to the application of water and fertilizers (Shochat et al. 2006).

There are a limited number of studies that make comparisons between mycorrhizal communities in urban areas with the surrounding natural ecosystems. When Baxter et al. (1999) compared ectomycorrhizal communities in *Quercus* (oaks) in rural and urban areas in the New York City area, they found fewer ectomycorrhizal types on mature oaks in urban soils in comparison to rural soils, although the number of ectomycorrhizal types/cm fine roots did not differ between urban and rural soils. Ectomycorrhizal colonization in oak trees was also lower in urban areas compared to rural areas. A study by Stabler et al. (2001) compared AMF diversity in urban sites in Phoenix, Arizona with sites in an adjacent desert remnant. They found differences in the number of AMF species detected and found some differences in species composition between urban and desert sites. Colonization of roots by AMF was greater in trees sampled from desert sites in comparison to urban sites. Cousins et al. (2003) compared AMF diversity at 20 sampling sites in the Phoenix metropolitan area and the surrounding desert and found that land use type, land use history, and vegetation type affected AM fungal community structure. Although mean species richness was not significantly different among land-use categories (urban, agriculture, desert), spore densities were higher at the desert sites than at the agriculture and urban sites. Since the focus of both the Stabler et al. (2001) and the Cousins et al. (2003) studies was on characterizing urban AMF communities, only a limited number of desert sites ($n = 3$) were examined.

In this current study, AM fungal diversity was assessed at additional sites in the Phoenix metropolitan area and compared to diversity at sites in the surrounding Sonoran desert. Several drivers have been proposed for alterations in animal community structure in urban ecosystems including altered habitat structure through the replacement of indigenous vegetation with managed vegetation in landscapes (McKinney 2002; Shochat et al. 2006), increased plant productivity due to inputs such as irrigation and fertilizers (McKinney 2002; Cook and Faeth 2006), competitive exclusion associated with the increase in habitat productivity (Shochat et al. 2006), temporal changes in habitat structure, the availability of resources such as food and water due to altered local climates and buffered seasonality (Shochat et al. 2006), and fragmentation of habitats and high spatial heterogeneity resulting from varying land use (Faeth et al. 2005). Because the study of Cousins et al. (2003) suggested that there might be differences between AMF communities associated with indigenous and non-indigenous plants, we decided to investigate if the introduction of non-indigenous landscape plants might be associated with changes in AMF community structure in urban ecosystems by selecting sites with non-indigenous, landscape trees and shrubs and comparing them with urban sites with indigenous trees and shrubs.

The research questions for this project were: (1) Does AMF diversity and species composition differ between urban areas and the surrounding desert? (2) Are there any differences in AM fungal diversity and species composition associated with indigenous plants and non-indigenous plants growing in urban areas? (3) Are there any differences in AM fungal diversity and species composition between indigenous plants growing in urban areas and in the surrounding desert? We hypothesized that the total number of AMF species detected and AMF species richness would be smaller at urban sites in comparison to desert sites and that a greater number of AMF species would be associated with indigenous plants in comparison to non-indigenous plants. We were also interested in determining if AMF species other than those commonly reported in the Sonoran desert (Stutz et al. 2000) were present at urban sites and if shifts in rank frequency occurred with urbanization.

14.2 Materials and Methods

Soil samples were collected from 30 sites in Maricopa County, Arizona, USA. The county is comprised of a land area of roughly 6,400 km^2 that includes the Phoenix metropolitan area. Maricopa County is located in the northeast portion of the Sonoran Desert and lies at the junction of the Lower Colorado River Valley and Arizona Upland Subdivisions (Turner and Brown 1994). Mean annual rainfall for the region is 180 mm and annual mean temperatures range from a maximum of 41 C in July to a minimum of 5.1 C in January. The city of Phoenix and the 24 surrounding municipalities make up one of fastest-growing metropolitan areas in the United States with population growth more than 40% since 1990 to a current population over 3.5 million people (Grimm and Redman 2004). Rapid expansion and conversion of the surrounding Sonoran Desert for anthropogenic use has created a heterogeneous mosaic of land-use types (Hope et al. 2003; Grimm and Redman 2004).

The sites used in this study were a subset of 204 sites that are part of the Central Arizona – Phoenix Long-Term Ecological Research (CAP LTER) Survey 200 (Cousins et al. 2003; Grimm and Redman 2004) in which spatially dispersed and unbiased sampling sites were selected using a dual-density, randomized, tessellation-stratified design (Hope et al. 2003). All sites were categorized by land-use: desert, urban-residential, urban non-residential, and agriculture (Hope et al. 2003). We selected a subset of these 204 sites based on land-use category and vegetation with ten sites categorized as urban with indigenous plants, ten sites categorized as urban with non-indigenous plants, and ten sites categorized as desert. Indigenous vegetation was defined in this study as vegetation traditionally found in the Sonoran Desert surrounding Phoenix. All but three of the urban sampling sites were located in residential landscapes. One of the urban sites with indigenous plants was located in a transportation/freeway corridor and a second was located in a vacant lot with indigenous vegetation. The third urban site with non-indigenous plants was located in the landscaped area of a church.

In Spring 2000 and 2003, soil samples were collected using a metal trowel at the 30 GPS-located sampling sites from the rhizosphere of the three woody plants (trees or shrubs) located nearest to the predetermined plot-center of each site. Samples were collected from 20 non-indigenous genera, all common landscape plants and most previously reported as forming arbuscular mycorrhizal communities, including *Ulmus* (elms), *Nerium* (oleanders), and several genera of palms. A limited number of plants from genera commonly reported as ectomycorrhizal such as *Pinus* (pines) were included due to the sampling protocol. Samples were collected from nine indigenous genera including *Atriplex* (salt-bush), *Encelia* (brittlebush), *Larrea* (creosotebush), and *Parkinsonia* (paloverde). Each soil sample was placed in a self-sealing plastic bag and stored at 4 C until analyzed. Because of low spore numbers in soil, soil samples were used to establish pot cultures to obtain AMF spores for identification. Two generations of pot cultures were grown in a greenhouse following the method of Cousins et al. (2003) with *Sorghum bicolor* (sudan grass) as the host plant. Pot cultures of field soils have been successfully used to induce sporulation of cryptic AMF species through the colonization of the trap plants by dormant propagules (spores and external hyphal segments) in soil allowing for a more complete assessment of diversity than a one-time sample of spores from field soils (Bever et al. 1996; Stutz and Morton 1996; Brundrett et al. 1999). Spores were collected from pot cultures by wet sieving and decanting, followed by sucrose gradient centrifugation (Daniels and Skipper 1982). Spores were observed using a stereomicroscope, and distinct morphotypes were mounted on slides in polyvinyl alcohol-lactic acid-glycerol (PVLG) (Koske and Tessier 1983) and PVLG mixed 1:1 (v/v) with Melzer's reagent. Identification was based on spore morphology observed using the light stereomicroscope and compound microscope and compared to voucher specimens and to descriptions from the International Culture Collection of Arbuscular and Vesicular-Mycorrhizal Fungi (INVAM) web page (http://invam.caf.wvu.edu/). Undescribed species of *Glomus* were referenced according to INVAM accession codes of the trap cultures from which the unique spore morphotypes were first detected.

The mean number of species per site and the frequency of occurrence of AMF species were determined (Cousins et al. 2003). Rank frequency differences of species richness per site and number of AMF species per plant were performed. A two-sample *t* test was performed to test the means of the urban and desert treatments and a one-way analysis of variance (ANOVA) was performed to test for significance of mean number of species between the three plant/land-use type categories. EstimateS (Colwell et al. 2004; http://viceroy.eeb.uconn.edu/estimates) was used to generate species sampling effort curves to determine the total number of AMF species in urban and desert communities. Relative frequency (the number of times a species was detected/total number of samples collected × 100) was used as an indication of species abundance because spore numbers were low in the soil and trap cultures were used to detect AMF. Because host plants are known to impact AM fungal communities (Johnson et al. 1992; Johnson et al. 2003), comparisons were also made between the AM fungal communities associated with the most common plant in the study, the indigenous plant *Larrea tridentata* ($n = 12$ at urban sites and $n = 15$ at desert sites). Data was analyzed using JMP 5.0.1 statistical software SAS Institute (2002) JMP. Version 5.0.1a. SAS Institute, Cary, NC.

14.3 Results

A total of 21 species of AMF were detected at 30 study sites in the Phoenix metropolitan area and the surrounding Sonoran Desert (Table 14.1) including three undescribed *Glomus* morphotypes (*Glomus* sp. AZ112, *Glomus* sp. AZ117 and *Glomus* sp. AZ123). There were 19 species of AM fungi detected at the desert sites and 17 species detected at the urban sites with 16 species detected at sites where samples were collected from indigenous vegetation and 16 detected at sites where samples were collected from non-indigenous vegetation (Table 14.1). Sampling effort curves, used to determine the total number of AM fungal species in urban and desert communities, suggests that most species from the urban sites were detected during this study (Fig. 14.1). Additional species could be detected at the desert sites if more samples were collected.

Table 14.1 AMF species detected in trap cultures started with soil collected at urban and desert sites located within the Phoenix, Arizona, USA, metropolitan area and the surrounding Sonoran Desert. Urban/indigenous, Urban/non-indigenous, and Desert/indigenous designate land-use and vegetation type. Species presence is notated with an X and species absence is notated with a dash (–)

AM fungal species	Authority	Urban/ indigenous	Urban/ non-indigenous	Desert/ indigenous
Archaeospora trappei	Ames & Linderman) Morton & Redecker emend. Spain	X	X	X
Acaulospora morrowiae	Spain & Schenck	–	X	–
Entrophospora infrequens	(Hall) Ames & Schneider	X	X	X
Diversispora spurca	(Pfeiff., Walker & Bloss) Walker & Schüßler	X	X	X
Glomus claroideum	(Schenck & Smith emend Walker & Vestberg	X	X	X
G. coremioides	(Berk. & Broome) Redecker & Morton	–	–	X
G. eburneum	Kennedy, Stutz & Morton	X	X	X
G. etunicatum	Becker & Gerd.	X	X	X
G. fasciculatum	(Thaxter) Gerd. & Trappe emend. Walker & Koske	X	–	X
G. intraradices	Schenck & Smith	X	X	X
G. luteum	Kennedy, Stutz & Morton	X	X	X
G. macrocarpum	Tulasne & Tulasne	X	X	–
G. microaggregatum	Koske, Gemma & Olexia	X	X	X
G. mosseae	(Nicolson & Gerd.) Gerd. & Trappe	X	X	X
G. sinuosum	(Gerd. & Bakshi) Almeida & Schenck	X	X	X
G. tortuosum	Schenck & Smith	–	–	X
G. versiforme	(Karston) Berch	–	–	X
G. sp. AZ112	Undescribed	X	X	X
G. sp. AZ117	Undescribed	–	–	X
G. sp. AZ123	Undescribed	X	X	X
Paraglomus occultum	(Walker) Morton & Redecker	X	X	X
	Total number of species detected	16	16	19

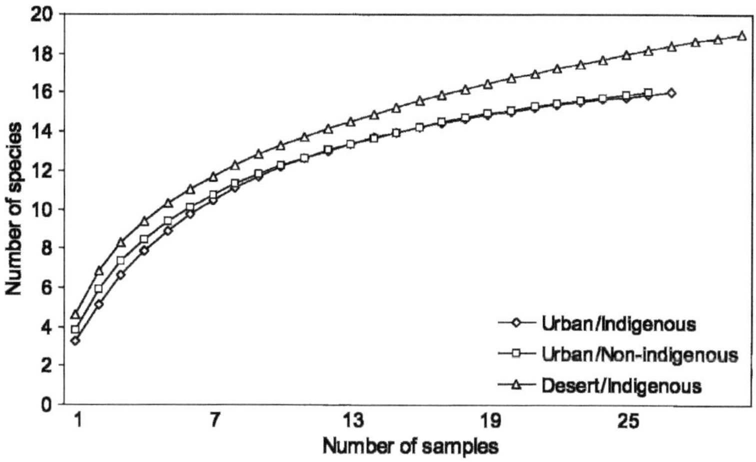

Fig. 14.1 Sampling effort curve for each land-use type/vegetation category located in the Phoenix, Arizona metropolitan area (*urban*) and the surrounding Sonoran Desert (*desert*)

Table 14.2 Sorenson's index of similarity for AMF species composition detected at sampling sites in the Phoenix, Arizona metropolitan area and the surrounding Sonoran Desert. Values closer to 1.00 represent species compositions that are more similar between land use and vegetation types

	Urban/non-indigenous	Desert/indigenous
Urban/indigenous	0.94	0.86
Urban/non-indigenous		0.80

There was an overlap of AM fungal species composition between the Phoenix urban and desert sites (Table 14.1) with about 70% of the species detected at both urban and desert locations. Four *Glomus* species (*G. coremoides*, *G. tortuosum*, *G. versiforme*, and *G.* sp. AZ117) were only detected at desert sites. Two species (*Acaulospora morrowiae* and *G. macrocarpum*) were only detected at urban sites. A Sorenson similarity coefficient analysis also confirmed the similarity of urban and desert sites to one another (Table 14.2). Although the urban sites had the greatest similarity to one another, the urban and desert sites had similarity index numbers greater than 0.80.

The number of species detected per plant differed between desert and urban sites ($F = 5.874$, $df = 2$, $P = 0.004$) with a greater overall mean number of AMF species detected per plant at the desert sites (4.63 ± 0.4) than at the urban sites with indigenous plants (2.9 ± 0.4) and urban sites with non-indigenous plants (3.3 ± 0.4). Because of the sampling protocol, soil was collected from a small number ($n = 6$) of landscape plants typically thought to be ectomycorrhizal such as pine and juniper. The number of species detected for these plants (3.0 ± 1.2) was similar to other non-indigenous plants. Differences in species richness (number of AM fungi detected per site) were observed ($F = 3.05$, $df = 2$, $P = 0.064$) between desert sites (8.10 ± 0.8), urban sites with indigenous plants (5.6 ± 0.7), and urban sites with non-indigenous plants (6.5 ± 0.7).

G. intraradices was the most frequently detected species and was detected at every site. Three other species, *G. mosseae*, *Diversispora spurca*, and *G. microaggregatum*, were detected at more than half of the sites. At the desert sites, seven species were found at more than half of the sites with *G. intraradices*, *G. microaggregatum*, and *G. mosseae* found at all the sites (Fig. 14.2). *D. spurca*, *Glomus* sp. AZ112, and *G. luteum* were found at seven or more sites. At the urban sites with non-indigenous plants, *G. intraradices* and *G. mosseae* were detected at all the sites. Five additional species, including *G. microaggregatum*, *D. spurca*, *G. eburneum*, *Paraglomus occultum*, and *G. claroideum* were found at five or more sites. At urban sites with indigenous plants, only four species, *G. intraradices*, *G. mosseae*, *G. eburneum*, and *Glomus* sp. AZ123, were detected at five or more sites. *Entrophospora infrequens* and *Glomus* sp. AZ123 were primarily detected at sites where indigenous plants were sampled.

When relative frequency (a ratio of the number of times a species was detected to the number of samples collected) was calculated, differences were observed for many AMF species between the desert and urban areas (Fig. 14.3). The relative frequency of *G. mosseae* and *G. microaggregatum* was greater than 70% when samples were collected from desert areas, while these species were detected in less than 50% of the samples collected from urban areas, with *G. microaggregatum* detected in less than 5% of the samples collected from indigenous plants in urban

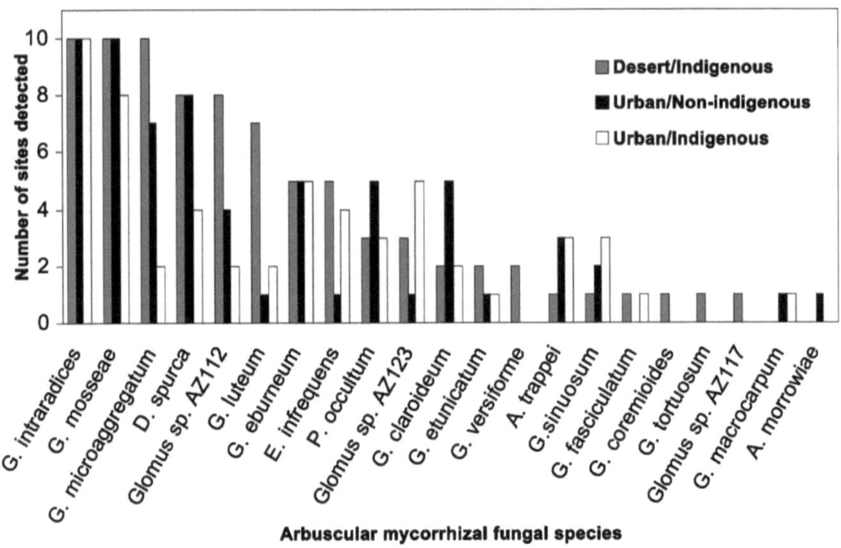

Fig. 14.2 Number of sites in which each AMF species was detected for each land-use/vegetation type (n = 10) located in the Phoenix, Arizona metropolitan area (*urban*) and the surrounding Sonoran Desert (*desert*)

areas. *G. luteum* and *G.* sp. AZ112 also had greater relative frequencies (40%) when samples were collected from desert areas in comparison to those collected from urban areas (less than 10%). In contrast, several rarer species including *P. occultum*, *Archeospora trappei*, and *G. sinuosum* had slightly greater relative frequencies in samples collected from urban areas in comparison to samples collected from the desert areas surrounding Phoenix.

When comparisons were made between the AMF communities associated with the most common plant in the study, the indigenous species *L. tridentata*, the mean number of AMF associated with plants growing at desert sites (4.1 ± 0.5 AMF species/plant) differed ($F = 3.83$, $df = 1$, $P = 0.062$) from the mean number of AMF species at urban sites (2.6 ± 0.6 AMF species/plant). The relative frequencies of AMF species associated with *L. tridentata* at desert sites were slightly different from those of AM fungi associated with *L.* tridentata at urban sites (Fig. 14.4). *G. intraradices* and *G. mosseae* had relative frequencies of greater than 90% when samples were collected from plants in desert areas surrounding Phoenix, while

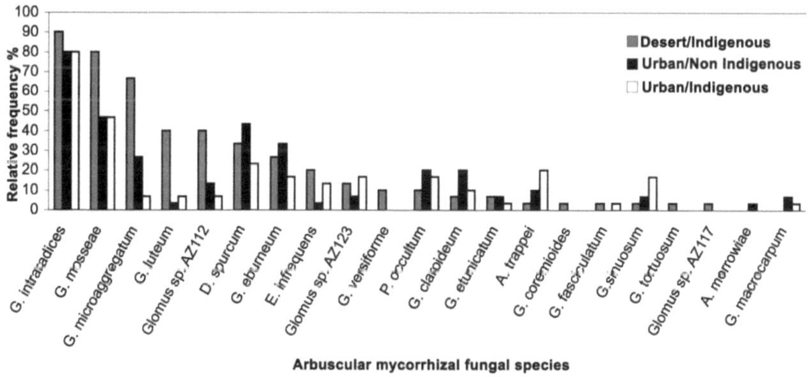

Fig. 14.3 Relative frequency of AMF for each land-use/vegetation type ($n = 30$) located in the Phoenix, Arizona metropolitan area (*urban*) and the surrounding Sonoran Desert (*desert*)

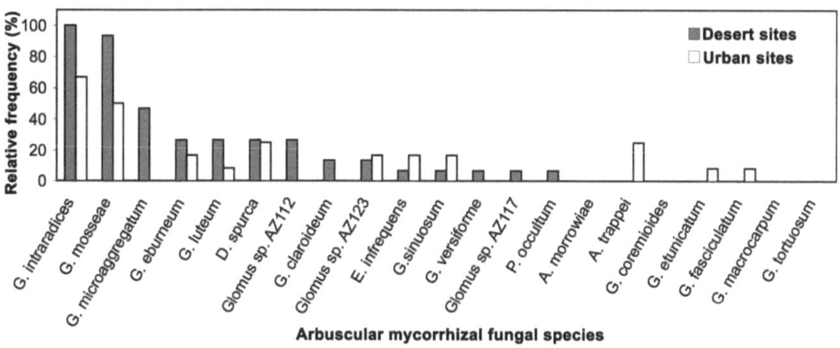

Fig. 14.4 Relative frequency of AMF species detected in soil samples collected from the rhizosphere of *L. tridentata* plants at urban and desert sites ($n = 12$ for urban, $n = 15$ for desert plants)

relative frequencies when samples were collected in urban areas were around 50%. *G. microaggregatum* and rarer species *G.* sp. AZ112, *G. claroideum*, *G. versiforme*, *G.* sp. AZ117, and *P. occultum* were only detected in samples collected at desert sites. Some rare species (*A. trappei*, *G. etunicatum*, and *G. fasciculatum*) were only detected in samples from urban sites.

14.4 Discussion

Several differences were observed in AMF community structure between urban and desert areas. The total number of AMF species detected, the mean number of AMF species per plant, and the mean number of species per site were greater in the desert sites in comparison to urban sites. These results were similar to the decreases in species richness that occur with urbanization for mammals, birds, butterflies, and lizards (McKinney 2002). These results did differ from the previous studies of Stabler et al. (2001) and Cousins et al. (2003) who did not find a significant difference in species richness between urban and desert sites in the Phoenix metropolitan area. One possible explanation for the contradictory findings of these studies and the current one may be the low number of sites sampled in the previous studies ($n = 3$ for desert sites). In addition, all the study sites of Stabler et al. (2001) and most of Cousins et al. (2003) were located in desert remnant parks. In the current study, most sampling sites were from the surrounding desert. Recent studies of ground arthropod communities in Phoenix have shown that remnant desert parks exhibit differences in species richness, composition, and trophic structure compared with the deserts surrounding Phoenix (Faeth et al. 2005). Future research is needed to ascertain if AMF communities of desert remnants differ from those in the surrounding desert.

Despite the differences between AMF communities from urban and desert sites in species richness, there was a significant overlap in the AMF species composition between these areas. Like other studies of arid environments (Jacobson 1997; Stutz et al. 2000; Azcón-Aguilar et al. 2003; Cousins et al. 2003; Tao and Zhiwei 2005; Beauchamp et al. 2006; Uhlmann et al. 2006), the AMF community at both urban and desert sites was dominated by members of the Glomaceae. Like previous reports of AMF diversity in urban soils in the Phoenix metropolitan area (Stutz and Martin 1998; Stabler et al. 2001; Cousins et al. 2003; Whitcomb and Stutz 2007), we did not commonly detect AMF species other than those commonly reported in the Sonoran desert (Stutz et al. 2000). The rank frequency of AMF species also varied between desert and urban sites with several frequently detected species from desert areas (detected in 40% of collected samples) detected in less than 10% of samples from urban areas. This result was similar to differences in rank frequency between desert and urban sites reported for birds, arthropods, and spiders (Faeth et al. 2005).

We did not find a significant difference between urban sites with indigenous and non-indigenous plants with respect to the total number of AM fungal species detected, the mean number of AM fungal species per plant, and the mean number of species per site. These results were unexpected because Cousins et al. (2003) suggest species richness in the Phoenix urban area was affected by the plants from which the species were collected. AMF species composition was also similar between urban sites with indigenous and non-indigenous plants. In fact, the Sorenson's similarity coefficients between the AM fungal communities associated with the vegetation types found in the urban sites were the most similar. Indigenous plants growing in urban sites had both a lower total number of AM fungal species and mean number of AM fungal species per plant than indigenous plants growing in the surrounding desert. This result was also unexpected because, in the study of Cousins et al. (2003), species richness was highest at sites where samples were collected from native vegetation regardless of whether plants were growing at urban or desert sites. As mentioned above, the differences between that study and the current study could be due to the lower number of study sites and differences in the location of sites classified as "desert" (desert remnant parks versus desert sites from areas surrounding Phoenix). In this current study, when only one indigenous plant was examined, *L. tridentata*, there was still a greater number of AM fungi associated with this plant at the desert sites than at the urban sites. This suggests that different plant species do not account for the differences observed.

Although the Sorenson's similarity coefficient showed that AM fungal communities between the indigenous plants at the urban site and desert site were similar, several species including *G. mosseae*, *G. microaggregatum*, *G. luteum*, and *G.* sp. AZ112 were less frequently associated with indigenous plants at urban sites than these types of plants at desert sites. Similar results were found when comparing the relative frequencies of these fungi associated with *L. tridentata* growing in desert and urban locations. Cousins et al. (2003) also found that *G. mosseae* was detected at over 50% of the desert sites but less frequently at the urban sites. The indigenous status of plants does not seem to be the driving force effecting differences in AMF species composition between the urban and desert sites. *G. intraradices* and *G. mosseae* were frequently detected at both urban and desert sites regardless of plant species and were also associated with *Larrea* at both the urban and desert sites. The decrease in detection of *G. microaggregatum* at the urban sites suggests that there is something other than indigenous status of the host influencing AM fungal species composition. Some rarer AMF (*D. spurca* and *G.* sp. AZ123) were detected with equal frequencies in soil collected from *L. tridentata* between the urban and desert sites, which may indicate a host preference for these species. Comparing AM fungal composition across all sites, *P. occultum* frequency was greater at the urban sites compared to the desert yet was not detected to be associated with urban *Larrea* plants. *E. infrequens* was detected with greater frequency in the desert plants compared to the urban, yet for *Larrea* plants, *E. infrequens* was detected with greater frequency at urban locations.

In conclusion, several differences were observed in AMF community structure between urban and desert areas. Since this study was limited to one sampling time, it is possible that additional sampling times and locations could result in a more complete picture of AMF diversity. Results of this study also seem to indicate that differences in environmental conditions between the Phoenix urban area and the surrounding desert had a much stronger influence on AMF diversity than the influence of indigenous and non-indigenous plants. The soil disturbance caused by the process of urbanization could be a reasonable explanation for the differences observed. An effect of disturbance may not be detected in this study, because the sites were not selected to have a range of time since development. A future study that examines AMF communities using space for time substitution design might tease out the effects of development. A preliminary report by Stutz and Martin (1998) found that AMF species richness increased with time since development. Increases in soil N due to atmospheric N deposition in urban areas and application of fertilizers as part of horticultural practice (Kaye et al. 2006) could also be factors influencing the differences in AMF communities observed between urban and desert sites. Simulation experiments exploring different amounts of N enrichment have been shown to be associated with decreases in species richness and shifts in AMF species composition (Johnson 1993; Egerton-Warburton and Allen 2000; Johnson et al. 2005) and to alter the interaction between host plants (Johnson et al. 2005). Future manipulative experiments would be necessary to ascertain the drivers that may be responsible for the differences in AMF communities between urban and desert areas.

Acknowledgments This material is based upon work supported by the National Science Foundation under Grant No. DEB-0423704, Central Arizona – Phoenix Long-Term Ecological Research (CAP LTER). Any opinions, findings, and conclusions or recommendation expressed in this material are those of the author(s) and do not necessarily reflect the views of the National Science Foundation (NSF).

References

Abbott LK, Gazey C (1994) An ecological view of the formation of VA mycorrhizas. Plant Soil 159:69–78

Abbott LK, Robson AD (1991) Factors influencing the occurrence of vesicular-arbuscular mycorrhizas. Agric Ecosyst Environ 35:121–150

Austin AT (2004) The human footprint in ecology – past, present and future. New Phytol 164: 419–422

Azcón-Aguilar C, Palenzuela J, Roldán A, Bautista S, Vallejo R, Barea JM (2003) Analysis of the mycorrhizal potential in the rhizosphere of representative plant species from desertification-threatened Mediterranean shrublands. Appl Soil Ecol 22:29–37

Baxter JW, Pickett STA, Carreiro MM, Dighton J (1999) Ectomycorrhizal diversity and community structure in oak forest stands exposed to contrasting anthropogenic impacts. Can J Bot 77: 771–782.

Beauchamp VB, Stromberg JC, Stutz JC (2006) Arbuscular mycorrhizal fungi associated with *Populus-Salix* stands in a semi arid riparian ecosystem. New Phytol 170:369–380

Bever JD, Morton JB, Antonovics J, Schultz PA (1996) Host-dependent sporulation and species diversity of arbuscular mycorrhizal fungi in a mown grassland. J Ecol 84:71–82

Brundrett MC, Abbott LK, Jasper DA (1999) Glomalean mycorrhizal fungi from tropical Australia I. Comparison of the effectiveness and specificity of different isolation procedures.. Mycorrhiza 8:305–314

Colwell RK, Mao CX, Chang J (2004) Interpolating, extrapolating, and comparing incidence-based species accumulation curves. Ecology 85:2717–2727

Cook WM, Faeth SH (2006) Irrigation and land use drive ground arthropod community patterns in urban desert. Environ Entomol 35:1532–1540

Cousins JR, Hope D, Gries C, Stutz JC (2003) Preliminary assessment of arbuscular mycorrhizal fungal diversity and community structure in an urban ecosystem. Mycorrhiza 13:319–326.

Daniels BA, Skipper HD (1982) Methods for the recovery and quantitative estimation of propagules from soil. In: Schenck NC (ed) Methods and principles of mycorrhizal research. American Phytopathological Society, St. Paul, MN, pp 29–35

Egerton-Warburton LM, Allen EB (2000) Shifts in arbuscular mycorrhizal fungal communities along an anthropogenic nitrogen deposition gradient. Ecol Appl 10:484–496

Faeth SH, Warren PS, Shochat E, Marussich WA (2005) Trophic dynamics in urban communities. BioScience 55:399–407

Grimm NB, Redman CL (2004) Approaches to the study of urban ecosystems: The case for central-Phoenix. Urban Ecosyst 7:199–213

Grimm NB, Grove JM, Pickett ST, Redman CL (2000) Integrated approaches to long-term studies of urban ecological systems. BioScience 50:571–584

Hope D, Gries C, Zhu W, Fagan WF, Redman CL, Grimm NB, Neson AL, Martin C, Kinzig A (2003) Socioeconomics drive urban plant diversity. Proc Natl Acad Sci USA 100:8788–8792

Jacobson KM (1997) Moisture and substrate stability determine VA-mycorrhizal fungal community distribution and structure in an arid grassland. J Arid Environ 35:59–75

Johnson D, Vandenkoornhuyse PJ, Leake JR, Gilbert L, Booth RE, Grime JP, Young JPW, Read DJ (2003) Plant communities affect arbuscular mycorrhizal fungal diversity and community composition in grassland microcosms. New Phytol 161:503–515

Johnson NC (1993) Can fertilization of soil select less mutualistic mycorrhizae? Ecol Appl 3:749–757

Johnson NC, Tilman D, Wedin D (1992) Plant and soil controls on mycorrhizal fungal communities. Ecology 73:2034–2042

Johnson NC, Wolf J, Reyes MA, Panter A, Koch GW, Redman A (2005) Species of plants and associated arbuscular mycorrhizal fungi mediate mycorrhizal responses to CO2 enrichment. Global Change Biol 11:1156–1166

Kaye JP, Groffman PM, Grimm NB, Baker LA, Pouyat RV (2006) A distinct urban biogeochemistry? Trends Ecol Evol 21:192–199

Koske RE, Tessier B (1983) A convenient, permanent slide mounting medium. Mycol Soc Am Newsl 34:59

McIntyre NE (2000) Ecology of urban arthropods: a review and a call to action. Ann Entomol Soc Am 93:825–835

McKinney ML (2002) Urbanization, biodiversity, and conservation. BioScience 52:883–890

Merryweather J, Fitter A (1998) The arbuscular mycorrhizal fungi of *Hyacinthoides non-scripta* I. Diversity of fungal taxa. New Phytol 138:117–129

Oehl F, Sieverding E, Ineichen K, Mäder P, Boller T, Wiemken A (2003) Impact of land use intensity on the species diversity of arbuscular mycorrhizal fungi in agroecosystems of central Europe. Appl Environ Microbiol 69:2816–2824

Pickett ST, Cadenasso ML, Grove JM, Nilon CH, Pouyat RV, Zipperer WC, Costanza R (2001) Urban ecological systems: Linking terrestrial, ecological, physical, and socioeconomic components of metropolitan areas. Ann Rev Ecol System 32:127–157

SAS Institute (2002) JMP. Version 5.0.1a. SAS Institute, Cary, NC.

Shochat E, Warren PS, Faeth SH, McIntyre NE, Hope D (2006) From patterns to emerging processes in mechanistic urban ecology. Trends Ecol Evol 21:186–191

Smith SE, Read DJ (1997) Mycorrhizal symbiosis, 2nd Edition. Academic, San Diego, CA

Stabler LB, Martin CA, Stutz JC (2001) Effect of urban expansion on arbuscular mycorrhizal fungal mediation of landscape tree growth. J Arbor 27:193–201

Stutz JC, Martin C (1998) Arbuscular mycorrhizal fungal diversity associated with ash trees in urban landscapes in Arizona. Phytopathology 88:586

Stutz JC, Morton JB (1996) Successive pot cultures reveal high species richness of arbuscular mycorrhizal fungi in arid ecosystems. Can J Bot 74:1883–1889

Stutz JC, Copeman R, Martin CA, Morton JB (2000) Patterns of species composition and distribution of arbuscular mycorrhizal fungi in arid regions of southwestern North America and Namibia, Africa. Can J Bot 78:237–245

Szlavecz KS, Placella A, Pouyat RV, Groffman PM, Csuzdi C, Yesilonis I (2006) Invasive earthworm species and nitrogen cycling in remnant forest patches. Appl Soil Ecol 32:54–62

Tao L, Zhiwei Z (2005) Arbuscular mycorrhizas in a hot and arid ecosystem in southwest China. Appl Soil Ecol 29:135–141

Turner RM, Brown DE (1994) Sonoran desert scrub. In: Brown DE (ed) Biotic communities: Southwestern United States and northwestern Mexico. University of Utah Press, Salt Lake City, Utah, pp 181–222

Uhlmann E, Görke C, Petersen A, Oberwinkler F (2006) Arbuscular mycorrhizae from arid parts of Namibia. J Arid Environ 64:221–237

United Nations Population Division (1997) Urban and rural areas 1950–2030 (the 1996 revision). New York, United Nations

Vitousek PM, Mooney HA, Lubchenco J, Melillo JM (1997) Human domination of the earth's ecosystems. Science 277:494–499

Whitcomb S, Stutz JC (2007) Assessing diversity of arbuscular mycorrhizal fungi in a local community: role of sampling effort and spatial heterogeneity. Mycorrhiza 17: 429–437

Zhang Y, Guo L-D, Liu R-J (2004) Survey of arbuscular mycorrhizal fungi in deforested and natural forestland in the subtropical region of Dujiangyan, southwest China. Plant Soil 261:257–263

Chapter 15
Desert Truffle Cultivation in Semiarid Mediterranean Areas

Asunción Morte, Mar Zamora, Almudena Gutiérrez, and Mario Honrubia

Abstract This chapter focusses on showing some taxonomic and ecological characters of the main desert truffle species in Spain as well as the main biotechnological procedures followed to produce mycorrhizal plants with some of these fungal species. The full micropropagation protocol of the host plant *Helianthemum violaceum* is described for the first time. Five different ways of producing mycorrhizal synthesis between desert truffles and the *Helianthemum* species are reported, according to the type of fungal inoculum, plant source and culture conditions used. Finally, a management protocol for the established desert truffle plantations is discussed. Irrigation, one of the most important factors considered for successful cultivation, should be applied at the end of the summer during dry years when rainfall is less than 150 mm and for a second time at the beginning of the fruiting season in very dry years.

15.1 Introduction

The term 'desert truffles' comprises species of different hypogeous Ascomycetes genera, such as *Terfezia, Balsamia, Delastreopsis, Delastria, Leucangium, Mattirolomyces, Phaeangium, Picoa, Tirmania* and some *Tuber* species. The name 'desert truffles' matches the nature of its distribution, which is typical of countries or territories with arid and semiarid conditions (Honrubia et al. 1992). All these fungi require localised outbreaks of rain to provide soil with a water supply for them to develop and fructify.

Among ascomycetous truffles, several genera have an excellent record as edible fungi, two of which are of considerable economic importance: *Terfezia* and *Tuber*. Of these two genera, only a few species of *Tuber* have been cultivated commercially

A. Morte (✉), M. Zamora, A. Gutiérrez, and M. Honrubia
Departamento de Biología Vegetal (Botánica),
Facultad de Biología, Universidad de Murcia,
Campus de Espinardo, 30100, Murcia, Spain
e-mail: amorte@um.es

C. Azcón-Aguilar et al. (eds.), *Mycorrhizas - Functional Processes and Ecological Impact*,
DOI: 10/1007-978-3-540-87978-7_15,

until now. Recently, however, biotechnological methods to produce both fungal inoculum and mycorrhizal plants, as well as new plantation management techniques, have been developed to cultivate some species of the *Terfezia* genus (Morte et al. 2008). The name *Terfezia* comes from the word 'Terfass' ('Terfess'), the name given to these hypogeous fungi in Arabic countries, although they are also known as 'sand truffles' (Khabar et al. 2001; Honrubia et al. 2003). Interest in desert truffles is not merely culinary or gastronomic, and not even purely commercial, since truffles also offer a high nutritional-dietary value (Murcia et al. 2002, 2003). Moreover, they are a priori 'ecological' given that they do not require any chemical input to be produced and because of their potential for promoting rural development, if we take into account the possibility to establish these fungi as an alternative or complementary crop in a sustainable farming context (see Chapter 1 by Atkinson).

Desert truffles have been known since ancient times and have been associated with Mediterranean cultures since their origins. We now know they were traded by the Greeks and Romans alike, and were imported from Libya to be sold in various markets of their respective Empires (Honrubia et al. 2007). Nowadays, they continue to be marketed and consumed.

The first references to truffles in the Mediterranean region were by Theophrastus, Dioscorides and Pliny. The first refers to Terfezias from Asia Minor under the name of '*Mizy*' or '*Mison*', defined as a 'vegetable without a root,…whose origin is in thunder,… the more thunder, the more truffles that grow' (Delmas 1989; Reyna 2000). Ethnomycological studies on desert truffles (Cano 2003; Honrubia et al. 2003, 2007; Mandeel and Al-Laith 2007) have suggested that the cultivation of truffles could contribute to improving the rural economic development in the semiarid Mediterranean areas where they can be cultivated.

15.2 Biodiversity, Geographical Distribution and Ecology of Desert Truffles

The genus *Terfezia* includes 12 species (Kirk et al. 2001), of which only 3 (*T. arenaria* (Moris) Trappe, *T. claveryi* Chatin and *T. leptoderma* Tul. & Tul.) are commercially valued in Spain because of their gastronomic interest and their crop yields. Two other species, *T. boudieri* Chatin and *T. olbiensis* Nees, which are also harvested for consumption purposes, have a lower commercial impact given their limited presence and the fact that they are poorer in taste than the other species (Gutiérrez et al. 2004). The truffle variety *Picoa lefebvrei* (Pat.) Maire is widely distributed and is known to have excellent antioxidant properties (Murcia et al. 2002); however, is not marketed because of its small size.

The most important macroscopic characteristics that distinguish the most common *Terfezia* species in Spain are summarised in Table 15.1. *Terfezia boudieri* is a species close to *T. claveryi*, and some authors consider them synonyms. In any case, their macro-morphological nature, ecology, hosts, fruiting periods, etc. are extremely similar. The greatest morphometric difference is noted at a microscopic level and lies

Table 15.1 Main characteristics for recognising the different marketable *Terfezia* species in Spain

Terfezia species	pH of soil	*Peridium*	*Gleba*	Ornamentation of spores	ø spores (μm)
T. claveryi	Basic	Thick/dark	Pink	Reticule	17–24
T. boudieri	Basic	Thick/dark	Pink	Wart	17–23
T. olbiensis	Basic	Thin/pale	White	Short thorn	16–20
T. arenaria	Acid	Thick /dark	Pink	Truncated wart	23–30
T. leptoderma	Acid	Thin/pale	Green	Long thorn	17–24

in spore ornamentation. In *T. claveryi*, the spore presents a well-defined reticulum upon which slightly prominent warts might appear in accordance with the degree of maturity, whereas pronounced and sometimes truncated warts appear in the spores of *T. boudieri*, which resemble those of *T. arenaria*. This character is insufficient to separate both species because of a possible continuum between the ornamentation of one species and that of the other, given the presence of verrucose growths on the reticulum nodes of the *T. claveryi* spores. Nonetheless, molecular studies of gene sequencing have detected notable differences between populations of both species, but caution is still needed (see Chapter 10 by Morton). In short, for the time being, they ought to be kept separate in taxonomical terms (Gutiérrez 2001). At any rate, the most frequent and sought after species of *Terfezia* for consumption purposes in Spain (basic soils) is *T. claveryi*. In the market place, however, it is quite possible to find carpophores of *T. boudieri* mixed with those of *T. claveryi* since even the rural population cannot distinguish between the two species.

The geographical distribution of desert truffles is limited to arid and semiarid lands, mostly in countries around the Mediterranean sea such as central and southern Spain, Portugal, Italy, France, Hungary, Turkey, extending from Morocco to Egypt, Israel, the Arabian Peninsula, Iran, Iraq, Syria and Kuwait. Moreover, some desert truffle species have been found in South Africa (Botswana) (Marasas and Trappe 1973), North America (Trappe and Sundberg 1977) Japan and China (Wang, personal communication).

Generally, the regions where desert truffles grow have an annual rainfall ranging from 50 to 380 mm. The truffle season produces good yields if rainfall ranges from 70 to 120 mm in North African countries, and from 100 to 350 mm in countries of southern Europe. The distribution of this rainfall is as important as the quantity; that is, rain is necessary no later than the beginning of December in North African and Middle Eastern countries and no later than the beginning of October in the countries of south Europe (Morte et al. 2008).

Most desert truffles establish mycorrhizal symbiosis with species of the *Helianthemum* genus (relatives of the common rock rose cultivars) of the *Cistaceae* family. Thus, the distribution and ecology of desert truffles are related to those of their host plants. Below, we provided a short description of the characteristics of the main desert truffles collected in Spain:

- *Terfezia arenaria* (Moris) Trappe: This is an acidophilic species of the Mediterranean area, which has developed mycorrhizal associations with *Tuberaria guttata* (L.) Fourr. (Calonge et al. 1977; Khabar et al. 2001), an annual species of the Cistaceae family. It therefore grows in herbaceous grasslands, areas that are open, rarely mountainous with gentle slopes, sandy, wet and poor in organic matter. It is widely distributed throughout southern and western Spain. Its fruiting period is in spring during the flowering period of its host plant.

- *Terfezia claveryi* Chatin: This species is widely distributed around the Mediterranean region, from countries of North Africa to Asia Minor. In the Iberian Peninsula, it is also widespread in southern, southeastern and central areas, from sea level up to around 1,100 m a.s. l., always in soils that are basic, carbonated and clayey, or in sandy soils on the coast. *Terfezia claveryi* has also been found in the Canary Islands, specifically in Lanzarote (Honrubia et al. 2007). It was first reported under the name *Tirmania pinoyi* (Maire) Malençon by Rodríguez et al. (1988), and later by Calonge (1990) as *T. claveryi*. This species is associated with different rockrose species of the genus *Helianthemum*, including chamaephytes, hemicrytophytes and therophytes. Consequently, it is located in open, sunny scrubland, or in the meadows of mountain plains. Like the previous species, it is vernal, although its fruiting period starts once its host has finished flowering. Occasionally, its host's flowering periods may start earlier or be delayed, which results in changes in the carpophore production period.
- *Terfezia leptoderma* Tul. & Tul.: This is a typical Mediterranean, acidophilic species that appears in spring, and which is mainly associated with members of the rock rose family of *Tuberaria guttata*. This is a particularly common species in southwestern Spain, where it is highly appreciated. It has a similar ecology to that of *T. arenaria*. Its size, peridium and, particularly, its spore ornamentation at microscopic level, distinguish it from *T. arenaria*.
- *Terfezia olbiensis* Nees: This is a poorly known species which is relatively common in southeast areas of the Iberian Peninsula, where it is consumed indiscriminately with *T. claveryi*. It is collected in spring in open pine groves with rock rose srub.
- *Picoa lefebvrei* (Pat.) Maire: This species appears in spring. It shares its ecological niche with the rest of the basiphilic *Terfezias*. Its fruiting period precedes that of *T. claveryi* by several weeks. Its lower ecological needs permit it to occupy a more dispersed area, and its productivity is much greater than that of any of the other species of the *Terfezia* genus. Nonetheless, its small size and the pubescence in its peridium, which means it accumulates soil debris, lowers its commercial value. For this reason, only the larger-sized fungi are eaten since they have a pleasant taste and a high culinary value given their lipid content (Murcia et al. 2002).

In general, desert truffles grow much more prolifically than *Tuber* species. Sizable quantities of several species of wild *Terfezia* are collected and marketed in southern Europe, North Africa and other countries bordering the Mediterranean sea. However, natural areas where desert truffles grow naturally have gradually disappeared. Large areas of the coastal desert in Egypt and Libya were mined in World War II. More recently, in Kuwait, the effects of the 1990–1991 Gulf War have apparently ruined many truffle-gathering areas, whereas in Europe, the widespread construction that has taken place in these 'sunny' areas over the last years is the underlying cause (Morte et al. 2008). In Spain, the natural production of this species is not controlled and there are no official data, but only approximate, barely reliable estimates for some regions. The areas where *Terfezia* grow have diminished partly because of the transformation in land use in these areas, including reforestation projects (mainly with *Pinus halepensis* Miller) and also due to the fact that mountain vegetation is becoming denser as traditional practices such as grazing decrease (Cano 2003).

15.3 Biotechnological Aspects of Mycorrhizal Plant Production

All species of the host plant *Helianthemum* display erratic seed germination, and it is necessary to scarify seeds to increase germination rates (Pérez-García and González-Benito 2006). In addition, more than 50% of germinated seedlings normally die after 4–6 weeks during nursery production. For these reasons, micropropagation of these plant species is an interesting option for plant production since adult plants are produced directly. Micropropagation is also associated with mass production at a competitive price.

Among the *Helianthemum* species able to establish mycorrhizal symbioses with desert truffles, only *H. almeriense* has been successfully micropropagated (Morte and Honrubia 1992, 1997), and its distribution is restricted to southeast Spain and North Africa (Morocco). Recently, a protocol for the in vitro propagation of *H. violaceum* has also been described (Zamora et al. 2006). This species inhabits a larger territory than *H. almeriense*, not only in Spain but also in south Europe and North Africa. The full micropropagation protocol for *H. violaceum* is described as follows:

- Initial explants: Shoot tips and nodal segments of axenically germinated seeds. After scarification, mature seeds were surface-sterilised by a 30-min treatment in 20% commercial bleach solution, followed by three 5-min rinses in sterile distilled water, and were then transferred to basal media. The percentage of axenically germinated seed was 75%.
- Culture conditions: 22 ± 2 C, 40 μmol m^2 s Growlux fluorescent light and 16-h photoperiod.
- Culture medium: Murashige and Skoog (MS) (1962), with 0.8% agar, 3% sucrose, pH 5.8. This medium was selected because it is the most suitable for Cistaceae species (M'Kada et al. 1991; Morte and Honrubia 1992, 1997; Iriondo et al. 1995).
- Multiplication stage: the propagation system followed with these species was mainly the formation of axillary buds and the culture of nodal segments. The incorporation of kinetin to the medium caused the elongation of the explants, which allows the obtaining of several nodal segments. This was the same propagation system followed for *H. almeriense* (Morte and Honrubia 1992, 1997).

Kinetin was more effective than the other cytokinin tested (6-benzyladenine, BA) in the production of shoots (Table 15.2). The best multiplication rates were obtained on the MS medium without a plant growth regulator (2.39), or with 0.23 and 0.46 μM kinetin (2.55 and 2.60, respectively) for 4 weeks (Table 15.2). However, they were three times lower than those obtained for *H. almeriense*, with 0.46 μM kinetin (7.72) (Morte and Honrubia 1992, 1997). The concentration of 0.23 μM kinetin was selected for propagation because the presence of kinetin reduced variations in the multiplication rate during subcultures (Fig. 15.1). Moreover, explants showed the lowest percentages of chlorosis, hyperhydricity, tip necrosis and callus formation with this kinetin concentration (Table 15.3).

- Elongation stage: shoots elongated in the same multiplication medium, avoiding the need for a new subculture. This was due to the propagation system used: the formation of nodal segments, which resulted in explants of up to 10 cm long in the multiplication stage. Furthermore, the type of cytokinin used and its relative low concentration allowed the axillary shoots to reach an adequate length at the end of the multiplication stage for their direct use at the following rooting stage. In *H. violaceum*, therefore, the shoot multiplication rate is directly related to shoot elongation and to the number of nodal explants available in each species. These results agree with those obtained for *H. almeriense* (Morte and Honrubia 1992, 1997).
- Rooting stage: spontaneous rooting (25–75%) was observed during the simultaneous multiplication and elongation stages in culture media with a low level of cytokinins. This rooting percentage was improved to 100% with full MS medium and without plant growth regulators after 5 weeks. Contrary to *H. almeriense*, macronutrient dilution did not improve the number of roots per explant nor the rooting rate, but it did increase shoot tip necrosis for the two dilutions tested, as well as the chlorosis and callus formation in ¼ and ½ dilutions, respectively (Table 15.4).
- Weaning stage: plantlets were transferred to pots containing a peat-sand-vermiculite mixture (2:1:1, v/v), and gradually exposed to reduced relative humidity for 2–3

Table 15.2 Influence of 6-benzyladenine (BA) and kinetin on the number of microshoots produced from shoot tips and nodal segments of *H. violaceum* cultured in MS medium. Data in a column followed by the same letter are not significantly different (P 0.05) as determined by Duncan's test

Plant growth regulators (μM)		
Kinetin	BA	No of shoots per explant
–	–	2.39 ab
0.23	–	2.55 a
0.46	–	2.60 a
0.93	–	2.00 bc
2.32	–	1.60 c
4.60	–	0.84 d
–	0.22	0.81 d
–	0.89	0.52 de
–	2.22	Callus e
–	4.40	Callus e

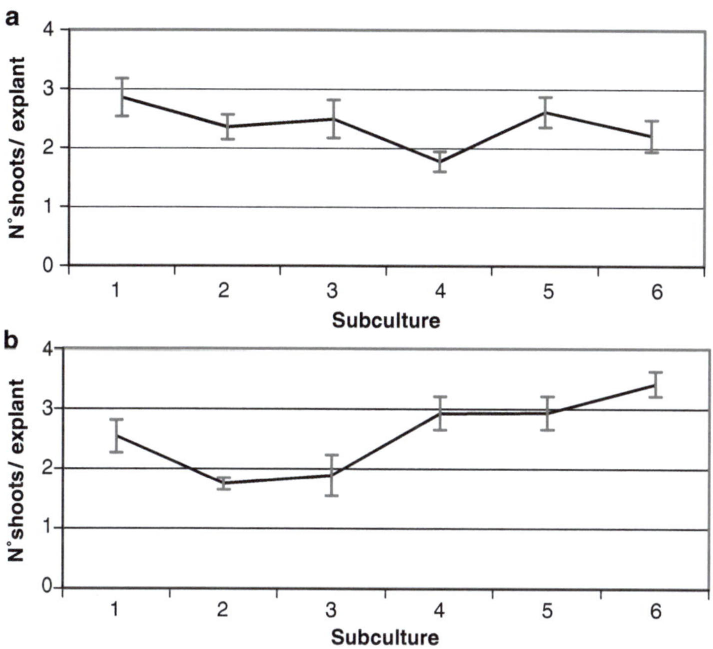

Fig. 15.1 Variation of the multiplication rate of *H elianthemum violaceum* explants cultured on the MS medium without plant growth regulators (**a**) and with 0.23 μM of kinetin (**b**), during six subcultures of 4 weeks each

Table 15.3 Influence of kinetin concentration on percentage of chlorosis, hyperhydricity, tip necrosis and callus formation of *H. violaceum*. Values are the means during the multiplication stage of six subcultures. Data in a column followed by the same letter are not significantly different (*P* 0.05) as determined by Duncan's test

Kinetin (μM)	Chlorosis (%)	Hyperhydricity (%)	Tip necrosis (%)	Callus formation (%)
0	30.6 ab*	3.1 ab	14.6 a	11.0 a
0.23	9.7 a	0.5 a	14.4 a	12.6 a
0.46	19.2 ab	0.9 a	19.3 ab	24.0 a
0.93	15.8 ab	3.5 ab	21.1 ab	16.4 a
2.32	44.9 b	7.5 b	34.4 bc	25.8 a
4.60	42.8 b	14.3 c	42.6 c	54.9 b

Table 15.4 Influence of the concentration of macronutrients of MS medium on in vitro rooting of *H. violaceum* explants without growth regulators after 5 weeks. $n = 210$ explants for each medium culture

Medium	No roots	Rooting (%)	Root length (cm)	Tip necrosis (%)	Chlorosis (%)	Callus formation (%)
MS0	6.5	100	2.8	23.3	30	3
1/2MS	5.5	96.6	3.1	66.7	30	0
1/4MS	4.7	80	1.2	70	96.7	6.7

weeks in the greenhouse. At the end of 1 month, approximately 95% of the plantlets survived.

The in vitro micropropagation protocols set up for the *Helianthemum* species (*H. almeriense* and *H. violaceum*) are quite rapid (about 10 weeks) because plant multiplication, elongation and rooting occur in the same subculture. Consequently, it is also a cheap protocol because only a small amount of plant growth regulators and labour are required. To obtain plants that are suitable for fungal inoculation from seeds germinated under nursery conditions takes at least 6 months. This traditional propagation method is cheaper than in vitro propagation. Both methods can be used for plant production, although it is advisable to use micropropagation when possible (Morte et al. 2008).

For the mycorrhizal synthesis, both seedlings and micropropagated plants of these *Helianthemum* species, together with spores and mycelium (from *T. claveryi*, *T. olbiensis* and *P. lefebvrei*), have been successfully used (Fig. 15.2). Five options have been designed not only to obtain continuous desert truffle mycorrhizal plant production, but also to avoid any limiting factor in this production (Fig. 15.2):

(1) Plant seedlings and mature spores: seedlings must be 5–6 months old before fungal inoculation to develop a good root system that is suitable for mycorrhizal establishment. These 6-month-old seedlings are inoculated with desert truffle spore suspensions and three further months are needed to get adequately mycorrhizal plants. Therefore, 9 months in total are required, the longest time to obtain mycorrhizal plants.
(2) Seedlings with a mycelial suspension obtained from fermentation in a bioreactor: the saving in time results from the rapid root colonisation by the mycelium (only 1.5–2 months). This procedure to grow the mycelium in bioreactors has been successful only for *T. olbiensis,* strain 111ET. The production of *T. olbiensis* mycelial inoculum was carried out in MMN liquid medium with a Braun BIOSTAT® B 5-L fermentor (Morte et al. 2004, 2008).
(3) Acclimatised micropropagated plants with the spore suspension: the 2-month reduction in time is due to the rapid plant micropropagation. After 4-month acclimatisation of the micropropagated plants, they have a suitable root system for mycorrhizal establishment. However, 3 months are necessary for spores to adequately colonise the *Helianthemum* roots.
(4) Micropropagated *Helianthemum* plants and pieces of agar with mycelium: on sterilised vermiculite watered with MH liquid medium (Morte and Honrubia 1994, 1995). Only 21–30 days are required to transfer the in vitro mycorrhizal plants to ex vitro conditions.
(5) Micropropagated *Helianthemum* plants with mycelial suspension obtained from fermentation in a bioreactor: 2 months are required to obtain in vitro rooted plantlets, whereas 1.5–2 months is needed for in vitro mycorrhization on the MH medium as described in (4).

In short, mycorrhizal synthesis between desert truffles and the *Helianthemum* species can be carried out in different ways according to the type of fungal inoculum (spores or mycelium), the plant source (seedlings or micropropagated plantlets) and

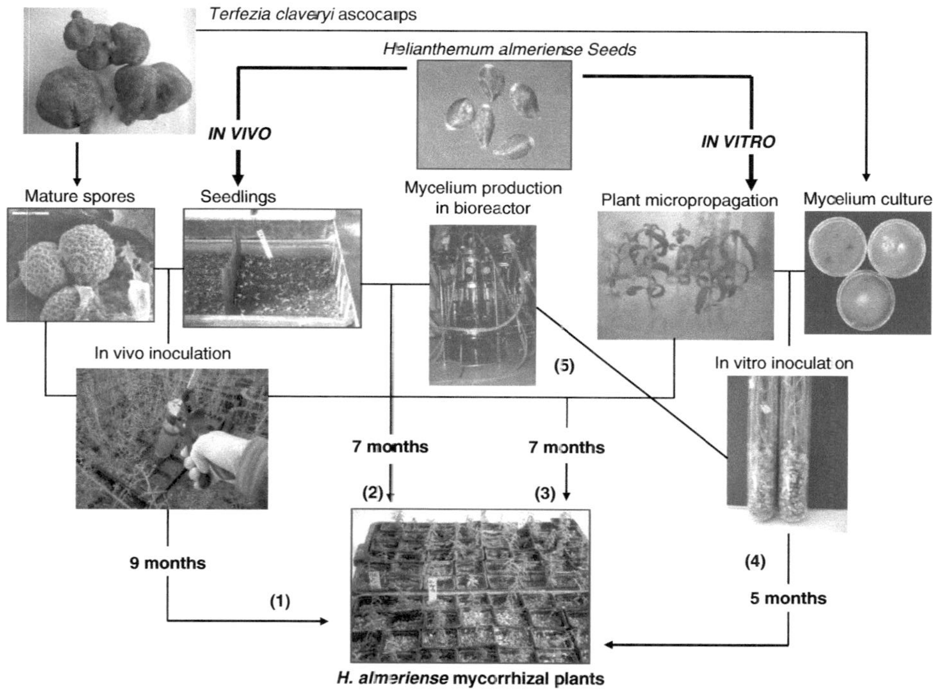

Fig. 15.2 Schema of the in vivo and in vitro methods of producing desert truffle mycorrhizal plants and the time required for each of them. (*1*) Plant seedlings and mature spores. (*2*) Seedlings with a mycelial suspension obtained from fermentation in a bioreactor. (*3*) Acclimatised micropropagated plants with the spore suspension. (*4*) Micropropagated plants and pieces of agar with mycelium. (*5*) Micropropagated plants with mycelial suspension obtained from fermentation in a bioreactor

the culture conditions (in vivo or in vitro) used (Fig. 15.2). The mycelial inoculum and micropropagated plants can be produced whenever required. However, an in vitro culture laboratory and specialised personnel are required, which might increase the final cost of the mycorrhizal plants (Morte et al. 2008).

Characterisation of the mycorrhiza formed by the *Helianthemum* plants with these desert truffles is extremely important to ensure high-quality mycorrhizal plants. For this reason, a morphological and molecular characterisation of the desert truffle mycorrhiza was carried out (Gutiérrez et al. 2003). Such characterisation is also important to evaluate the permanence of the mycorrhiza in field conditions.

15.4 Cultivation Management of the Desert Truffle *Terfezia claveryi*

Since 1999, when the first desert truffle plantation was established in the province of Murcia (Spain) (Fig. 15.3), we have developed a simple management plantation protocol using the experience acquired over the years that has allowed us to reduce the fructification time of the fungus from 23 to 12 months. In all plantations established, *H. almeriense* was selected as host plant mycorrhizal with *T. claveryi*. The management protocol mainly consists of planting in spring, one single irrigation during summer at the end of August of around 60–100 l m^2, and a yearly weeding after the third year of plantation (Morte et al. 2008). Soil fertilisation has not been necessary until now. In general, the soils used for plantations are characterised by a low content of organic carbon (0.9–3.9) and a C/N rate of 7–10. Following such management practices and given the prices they fetch in the open market, the cultivation of *T. claveryi* is profitable.

The fact that desert truffle fruiting depends on rainfall is one of the most important factors for successful cultivation. Irrigation is not necessary when rainfall is abundant (between 350 and 400 mm in Murcia) because the mycorrhizal association is well adapted to arid and semiarid climates (Morte et al. 2000). However, irrigation should be applied during dry years when rainfall is less than 150 mm, at the end of the summer (August/September), and for a second time at the beginning of the fruiting season (January/February) in very dry years. A drip irrigation system is recommended to save water, although watering by sprinklers is also effective. The amount of water to be applied should range between 60 and 100 l m^2, depending on the plant status. After the third year of plantation, one manual, annual and superficial weeding is necessary to avoid plant competition for water, and to maintain the open and sunny desert truffle ecosystem. This weeding should be carried out in the same season every year. In one of the experimental plantations (number 4) production decreased by 75% in 2007 because weeding was not carried out (Table 15.5). Species of the genus *Plantago* tend to cover most of the soil surface around the *H. almeriense* mycorrhizal plants. To facilitate this management, a plantation frame of 3 × 3 or 3 × 2 m should be used (Morte et al. 2008).

The application of such plantation management practices are necessary if desert truffle production is to be maintained because, without them, plantations lost their

Fig. 15.3 Evolution of a plantation of mycorrhizal *H. almeriense* plants with *T. claveryi* in Lorca (Murcia, Spain) (plantation number 1 in Table 15.5). (**a**) One-month-old mycorrhizal plant. (**b**) Plantation at six months after establishment. (**c**) Plantation at 1 year. (**d**) Plantation at 2 years. (**e**) *T. claveryi* ascocarp obtained 1 year after plantation establishment

Table 15.5 *Terfezia claveryi* production in different plantations in Spain (Lorca, Murcia). Data obtained until September 2007

Plantation	Number mycorrhizal plants	Plantation establishment date	Management	Truffle production (kg/ha)			
				Up to 2004	2005	2006	2007
1	60	May 1999	Yes	1,147.6	0.6	1.6	700
2	188	July 2001	No	4.8	–	–	25.8
3	83	March 2002	No	19.5	0.02	–	17.1
4	196	March 2002	Yes	29.4	600	120	29[a]
5	80	March 2002	No/Yes[b]	8.2	–	–	187.5
6	640	January 2003	No	–	–	–	3.6

[a]*No* No weeding.

[b]*Yes* Weeding from September 2006.

productivity after 2–3 years. Such practices become even more essential during years of severe drought, like 2005 and 2006 (Table 15.5). However, mycorrhizal plants are able to produce a 'base amount' of truffles with no management in rainy years like 2007. Furthermore, mycorrhizal plants quickly respond to management, and are able to increase truffle production in the following season (see plantation number 5; Table 15.5). Despite this, desert truffle production fluctuates in the same plantation from one year to the next, due to other environmental or soil constrains, such as temperature and relative humidity, or animal disturbance, which always influences any crop production in the field.

Acknowledgements This work was supported by the projects CGL2007-61175/BOS (Ministerio de Educación y Ciencia, Spain) and 03121/PI/05 (Fundación Séneca, Murcia, Spain). The authors are indebted to the farmers of the plantations (Francisco González, Alfonso and Julia) who carried out the field practices.

References

Cano F (2003) Conocimiento y aprovechamiento popular de un recurso alimenticio de los montes murcianos: Los hongos del género *Terfezia*. *Lactarius* 12:35–53

Calonge FD (1990) *Terfezia claveryi* Chatin (Ascomycotina) in the Canary Islands. *Bol Soc Micol Madrid* 15:193–196

Calonge FD, De La Torre M, Lawrynowicz M (1977) Contribución al estudio de los hongos hipogeos de España. *Anal Inst Bot Cavanilles* 34:15–31

Delmas J (1989) Familla des *Terfeziaceae*. In: Les Champignons et leur cultures. La Maison Rustique, Flammarion, París, pp 697–703

Gutiérrez A (2001) Caracterización, micorrización y cultivo en campo de las trufas de desierto. Doctoral Thesis. Universidad de Murcia

Gutiérrez A, Morte A, Honrubia M (2003) Morphological characterization of the mycorrhiza formed by *Helianthemum almeriense* Pau with *Terfezia claveryi* Chatin and *Picoa lefebvrei* (Pat.) Maire. *Mycorrhiza* 13:299–307

Gutiérrez A, Moreno G, Morte A, Honrubia M (2004) *Terfezia olbiensis* Ness in south-east of Spain: A different species from *T. leptoderma* Tul. Actes du Premier Symposium sur les Champignons hypoges du Basin Mediterraneen. p 102, Rabat, Morocco

Honrubia M, Cano A, Molina-Niñirola C (1992) Hypogeous fungi from southern Spanish semi-arid lands. *Persoonia* 14:647–653

Honrubia M, Morte A, Gutiérrez A, González F, Dieste C (2003) Las Turmas o Trufas de Desierto. In: Esteve-Selma MA, Lloréis-Pascual M, Martínez-Gallur C (eds) Los Recursos Naturales de la Región de Murcia. Un Análisis Interdisciplinar. Servicio de Publicaciones de la Universidad de Murcia, Spain, pp 277–279

Honrubia M, Morte A, Gutiérrez A (2007) Las Terfezias. Un cultivo para el desarrollo rural en regiones áridas y semi-áridas. In: Reyna S (ed) Truficultura. Fundamentos y técnicas. Ediciones Mundi-Prensa, Madrid, pp 365–397

Iriondo JM, Moreno C, Pérez C (1995) Micropropagation of six rockrose (*Cistus*) species. *HortScience* 30:1080–1081

Khabar L, Najim L, Janex-Favre MC, Parguey-Leduc A (2001) Contribution à l'étude de la flore mycologique du Maroc. *Les truffes marocaines. Bull Soc Mycol Fr* 117:213–229

Kirk PM, Cannon PF, David JC, Stalpers JA (2001) Dictionary of the fungi, 9th edition. CABI Publishing, Wallingford, 653 pp

M'Kada JM, Dorion N, Bigot C (1991) In vitro propagation of *Cistus x purpureus* Lam. *Sci Hortic* 46:155–160

Mandeel QA, Al-Laith AA (2007) Ethnomycological aspects of the desert truffle among native Bahraini and non-Bahraini peoples of the Kingdom of Bahrain. *J Ethnopharmacol* 110:118–129

Marasas WFO, Trappe JM (1973) Notes on Southern African Tuberales. *Bothalia* 11:139–141

Morte A, Honrubia M (1992) In vitro propagation of *Helianthemum almeriense* Pau (Cistaceae). *Agronomie* 12:807–809

Morte A, Honrubia M (1994) Método para la micorrización in vitro de plántulas micropropagadas de *Helianthemum* con micelio de *Terfezia claveryi*. Patent n 9402430. Universidad de Murcia

Morte A, Honrubia M (1995) Improvement of mycorrhizal synthesis between micropropagated *Helianthemum almeriense* plantlets with *Terfezia claveryi* (desert truffle). In: Elliot TJ (ed) Science and cultivation of edible fungi, vol 2. Balkema, Rotterdam, pp 863–868

Morte A, Honrubia M (1997) Micropropagation of *Helianthemum almeriense*. In: Bajaj YPS (ed) Biotechnology in agriculture and forestry. High-tech and micropropagation, vol 40. Springer, Berlin/Heidelberg, pp 163–177

Morte A, Lovisolo C, Schubert A (2000) Effect of drought stress on growth and water relations of the mycorrhizal associations *Helianthemum almeriense-Terfezia claveryi*. *Mycorrhiza* 10:115–119

Morte A, Dieste C, Díaz G, Gutiérrez A, Navarro A, Honrubia M (2004) Production of *Terfezia olbiensis* mycelial inoculum in a bioreactor. Act 1er Symp Champignons Hypoges du Basin Mediterraneen. Rabat, Morocco, pp 146–149

Morte A, Gutiérrez A. Honrubia M (2008) Biotechnology and cultivation of desert truffles. In: Varma A (ed) Mycorrhiza: Biology, genetics, novel endophytes and biotechnology. Third edition. Springer, Germany, pp 467–483

Murashige T, Skoog F (1962) A revised medium for rapid growth and bioassays with tobacco tissue cultures. *Physiologia Plantarum* 15:437–497

Murcia MA, Matínez-Tomé M, Jiménez AM, Vera AM, Honrubia M, Parras P (2002) Antioxidant activity of edible fungi (truffles and mushrooms): Losses during industrial processing. *J Food Protect* 65:22–30

Murcia MA, Martínez-Tomé M, Vera A, Morte A, Gutiérrez A, Honrubia M, Jiménez AM (2003) Effect of industrial processing on desert truffles *Terfezia claveryi* and *Picoa juniperi* Vitt: proximate composition an fatty acids. *J Sci Food Agric* 83:535–541

Pérez-García F, González-Benito ME (2006) Seed germination of five *Helianthemum* species: Effect of temperature and presowing treatments. *J Arid Environ* 65:688–693

Reyna S (2000) Trufa, truficultura y selvicultura trufera. Ediciones Mundi Prensa, Madrid

Rodríguez L, Beltrán E, Bañares A, Gonzalez MD (1988) Adiciones a la flora micológica canaria VI. *Documents Mycol* 17:65–72

Trappe JM, Sundberg WJ (1977) *Terfezia gigantea* (Tuberales) in North America. *Mycologia* 69:433–437

Zamora M, Morte A, Gutiérrez A, Honrubia M (2006) *Helianthemum violaceum* Pers., a new host plant for mycorrhizal desert truffle plant production. 5th Inter Conf Mycorrhiza, Granada, Spain, pp. 223

Index

MIX
Papier aus verantwortungsvollen Quellen
Paper from responsible sources
FSC® C105338
FSC
www.fsc.org

If you have any concerns about our products,
you can contact us on
ProductSafety@springernature.com

In case Publisher is established outside the EU,
the EU authorized representative is:
Springer Nature Customer Service Center GmbH
Europaplatz 3, 69115 Heidelberg, Germany

Printed by Libri Plureos GmbH
in Hamburg, Germany